WITHDRAWN

Intermediate Mathematics of Electromagnetics

PRENTICE-HALL ELECTRICAL ENGINEERING SERIES

Intermediate Mathematics of Electromagnetics

DONALD C. STINSON

Hughes Aircraft Company

PRENTICE-HALL, INC., Englewood Cliffs, New Jersey

Library of Congress Cataloging in Publication Data

STINSON, DONALD C.
  Intermediate mathematics of electromagnetics
  Includes bibliographical references and index.
  1. Electromagnetism—Mathematics. I. Title.
QC760.S74   537'.01'51   75-34206
ISBN 0-13-470633-1

10   9   8   7   6   5   4   3   2   1

Printed in the United States of America

PRENTICE-HALL INTERNATIONAL, INC., *London*
PRENTICE-HALL OF AUSTRALIA PTY. LIMITED, *Sydney*
PRENTICE-HALL OF CANADA, LTD., *Toronto*
PRENTICE-HALL OF INDIA PRIVATE LIMITED, *New Delhi*
PRENTICE-HALL OF JAPAN, INC., *Tokyo*
PRENTICE-HALL OF SOUTH-EAST ASIA PRIVATE LIMITED, *Singapore*

Contents

Preface

This book introduces much of the mathematics relevant to electromagnetics and then applies that mathematics to some typical problems. Orthogonal functions, Green's functions, and Fourier transforms are presented in the first three chapters. The equations of electromagnetics in simple and non-simple media are introduced in Chapter 4. The normal modes and Green's functions for Laplace's and Helmholtz's equations in cartesian, cylindrical, and spherical coordinate systems are presented in Chapter 5. Finally, wave propagation in simple and non-simple media in unbounded and bounded space is treated in the last two chapters.

The problems are an integral part of the text and expand upon the material in each chapter. The answers to all of the problems are given so that the student can determine immediately the correctness of a solution. In addition, many of the problems in the first part of the book are chosen so that they later appear as parts of the solutions to more complicated problems in the latter part of the book.

The material in this text was developed at the University of Arizona and was originally used to supplement several of the graduate courses in electromagnetics. However, the material is suitable for a beginning graduate course in electromagnetics as well as a specialized course for seniors. It was presented to both seniors and beginning graduate students in electrical engineering at the

University of Texas at Arlington and was found to be satisfactory except that the pace was considerably slower with the seniors. The book is also suitable for independent study by students as well as being useful as a reference or supplement with courses in electromagnetics and applied mathematics.

In conclusion, the author wishes to express his appreciation to Dr. Andrew E. Salis of the University of Texas at Arlington for the opportunity to use this material in several courses.

D.C.S.

Intermediate Mathematics of Electromagnetics

1

Orthogonal Functions and Introductory Mathematics

1.1 Introduction to Orthogonal Functions and Fourier Series

The concept of orthogonal functions is necessary for the solution of most problems in applied physics and engineering. These functions appear as solutions of the Sturm-Liouville equation which is one of the most basic equations in engineering. This equation will be discussed later after a few introductory remarks are made about Fourier series. The concept involved in the orthogonality of functions is but an extension of the concept of the orthogonality of vectors in n-dimensional space. For instance, let us define a vector

$$\mathbf{f} = \sum_{i=1}^{n} x_i \mathbf{a}_i$$

where $\mathbf{a}_i$ is the usual unit vector in the i-direction and x_i is the amplitude of the vector component of $\mathbf{f}$ in the i-direction. In three-dimensional space, n is 3, etc. If one wishes to evaluate x_j, this can be accomplished quite readily because of the orthogonality of the chosen coordinate system with the concomitant orthogonality of the unit vectors. Thus,

$$\mathbf{f} \cdot \mathbf{a}_j = \sum_{i=1}^{n} x_i \mathbf{a}_i \cdot \mathbf{a}_j = \sum_{i=1}^{n} x_i \delta_{ij} = x_j, \qquad \delta_{ij} = \begin{cases} 1, i = j \\ 0, i \neq j \end{cases}$$

where δ_{ij} is Kronecker's delta. This example illustrates the simplicity with which we can evaluate the amplitude of a particular vector component. One other item we shall need later is the norm N of the vector $\mathbf{f}$. A knowledge of the value of N allows us to normalize $\mathbf{f}$, i.e., find the unit vector $\mathbf{a}_f$ that is in the direction of $\mathbf{f}$. As we already know

$$N = \mathbf{f} \cdot \mathbf{f} = \sum_{i=1}^{n} x_i \mathbf{a}_i \cdot \sum_{j=1}^{n} x_j \mathbf{a}_j = \sum_{\substack{i=1 \\ j=1}}^{n} x_i x_j \delta_{ij} = \sum_{i=1}^{n} x_i^2$$

so that we can evaluate the desired unit vector as

$$\mathbf{a}_f = \frac{\mathbf{f}}{\sqrt{N}}$$

All of the aforementioned ideas involving vectors are applicable to functions except that one of the summations is replaced by an integral. Thus, if we have a function $f(x)$ specified in the interval (a, b), we can expand it in a set of orthogonal functions $u_i(x)$ as follows:

$$f(x) = \sum_{i=1}^{\infty} c_i u_i(x) \tag{1-1}$$

The c_i are analogous to the x_i and the $u_i(x)$ are analogous to the $\mathbf{a}_i$. However, the summation now extends over an infinite number of integers and the orthogonality of the $u_i(x)$ is expressed as follows:

$$N\delta_{ij} = \int_a^b u_i(x)u_j(x)\,dx \tag{1-2}$$

where N is the norm of $u_i(x)$. Thus, we evaluate c_j in a very similar manner to that used to evaluate x_j:

$$\int_a^b f(x)u_j(x)\,dx = \int_a^b \sum_{i=1}^{\infty} c_i u_i(x)u_j(x)\,dx = c_j N$$

and letting $i \longrightarrow j$,

$$c_i = \frac{1}{N} \int_a^b f(x)u_i(x)\,dx \tag{1-3}$$

We shall see later that the integrand in (1-2) should include a weighting function $w(x)$. We have replaced $w(x)$ by unity here in order to avoid unnecessary complications.

Let us now proceed to Fourier's problem[1] as a specific example of these introductory concepts. His problem was the approximation of a function

---

[1] A. Sommerfeld, *Partial Differential Equations in Physics*, Academic Press, Inc., New York, 1949, p. 2.

$f(x)$ specified in the interval $(-\pi, \pi)$ by a sum of $2n + 1$ sinusoidal terms as follows:

$$S_n(x) = \sum_{k=0}^{n} A_k \cos kx + \sum_{k=1}^{n} B_k \sin kx$$

The question that had to be resolved was how to choose the unknown coefficients A_k and B_k. The sinusoids $\cos kx$ and $\sin kx$ are the orthogonal functions here and satisfy the relations

$$\int_{-\pi}^{\pi} \cos kx \cos \ell x \, dx = \frac{2\pi \delta_{k\ell}}{\epsilon_k}, \qquad \epsilon_k = \begin{cases} 1, k = 0 \\ 2, k > 0 \end{cases}$$

$$\int_{-\pi}^{\pi} \sin kx \sin \ell x \, dx = \pi \delta_{k\ell}$$

$$\int_{-\pi}^{\pi} \sin kx \cos \ell x \, dx = 0$$

where ϵ_k is Neumann's number. To answer this question, Fourier defined an error term

$$\Delta_n(x) = f(x) - S_n(x)$$

and considered the minimization of the mean square error,

$$M = \frac{1}{2\pi} \int_{-\pi}^{\pi} \Delta_n^2(x) \, dx$$

When this is done, one obtains the relations

$$A_k = \frac{\epsilon_k}{2\pi} \int_{-\pi}^{\pi} f(x) \cos kx \, dx$$

$$B_k = \frac{1}{\pi} \int_{-\pi}^{\pi} f(x) \sin kx \, dx$$

(1-4)

The steps leading to (1-4) are left as a problem for the student. If we consider the error term, $\Delta_n(x)$, along with the relations (1-4), we find that the error term vanishes as $n \to \infty$ and we obtain the relation

$$f(x) = \sum_{k=0}^{\infty} (A_k \cos kx + B_k \sin kx)$$

(1-5)

where the coefficients A_k, B_k are given by (1-4). By comparing (1-5) with (1-1) and (1-4) with (1-3), we can see the similarity between the familiar Fourier series expansion of a function and the general orthogonal function expansion of a function. For this particular example, two sets of orthogonal functions, i.e., sin kx and cos kx, were required in the Fourier series expansion (1-5). This led to the two sets of coefficients A_k and B_k given by (1-4). We see immediately that our assumption that we can expand $f(x)$ in terms of one set of orthogonal functions $u_i(x)$ as in (1-1) is not sufficiently general since the expansion required two sets of orthogonal functions in this particular case. We now wish to consider the problem of whether an arbitrary function can be expanded in a series of arbitrary orthogonal functions. However, the first question that must be answered is how does one determine whether a group of functions is orthogonal. The general theory that allows one to determine a set of orthogonal functions can be developed from a study of the Sturm-Liouville equation, which will be considered next.

1.2   Sturm-Liouville Equation

Most of the useful ordinary differential equations of physics, chemistry, and engineering are second order and can be represented by the general form

$$L(u) + \lambda wu = 0 \qquad (1\text{-}6)$$

where the differential operator L is defined as

$$L(u) = (pu')' - qu \qquad (1\text{-}7)$$

The quantities p, q,[2] and w ($w \geq 0$) are functions of x in general but λ is a constant. We shall see later that the function $u(x)$ is an eigenfunction and that the constant λ is its corresponding eigenvalue. The prime in (1-7) means the derivative with respect to x which is not necessarily the argument of the $u(x)$. Let us digress for a moment and discuss the concept of an eigenfunction. The usual linear ordinary differential equation that is considered in a lower division mathematics course has constant coefficients so that the solution consists of several terms with a number of arbitrary constants. These arbitrary constants can be determined for a specific problem by giving specific boundary conditions. However, as one considers the common differential equations encountered in the applied mathematics area, one finds that the coefficients are no longer constants and that the general solutions are often found in the form of an infinite series. Moreover, one finds that the application of specific boundary conditions for a specific problem does not result in a specific (nonzero) solution unless some of the constants in the solution take on specific values. These specific, or characteristic, values are called eigenvalues and the specific solutions are called eigenfunctions. In order to illustrate these ideas, let us consider as a simple example a vibrating string

---

[2]The sign of q is sometimes taken as positive rather than negative. See G. Arfken, *Mathematical Methods for Physicists*, Academic Press, Inc., New York, 1966, p. 331.

of length L, linear mass density m, and tension T. If we let x be a variable along the length of the undisturbed string and $u(x, t)$ be the disturbed displacement at the point x and at the time t, the equation of motion for u is

$$m\frac{\partial^2 u}{\partial t^2} = T\frac{\partial^2 u}{\partial x^2} \qquad (1\text{-}8)$$

This is a linear second order partial differential equation and the solution may be found by the method of separation of variables. However, the equation was originally solved by d'Alembert (1747) by introducing the change of variables

$$\xi = x + \alpha_1 t, \qquad \eta = x + \alpha_2 t$$

whence (1-8) becomes

$$\frac{\partial^2 u}{\partial \xi \, \partial \eta} = 0 \qquad (1\text{-}9)$$

provided $\alpha_1 + \alpha_2 = 0$ and $\alpha_1\alpha_2 + (T/m) = 0$. The solution to (1-9) is

$$u = f(\xi) + g(\eta)$$
$$= f\left(x \pm \sqrt{\frac{T}{m}}\, t\right) + g\left(x \mp \sqrt{\frac{T}{m}}\, t\right) \qquad (1\text{-}10)$$

The solution (1-10) represents two traveling waves since we have not yet specified any conditions concerning the ends of the string. If we impose the boundary condition that the string is fixed at its ends $x = 0, L$, and use the method of separation of variables, we obtain as the general solution to (1-8)

$$u(x, t) = [A \cos \beta t + B \sin \beta t]\left[C \cos\frac{\beta x}{\alpha} + D \sin\frac{\beta x}{\alpha}\right]$$

where $\alpha = \sqrt{T/m}$. The boundary condition that $u(0, t) = 0$ causes C to vanish. However, the boundary condition that $u(L, t) = 0$ cannot be satisfied by allowing D to vanish since then our solution would be trivial. The only remaining choice is to let $\sin(\beta L/\alpha) = 0$ which means then that

$$\frac{\beta L}{\alpha} = m\pi, \qquad m = 1, 2, \ldots$$

Thus, we find that in contrast to familiar ordinary differential equations where the boundary conditions are satisfied by choosing specific values for the constants A, B, C, D, the solution $u(x, t)$ cannot be specified unless the parameters α and β take on specific values. For each specific value of α and β, we obtain one specific solution to (1-8). The most general solution now is an infinite series of terms as follows:

$$u(x, t) = \sum_{m=1}^{\infty} \left[A_m \cos \frac{m\pi\alpha}{L} t + B_m \sin \frac{m\pi\alpha}{L} t \right] \sin \frac{m\pi x}{L}$$

The general solution to (1-8) is the above series of eigenfunctions with an accompanying group of eigenvalues, $m\pi/L$. It also follows that solutions to (1-8) for values of α and β that do not correspond to the eigenvalues, $m\pi/L$, must be trivial (zero) solutions.[3]

Let us now return to the general equation (1-6) and show that the functions $u(x)$ are orthogonal, i.e., possess that property indicated in (1-2). We will also find that our definition of orthogonality in (1-2) is not sufficiently general to encompass all functions that satisty the Sturm-Liouville equation. Since the functions $u(x)$ satisfy the Sturm-Liouville equation, (1-6), then two specific eigenfunctions $u_i(x)$ and $u_j(x)$ with their corresponding eigenvalues λ_i and λ_j, respectively, satisfy (1-6), thus

$$L(u_i) + \lambda_i w u_i = 0$$
$$L(u_j) + \lambda_j w u_j = 0 \qquad (1\text{-}11)$$

We now apply a standard procedure, which is also used to derive Green's two identities in scalar or vector form, by multiplying the first equation by u_j, the second by $-u_i$, adding the two, and then integrating over the interval (a, b) to get

$$\int_a^b [u_j L(u_i) - u_i L(u_j)] \, dx = (\lambda_j - \lambda_i) \int_a^b w u_i u_j \, dx = [p(u_j u_i' - u_i u_j')]_a^b \qquad (1\text{-}12)$$

The third expression above results when one substitutes $L(u_i)$ and $L(u_j)$ from (1-7) into the first expression. The third expression vanishes because of the general boundary conditions

$$[p(u_j u_i' - u_i u_j')]_a^b = 0 \qquad (1\text{-}13)$$

When written in this form, one notes that the boundary conditions necessary for orthogonality can be satisfied by a variety of conditions, e.g.,

$$p u_j u_i']_a = p u_i u_j']_b$$
$$p(a) = p(b) = 0$$
$$p(a) = 0 = u_i(b)$$
$$u_j(a) = u_i(b) = 0$$
$$u_i'(a) = u_j'(b) = 0$$

---

[3]H. Sagan, *Boundary and Eigenvalue Problems in Mathematical Physics*, John Wiley & Sons, Inc., New York, 1961, Chap. 5.

When these boundary conditions are met, we obtain the expression of orthogonality

$$\int_a^b wu_i u_j \, dx = 0(\lambda_j - \lambda_i)^{-1}$$

As long as $\lambda_i \neq \lambda_j$, the right-hand side of the above expression vanishes. In most cases of interest, when $u_i \neq u_j$ then $\lambda_i \neq \lambda_j$ and the condition of orthogonality is valid. However, two exceptional cases of interest often occur where $u_i \neq u_j$ but $\lambda_i = \lambda_j$. In the first case, u_i and u_j are degenerate solutions of (1-6) because their eigenvalues are identical. In the second case, u_i and u_j are two linearly independent solutions of (1-6) for the same eigenvalue λ_i. This latter case leads to an expression involving the Wronskian.

If we consider the first case, where the eigenfunctions are degenerate, we find that the situation can be encountered in electromagnetic problems involving more than one dimension. For instance, the eigenfunctions, $u_{12} = \sin(\pi x/a) \sin(2\pi y/b)$ and $u_{21} = \sin(2\pi x/a) \sin(\pi y/b)$ have corresponding eigenvalues

$$\lambda_{12} = \left[\left(\frac{\pi}{a} \right)^2 + \left(\frac{2\pi}{b} \right)^2 \right]^{1/2} \quad \text{and} \quad \lambda_{21} = \left[\left(\frac{2\pi}{a} \right)^2 + \left(\frac{\pi}{b} \right)^2 \right]^{1/2}$$

respectively. In general, the eigenvalues λ_{12} and λ_{21} are different unless $a = b$. When $a = b$, the eigenvalues are identical, yet $u_{12} \neq u_{21}$. Thus, the eigenfunctions are degenerate. In this case, we see that the eigenfunctions u_{12} and u_{21} are orthogonal whether λ_{12} is different from λ_{21} or not. Since the physical nature of the problem does not change wildly when the boundaries or the boundary conditions are changed slightly, we see that the difficulties introduced by degeneracy above are more mathematical deficiencies rather than some extraordinary physical phenomenon. However, the physical problem of degeneracy is quite common in the real world and often occurs in classical[4] and quantum mechanical problems[5] as well as in electromagnetic problems.[6] The situation then is that several specific states or modes may possess the same energy or occupy the same energy level.

The other exceptional case of interest occurs when $u_i \neq u_j$ because they are two linearly independent solutions of (1-6) for the same eigenvalue λ_i. In other words, u_i is not a general solution of (1-6) but one of the two linearly independent solutions U. If we let the other linearly independent solution

[4]R. Wangsness, *Introduction to Theoretical Physics*, John Wiley & Sons, Inc., New York, 1963, p. 193.

[5]R. Wangsness, *Introductory Topics in Theoretical Physics*, John Wiley & Sons, Inc., New York, 1963, p. 232.

[6]Wangsness, *Introduction to Theoretical Physics*, p. 314.

be V, we obtain

$$\int_a^b wUV\,dx = \frac{0}{0}$$

Although, this integral is indeterminate, and of not enough interest to motivate its evaluation, let us consider the first and third expressions in (1-12) when $u_i = U$ and $u_j = V$ and the limits are ignored. Thus,

$$\int [VL(U) - UL(V)]\,dx = p(VU' - UV') = pW(V, U)$$

where $W(V, U)$ is the Wronskian[7] of V and U. If the Wronskian vanishes, U and V are not linearly independent solutions of (1-6). If we substitute the expressions for $L(V)$ and $L(U)$ from (1-7) into the left-hand side of the above integral, we obtain

$$\int [p(VU' - UV')]'\,dx = \int [pW(V, U)]'\,dx = pW(V, U) \neq 0$$

Unfortunately, this expression involving the Wronskian does not help us since we do not know the right-hand side. In fact, one might have expected the right-hand side to vanish since $\lambda_i = \lambda_j$ in the second expression in (1-11). However, this difficulty is caused by our implicit assumption that $\lambda_i \neq \lambda_j$ in (1-11). When $\lambda_i = \lambda_j$ in (1-11) the second expression in (1-12) does not appear since we obtain, for $u_i = U$ and $u_j = V$,

$$VL(U) - UL(V) = 0 = [pW(V, U)]'$$

and thus,

$$pW(V, U) = c \tag{1-14}$$

where c is a constant with respect to x. The reason we mention the difficulty encountered above is to emphasize the fact that the Sturm-Liouville equation is not particularly useful when evaluating the Wronskian or when evaluating the norm. The evaluation of the norm is very similar to the case we just discussed except that now either $u_i = u_j = U = V$ or $u_i = u_j = AU + BV$. In this case we find that the expression for the norm, (1-2), is not sufficiently general since the second expression in (1-12) gives us the correct general expression for the norm

$$\int_a^b wu_i^2\,dx = N$$

---

[7]H. Margenau and G. Murphy, *The Mathematics of Physics and Chemistry*, D. Van Nostrand Company, Inc., Princeton, N.J., 1943, p. 130.

and the general orthogonality expression

$$\int_a^b w u_i u_j \, dx = N \delta_{ij} \tag{1-15}$$

If we consider the first expression in (1-12) again, we find that

$$\int_a^b u_j L(u_i) \, dx = \int_a^b u_i L(u_j) \, dx, \qquad i \neq j$$

and the operator L is called a Hermitian operator. It is also possible for the eigenfunctions u_i and u_j to be complex, in which case the expression above and (1-11) must be modified. However, the eigenvalues λ_i and λ_j remain real. Further details concerning self-adjoint operators and Hermitian operators may be found elsewhere.[8]

1.3 Orthogonal Functions

In this section we consider some of the orthogonal functions that appear frequently in electromagnetic theory. It is assumed that the student is familiar with these functions, the various differential equations that they satisfy, and the general procedure for solving the equations. Thus, the functions can be introduced in terms of their generating functions. This alternate representation for the functions is usually very advantageous when deriving recurrence relations and other useful identities.

The first functions that we will discuss are the Bessel functions. These functions occur in problems involving heat flow, fluid flow, and wave motion provided solutions are being sought in the cylindrical coordinate system. The thing that is common to all of the above problems is that the spatial dependence of the function can be expressed in terms of the Laplacian. As a result, one can state that the Laplacian in cylindrical coordinates can be separated into ordinary differential equations, one of which is generally Bessel's equation. However, Bessel's equation also occurs in connection with other problems,[9] many of which still involve the Laplacian in different coordinate systems.[10] On the other hand, there are a great many ordinary differential equations that can be transformed by a change of variables into Bessel's equation so that their solutions can be expressed in terms of Bessel func-

---

[8] *Ibid.*, pp. 255, 328. Also, see Arfken, *Mathematical Methods for Physicists*, Chap. 9.

[9] H. Reddick and F. Miller, *Advanced Mathematics for Engineers*, 2nd ed., John Wiley & Sons, Inc., New York, 1947.

[10] Arfken, *Mathematical Methods for Physicists*, Chap. 8.

tions.[11] This procedure is advantageous because the Bessel functions are quite well tabulated.[12]

The next functions that are of interest are the Legendre functions. They occur generally when the Laplacian is separated in the spherical coordinate system as well as in some other coordinate systems. One also finds in the spherical coordinate system that the wave equation (or Helmholtz equation) separates into three ordinary differential equations, one of them in general being Legendre's equation of a more general form and the other in general being Bessel's equation of a more specific form. The more general form of Legendre's equation is called the associated Legendre equation and the more specific form of Bessel's equation is called the spherical Bessel's equation. This latter equation has solutions that are called spherical Bessel functions and can be expressed in terms of the usual (cylindrical) Bessel functions of order $n + \frac{1}{2}$. Moreover, when n is an integer, the spherical Bessel functions can be expressed in a closed form in terms of the cosine and sine functions.[13]

The tesseral or spherical harmonics[14] are two-dimensional functions that can be used to expand an arbitrary two-dimensional function on the surface of a sphere. They result from the separation of the Laplacian in spherical coordinates and represent the θ- and ϕ-dependent portions. In many cases, one is interested in expanding the arbitrary two-dimensional function over an interval that is less than the entire area of the sphere. In this case, it is necessary that the spherical harmonics be orthogonal over a shorter interval so that spherical harmonics of nonintegral order must be introduced. The situation is quite similar to that of the Fourier expansion of a function over the interval $(0, 2\pi)$ and the later generalization[15] to the interval $(0, L)$. That is, the functions $\sin nx$ and $\cos nx$ are orthogonal over the interval $(0, 2\pi)$, and the functions $\sin(n\pi x/L)$ and $\cos(n\pi x/L)$ are orthogonal over the interval $(0, L)$. Unfortunately, with spherical harmonics of nonintegral (noninteger) order, the orthogonality relations become considerably more complicated than those for the spherical harmonics of integral order.

The Mathieu functions are included because they occur later in Chap. 6 in connection with a problem involving a rectangular waveguide completely filled with a dielectric whose dielectric constant varies sinusoidally in the direction of the waveguide axis. However, the section on Mathieu functions can be omitted without disturbing the continuity of the presentation. Mathieu

---

[11]Sagan, *Boundary and Eigenvalue Problems in Mathematical Physics*, p. 205.

[12]E. Jahnke and F. Emde, *Tables of Functions with Formulae and Curves*, 4th ed., Dover Publications, New York, 1945.

[13]Sommerfeld, *Partial Differential Equations in Physics*, p. 113.

[14]*Ibid.*, p. 130.

[15]R. Churchill, *Fourier Series and Boundary Value Problems*, McGraw-Hill Book Company, Inc., New York, 1941, Chap. 4.

functions also occur in connection with wave propagation in a waveguide of elliptic[16] cross section. However, this topic will not be considered in this text.

The orthogonal functions that are of interest to us all satisfy the Sturm-Liouville equation (1-6) and so are of order two. As a result, the general solution consists of two linearly independent functions each of which satisfy the same equation. The test for linear independence once two functions are found requires that the Wronskian[17] does not vanish. This test is applicable as long as the functions are differentiable once. For ordinary differential equations of order n the functions must be differentiable $n - 1$ times in the interval of interest. This test is sufficiently general for our purposes although it will not be necessary to use it since we are concerned with well known functions. However, it is useful to know the Wronskian of many combinations of the functions that satisfy the different differential equations since this allows one to simplify the solutions to many problems. For our purposes, then, we can define the Wronskian for a second order ordinary differential equation as

$$W(y_1, y_2) = \begin{vmatrix} y_1 & y_2 \\ y_1' & y_2' \end{vmatrix} \tag{1-16}$$

where $y_1(x)$ and $y_2(x)$ are solutions to the differential equation. If y_1 and y_2 are linearly independent solutions, then

$$W(y_1, y_2) \neq 0$$

In fact, we have already shown in (1-14) that the Wronskian for two linearly independent solutions of the Sturm-Liouville equation[18] is a constant divided by $p(x)$.

1.4 Bessel Functions

The generating function for the Bessel function of order n is

$$\exp\left[\left(\frac{x}{2}\right)\left(t - \frac{1}{t}\right)\right] = \sum_{n=-\infty}^{\infty} J_n(x)t^n, \quad t \neq 0 \tag{1-17}$$

---

[16]J. Stratton, *Electromagnetic Theory*, McGraw-Hill Book Company, Inc., New York, 1941, Chap. 6.

[17]Margenau and Murphy, *The Mathematics of Physics and Chemistry*, Sec. 3.13. Also, see Arfken, *Mathematical Methods for Physicists*, Sec. 8.5.

[18]G. Duff and D. Naylor, *Differential Equations of Applied Mathematics*, John Wiley & Sons, Inc., New York, 1966, p. 46.

where n is a real integer and x may be complex. Bessel's equation is

$$x^2 J_n''(x) + x J_n'(x) + (x^2 - n^2) J_n(x) = 0 \qquad (1\text{-}18)$$

where the primes indicate derivatives with respect to the argument (x here). If we make the substitution $t = e^{j\phi}$ in (1-17), we obtain

$$e^{jx \sin \phi} = \sum_{n=-\infty}^{\infty} J_n(x) e^{jn\phi} \qquad (1\text{-}19)$$

Since the functions $e^{jn\phi}$ are orthogonal to the functions $e^{-jm\phi}$ in the interval $(0, 2\pi)$, let us multiply both sides of (1-19) by $e^{-jm\phi}$ and integrate over the interval $(0, 2\pi)$ to find the expression for $J_n(x)$. We obtain

$$J_n(x) = \frac{1}{2\pi} \int_0^{2\pi} e^{j(x \sin \phi - n\phi)} \, d\phi \qquad (1\text{-}20)$$

The function $J_n(x)$ is a real quantity (for x real) so the expression (1-20) can be written in the form

$$J_n(x) = \frac{1}{2\pi} \int_0^{2\pi} \cos(x \sin \phi - n\phi) \, d\phi \qquad (1\text{-}21)$$

$$0 = \frac{1}{2\pi} \int_0^{2\pi} \sin(x \sin \phi - n\phi) \, d\phi$$

where (1-21) is in Bessel's original form.

Since Bessel's equation, (1-18), is a second order differential equation, it must possess two linearly independent solutions. The solution we have discussed in (1-17) is usually called a Bessel function and the other independent solution is called a Neumann function or Bessel function of the second kind and given the symbol $N_n(x)$. The Neumann function is not easily obtainable from the differential equation (1-18) and is usually used only when n is an integer. When n is not an integer as we assumed in (1-17), (1-18) can be solved in the usual fashion (method of Frobenius) and $J_n(x)$ and $J_{-n}(x)$ are the two independent solutions. Since various combinations of $J_n(x)$ and $N_n(x)$ are also solutions to (1-18), it follows that the two functions defined below are satisfactory functions:

$$\begin{aligned} H_n^{(1)}(x) &= J_n(x) + jN_n(x) \\ H_n^{(2)}(x) &= J_n(x) - jN_n(x) \end{aligned} \qquad (1\text{-}22)$$

These two functions, $H_n^{(1)}(x)$ and $H_n^{(2)}(x)$, are called Hankel functions of the first and second kinds, respectively, and are quite useful in exterior boundary value problems.

There is no simple relation between a Bessel function of order n and either its derivative or any other Bessel function. Thus, one needs recurrence relations for Bessel functions of different orders and the derivative of a Bessel function. These relations can be obtained from the generating function (1-17). The relation involving three different orders is found by differentiating (1-17) with respect to t and comparing like powers of t. The relation involving the derivative is found by differentiating (1-17) with respect to x and comparing like powers of t. The results are

$$\frac{2n}{x} J_n(x) = J_{n-1}(x) + J_{n+1}(x)$$
$$2J_n'(x) = J_{n-1}(x) - J_{n+1}(x) \tag{1-23}$$

The orthogonality integral for Bessel functions is necessary when one is expanding a function in a Fourier-Bessel series, i.e., an orthogonal-function expansion indicated in (1-1). If one substitutes $x = \alpha x$ in (1-18), the equation is put into the form of (1-6). Thus, $w = x$ and the orthogonality integral[19] in the form of (1-15) for Bessel functions is

$$\int_a^b x B_n(\alpha x) B_m(\alpha x) \, dx = \delta_{mn} N$$

where

$$N = \left[\frac{x^2}{2} \{ B_n^2(\alpha x) - B_{n-1}(\alpha x) B_{n+1}(\alpha x) \} \right]_a^b$$
$$= \left[\frac{x^2}{2} \left\{ B_n'^2(\alpha x) + \left(1 - \frac{n^2}{\alpha^2 x^2} \right) B_n^2(\alpha x) \right\} \right]_a^b \tag{1-24}$$

In the above expression, $B_n(\alpha x)$ is any arbitrary solution to Bessel's equation, i.e., $J_n(\alpha x)$, $N_n(\alpha x)$, $H_n^{(1)}(\alpha x)$, $H_n^{(2)}(\alpha x)$, $J_n(\alpha x) + a_n H_n^{(2)}(\alpha x)$, etc. The recurrence relations in (1-23) are also valid with $J_n(x)$ replaced by $B_n(\alpha x)$ provided the prime means the derivative with respect to the argument, i.e.,

$$B_n'(\alpha x) = \frac{d}{d(\alpha x)} [B_n(\alpha x)] = \frac{1}{\alpha} \frac{d}{dx} [B_n(\alpha x)]$$

The Bessel functions are also orthogonal for $n = m$ and different eigenvalues α_i.[20] Thus, we can write that

$$\int_a^b x B_n(\alpha_i x) B_n(\alpha_j x) \, dx = \delta_{ij} N \tag{1-25}$$

---

[19] *Ibid.*, p. 308.

[20] R. Collin, *Field Theory of Guided Waves*, McGraw-Hill Book Company, Inc., New York, 1960, p. 196.

There are occasions when x is a complex quantity.[21] However, the only case of interest to us is when $x = jy$. In this case we obtain the modified Bessel functions[22] of the first and second kinds, respectively, defined as

$$I_n(y) = j^{-n} J_n(jy)$$

$$K_n(y) = \frac{\pi}{2} j^{n+1} H_n^{(1)}(jy)$$

and the differential equation that they both satisfy

$$y^2 I_n''(y) + y I_n'(y) - (y^2 + n^2) I_n(y) = 0 \qquad (1\text{-}26)$$

or

$$z^2 \frac{d^2 I_n(az)}{dz^2} + z \frac{dI_n(az)}{dz} - (a^2 z^2 + n^2) I_n(az) = 0$$

Various Wronskians of Bessel functions are found to be

$$J_n(x) N_n'(x) - J_n'(x) N_n(x) = J_n(x) N_{n-1}(x) - J_{n-1}(x) N_n(x) = \frac{2}{\pi x}$$

$$N_n(x) H_n^{(2)\prime}(x) - N_n'(x) H_n^{(2)}(x) = -\frac{2}{\pi x} \qquad (1\text{-}27)$$

$$J_n(x) H_n^{(2)\prime}(x) - J_n'(x) H_n^{(2)}(x) = \frac{2}{j\pi x}$$

Similar relations involving $H_n^{(1)}(x)$ can be found by noting that

$$H_n^{(1)}(x) = H_n^{(2)*}(x)$$

where the asterisk denotes a complex conjugate. Other useful relations are the following:

$$J_n(-x) = (-1)^n J_n(x)$$
$$N_n(-x) = (-1)^n [N_n(x) + 2j J_n(x)]$$
$$H_n^{(1)}(-x) = (-1)^{n+1} H_n^{(2)}(x)$$
$$H_n^{(2)}(-x) = (-1)^n [H_n^{(1)}(x) + 2 H_n^{(2)}(x)]$$

---

[21]S. Ramo, J. Whinnery, and T. Van Duzer, *Fields and Waves in Communication Electronics*, John Wiley & Sons, Inc., New York, 1965, p. 292.

[22]Arfken, *Mathematical Methods for Physicists*, Sec. 11.4.

1.5 Legendre Functions

The generating function for the Legendre functions of order n is

$$(1 - 2xt + t^2)^{-1/2} = \sum_{n=0}^{\infty} P_n(x)t^n \tag{1-28}$$

provided $|x| \leq 1$ and $|t| < 1$. If one expands the left side of (1-28) in a MacLaurin series about $t = 0$, one obtains

$$P_n(x) = \frac{1}{n!} \frac{\partial^n}{\partial t^n} (1 - 2xt + t^2)^{-1/2}]_{t=0}$$

It can be shown that $P_n(x)$ satisfies the second order ordinary differential equation known as Legendre's equation

$$(1 - x^2)y'' - 2xy' + n(n + 1)y = 0 \tag{1-29}$$

or

$$\frac{1}{\sin \theta} \frac{d}{d\theta} \left(\sin \theta \frac{dy}{d\theta} \right) + n(n + 1)y = 0$$

where $x = \cos \theta$.

An interesting electrostatic problem that illustrates Legendre functions consists of finding the potential of the two point charges shown in Fig. 1-1.

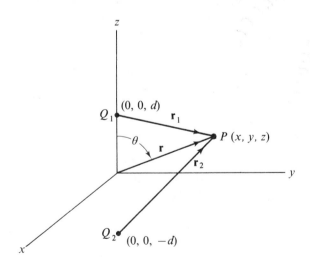

Fig. 1.1 Potential at point $P(x, y, z)$ of two charges Q_1 and Q_2 in terms of Legendre functions.

The potential at P is well-known from elementary courses in field theory as

$$\Phi_P = \frac{1}{4\pi\epsilon_0 r}\left(\frac{Q_1}{r_1} + \frac{Q_2}{r_2}\right)$$

Since $r_{1,2}^2 = r^2 + d^2 \mp 2rd \cos\theta$ from the cosine law, let us make the substitutions $x = \pm\cos\theta$ and $t = d/r < 1$. When we use these relations for r_1 and r_2 in the expression for Φ_P, we obtain

$$\Phi_P = \frac{1}{4\pi\epsilon_0 r}\sum_{n=0}^{\infty}\left(\frac{d}{r}\right)^n[Q_1 P_n(\cos\theta) + Q_2 P_n(-\cos\theta)], \qquad r > d$$

$$\Phi_P = \frac{1}{4\pi\epsilon_0 d}\sum_{n=0}^{\infty}\left(\frac{r}{d}\right)^n[Q_1 P_n(\cos\theta) + Q_2 P_n(-\cos\theta)], \qquad r < d$$

(1-30)

The other solution to Legendre's equation (1-29) is called a Legendre function of the second kind, denoted by $Q_n(x)$. These functions are not defined for $|x| = 1$ so do not appear in many physical problems. When n is an integer, as we have assumed thus far, the series representing $P_n(x)$ and $Q_n(x)$ have a finite number of terms. As a result, Legendre functions are usually called Legendre polynomials. However, this occurs only when n is an integer. The more involved cases for noninteger n are usually considered as special cases of solutions of the hypergeometric equation, also called Gauss' differential equation.

The recurrence relations for Legendre functions can be obtained from (1-28) in the same manner that the recurrence relations (1-23) for the Bessel functions were obtained from (1-17). One obtains

$$(n + 1)P_{n+1}(x) - (2n + 1)xP_n(x) + nP_{n-1}(x) = 0$$

$$P_1(x) - xP_0(x) = 0$$

and

$$P_n(x) = P'_{n+1}(x) + P'_{n-1}(x) - 2xP'_n(x)$$

$$P'_0(x) = 0$$

$$P'_1(x) = -P_0(x)$$

(1-31)

The orthogonality integral for Legendre functions is

$$\int_{-1}^{1} P_\ell(x)P_n(x)\,dx = \delta_{\ell n}N$$

where $N = 2/(2n + 1)$. Other relations that can be found from (1-18) are $P_n(-x) = (-1)^n P_n(x)$ and $P_n(\pm 1) = (\pm 1)^n$.

1.6 Tesseral or Spherical Harmonics

The problem of representing an arbitrary area on the surface of a sphere is just the problem of expanding $f(\theta, \phi)$ into two Fourier-type series of orthogonal functions. In this case the ϕ-dependent functions are the sinusoids and the θ-dependent functions are the associated Legendre functions. The associated Legendre functions satisfy the equation

$$(1 - x^2)z'' - 2xz' + \left[n(n + 1) - \frac{m^2}{1 - x^2}\right]z = 0$$

or   (1-32)

$$\frac{1}{\sin \theta}\frac{d}{d\theta}\left(\sin \theta\frac{dz}{d\theta}\right) + \left[n(n + 1) - \frac{m^2}{\sin^2 \theta}\right]z = 0$$

which reduces to (1-29) for $m = 0$. The solutions to (1-32) are the associated Legendre functions of the first and second kinds, $P_n^m(x)$ and $Q_n^m(x)$, respectively. These functions[23] are related to the usual Legendre functions as follows:

$$z = L_n^m = (-1)^m(1 - x^2)^{m/2}\frac{d^m y}{dx^m}$$

where y is $P_n(x)$ or $Q_n(x)$ and L_n^m (or z) is $P_n^m(x)$ or $Q_n^m(x)$, respectively.

As before, the functions $Q_n^m(x)$ are not well behaved at $|x| = 1$ so do not appear in many problems. Since both the θ- and ϕ-dependent functions satisfy the Sturm-Liouville equation, we know that it is possible to expand $f(\theta, \phi)$ as follows:

$$f(\theta, \phi) = \sum_{m=0}^{n} \sum_{n=0}^{\infty} (a_{mn}T_{mn}^e + b_{mn}T_{mn}^o)$$

where

$$T_{mn}^e(\theta, \phi) = P_n^m(\cos \theta) \cos m\phi$$
$$T_{mn}^o(\theta, \phi) = P_n^m(\cos \theta) \sin m\phi$$ (1-33)

and the orthogonality condition on the T_{mn}^i is

$$\int_0^{2\pi} d\phi \int_0^{\pi} \sin \theta \, d\theta \, T_{mn}^i(\theta, \phi)T_{pq}^j(\theta, \phi) = \frac{4\pi(n + m)!}{\epsilon_m(2n + 1)(n - m)!}\delta_{mp}\delta_{nq}\delta_{ij}$$

(1-34)

---

[23]R. Harrington, *Time-Harmonic Electromagnetic Fields*, McGraw-Hill Book Company, Inc., New York, 1961, Appendix E. Note that many authors do not include the factor $(-1)^m$ in their definition, e.g., see Sommerfeld, *Partial Differential Equations in Physics*, p. 128.

A similar orthogonality relation involving only the associated Legendre functions of the first kind is

$$\int_{-1}^{1} P_{\ell}^{m}(x) P_{n}^{m}(x) \, dx = \delta_{\ell n} N$$

where

$$N = \frac{2}{2n+1} \frac{(n+m)!}{(n-m)!}$$

An examination of the tesseral harmonics reveals the fact that they vanish at $\theta = 0$, π, and at $n - m$ other values of θ, giving $n - m$ parallels of latitude. The zeroes of the sinusoids produce m meridians. The zeroes of the tesseral harmonics of degree n and type m thus divide the sphere into spherical rectangles. As a result, one usually speaks of the tesseral harmonics as being the product of zonal and sectorial harmonics. The zonal harmonics are the θ-dependent functions and the sectorial harmonics are the ϕ-dependent functions. In general, both m and n may not be integers, in which case the preceding expressions must be modified. These cases will be considered in the next section.

One also finds that the spherical harmonics can be defined so that the ϕ-dependent functions in (1-33) are exponentials. This can be seen quite easily by forming the sum

$$T_{mn}^{e} + j T_{mn}^{o} = P_{n}^{m}(\cos \theta) e^{j m \phi}$$

This case is considered by others[24] who also extend the definition of the associated Legendre functions to include negative values for the index or order m. Various recurrence relations for the associated Legendre functions can be derived and are given in Sec. 1.7 for nonintegral values of n and m, changed to s and r, respectively. However, these recurrence relations[25] are also valid for integral values of s and r. More details are also available in Appendix C.4.

1.7 Spherical Harmonics of Nonintegral Degree and Order

From Sec. 1.6 the associated Legendre functions satisfy the equation

$$(1 - x^2)z'' - 2xz' + \left[n(n+1) - \frac{m^2}{1 - x^2} \right] z = 0 \qquad (1\text{-}32)$$

---

[24]Sommerfeld, *Partial Differential Equations in Physics*, pp. 128–29. Also, see Duff and Naylor, *Differential Equations of Applied Mathematics*, pp. 344–45.

[25]Harrington, *Time-Harmonic Electromagnetic Fields*, Appendix E.

which can also be written as

$$\frac{1}{\sin \theta} \frac{d}{d\theta} \left(\sin \theta \frac{dz}{d\theta} \right) + \left[n(n+1) - \frac{m^2}{\sin^2 \theta} \right] z = 0$$

when $x = \cos \theta$.

In general, $n(n+1)$ and m^2 are constants and n and m are not integers. However, when one expands an arbitrary function over the surface of a sphere in a set of orthogonal functions, one finds that n and m are integers. In this case, the orthogonal functions are the tesseral or spherical harmonics. This procedure was discussed in Sec. 1.6 and is valid as long as the entire surface of the sphere is considered, i.e., $|x| \leq 1$ or $0 \leq \theta \leq \pi$. On the other hand, there are problems that do not involve the entire surface of the sphere, whence one needs to consider the solutions of (1-32) for noninteger values of n and m. It is also occasionally necessary to discuss solutions to (1-32) for values of $x > 1$, for x complex, n complex, etc. We shall not discuss any of these cases here. However, let us discuss the problem encountered when one wishes to expand a function over the segment of spherical surface defined by $\theta = \theta_1$, $\theta = \theta_2$, $\phi = \phi_1$, $\phi = \phi_2$. When the function vanishes at $\phi = \phi_1$ and $\phi = \phi_2$, the ϕ-dependent solution will be of the form $\sin (p\pi/\alpha)(\phi - \phi_1)$ or $\sin (p\pi/\alpha)(\phi_2 - \phi)$, where $\alpha = \phi_2 - \phi_1$ with $p = 1, 2, 3, \ldots, \infty$. For example, see (5-61). When the normal derivative of the function vanishes on the boundary, the ϕ-dependent solution will be of the form $\cos (p\pi/\alpha)$ $(\phi - \phi_1)$ or $\cos (p\pi/\alpha)(\phi_2 - \phi)$ with p and α the same as above. It is also possible to have mixed boundary conditions, which we shall not consider. Now the constant m in (1-32) becomes $(p\pi/\alpha) = r$ and is not an integer. Note that integer values can still occur when $\alpha = q\pi$, $q = 1, 2, 3, \ldots$. However, in those cases image theory[26] can be used. Now (1-32) becomes

$$(1 - x^2)z'' - 2xz' + \left[s(s+1) - \frac{r^2}{1 - x^2} \right] z = 0 \qquad (1\text{-}32\text{a})$$

where we have replaced n by s, m by r, and $r = p\pi/\alpha$. The constants s and r are not integers. The two solutions to (1-32a) are still $P_s^r(x)$ and $Q_s^r(x)$ although $P_s^r(x)$ and $P_s^r(-x)$ are equally valid solutions. The function $P_s^r(x)$ is well behaved in the interval $-1 < x \leq 1$, i.e., it is singular at $x = -1$. The function $P_s^r(-x)$ is well behaved in the interval $-1 \leq x < 1$, i.e., it is singular at $x = 1$. The function $Q_s^r(x)$ is a combination of both $P_s^r(x)$ and $P_s^r(-x)$ and therefore well behaved in the interval $-1 < x < 1$, i.e., it is singular at $|x| = 1$. The recurrence relations in (1-31) are valid for n replaced by s. Several recurrence relations for associated Legendre functions are written below.

---

[26] W. Smythe, *Static and Dynamic Electricity*, 2nd ed., McGraw-Hill Book Company, Inc., New York, 1950, p. 158.

$$L_s^{r+1}(x) = \frac{-2rx}{\sqrt{1-x^2}}L_s^r(x) - (r+s)(r-s-1)L_s^{r-1}(x)$$

$$(1-x^2)\frac{dL_s^r(x)}{dx} = -(r+s)(r-s-1)\sqrt{1-x^2}L_s^{r-1}(x) + rxL_s^r(x)$$

$$\frac{dL_s^r(x)}{dx} = \frac{-rx}{1-x^2}L_s^r(x) - \frac{1}{\sqrt{1-x^2}}L_s^{r+1}(x)$$

$$(s-r+1)L_{s+1}^r(x) = (2s+1)xL_s^r(x) - (r+s)L_{s-1}^r(x)$$

$$(1-x^2)\frac{dL_s^r(x)}{dx} = -sxL_s^r(x) + (s+r)L_{s-1}^r(x)$$

for $r = 0$,

$$L_s^1(x) = -\sqrt{1-x^2}\,\frac{dL_s(x)}{dx}$$

In the above recurrence relations, $L_s^r(x)$ can be any general solution to (1-32a), i.e.,

$$L_s^r(x) = AP_s^r(x) + BQ_s^r(x)$$
$$L_s^r(x) = CP_s^r(x) + DP_s^r(-x)$$

(1-32b)

where A or B can vanish and C or D can vanish.

Since the associated Legendre functions usually appear multiplied by $e^{jr\phi}$ or $E\cos r\phi + F\sin r\phi$, one desired orthogonality relation will be of the form

$$\int_{x_1}^{x_2} L_s^r(x)L_t^r(x)\,dx = N_{1,2}\delta_{st}, \qquad |x_1|, |x_2| \neq 1, x_2 > x_1$$

where the orthogonality of the ϕ-dependent functions insures that the order r is the same on both associated Legendre functions in the integrand. The usual boundary conditions on the $L_s^r(x)$ are that $L_s^r(x_1) = L_s^r(x_2) = 0$ or that $dL_s^r(x_1)/dx_1 = dL_s^r(x_2)/dx_2 = 0$. If we consider the first boundary condition, it can be shown that the norm[27] N_1 is

$$N_1 = \frac{1}{2s+1}\left[(1-x^2)\frac{\partial L_s^r(x)}{\partial x}\frac{\partial L_s^r(x)}{\partial s}\right]_{x_1}^{x_2}, \qquad \text{when } L_s^r(x_1) = L_s^r(x_2) = 0$$

If we consider the same orthogonality condition for the other boundary condition, we find that the norm N_2 is

$$N_2 = -\frac{1}{2s+1}\left[(1-x^2)L_s^r(x)\frac{\partial^2 L_s^r(x)}{\partial x\,\partial s}\right]_{x_1}^{x_2}$$

[27]*Ibid.*, p. 156.

when

$$\frac{dL_s^r(x)}{dx}\bigg]_{x_1} = \frac{dL_s^r(x)}{dx}\bigg]_{x_2} = 0$$

Another orthogonality relation that is useful is of the form[28]

$$\int_{x_1}^{x_2}\left[(1-x^2)\frac{dL_s^r(x)}{dx}\frac{dL_t^r(x)}{dx} + \frac{r^2}{1-x^2}L_s^r(x)L_t^r(x)\right]dx = s(s+1)N_{1,2}\delta_{st}$$

It should be noted that the expressions for the norms above are valid when $x_1 = -1$ provided $L_s^r(x) = P_s^r(-x)$ and are valid when $x_2 = 1$ provided $L_s^r(x) = P_s^r(x)$. For completeness let us include the expression for the Wronskian of $P_s^r(x)$ and $Q_s^r(x)$, thus

$$W[P_s^r(x), Q_s^r(x)] \equiv P_s^r(x)\frac{dQ_s^r(x)}{dx} - \frac{dP_s^r(x)}{dx}Q_s^r(x)$$

$$= \frac{e^{j\pi r}2^{2r}\Gamma\left(1 + \dfrac{r+s}{2}\right)\Gamma\left(\dfrac{1+r+s}{2}\right)}{(1-x^2)\Gamma\left(1 + \dfrac{s-r}{2}\right)\Gamma\left(\dfrac{1+s-r}{2}\right)}$$

It is also possible to define $P_s^r(\pm x)$ in terms of a hypergeometric function[29] in cases where one needs explicit values for the functions.

If one desires the orthogonality integrals in terms of the angle θ rather than $x = \cos\theta$, it is sufficient to note that

$$\frac{\partial}{\partial x} = -\frac{1}{\sin\theta}\frac{\partial}{\partial\theta}$$

whence the orthogonality integrals become

$$\int_{\theta_1}^{\theta_2} L_s^r(\cos\theta)L_t^r(\cos\theta)\sin\theta\,d\theta = N_{1,2}\delta_{st}, \qquad \theta_2 > \theta_1, \theta_2 \neq \pi, \text{ and } \theta_1 \neq \pi$$

$$\int_{\theta_1}^{\theta_2}\left[\frac{dL_s^r(\cos\theta)}{d\theta}\frac{dL_t^r(\cos\theta)}{d\theta} + \frac{r^2}{\sin^2\theta}L_s^r(\cos\theta)L_t^r(\cos\theta)\right]\sin\theta\,d\theta$$

$$= s(s+1)N_{1,2}\delta_{st}$$

where

$$N_1 = -\frac{1}{2s+1}\left[\sin\theta\frac{\partial L_s^r(\cos\theta)}{\partial\theta}\frac{\partial L_s^r(\cos\theta)}{\partial s}\right]_{\theta_1}^{\theta_2}$$

[28]L. Bailin and S. Silver, "Exterior Electromagnetic Boundary Value Problems for Spheres and Cones," *Trans. IRE*, Vol. AP-4, pp. 516, January 1956.

[29]A. Erdélyi, W. Magnus, F. Oberhettinger, and F. Tricomi, *Higher Transcendental Functions*, Vol. I, McGraw-Hill Book Company, Inc., New York, 1955.

and

$$N_2 = \frac{1}{2s+1}\left[\sin\theta\, L_s^r(\cos\theta)\frac{\partial^2 L_s^r(\cos\theta)}{\partial\theta\,\partial s}\right]_{\theta_1}^{\theta_2}$$

A summary of relations involving Legendre functions is presented in Appendix C.4. Note that θ_1 and θ_2 above are interchanged from the θ_1 and θ_2 in Appendix C.4.10.

1.8 Spherical Bessel Functions

The spherical Bessel functions are derivable from the cylindrical Bessel functions described in Sec. 1.4. If we denote the spherical Bessel functions by $b_n(\alpha x)$, we find that

$$b_n(\alpha x) = \sqrt{\frac{\pi}{2\alpha x}}\, B_{n+(1/2)}(\alpha x) = (\alpha x)^{-1}\hat{B}_n(\alpha x)$$

with the accompanying ordinary differential equations

$$\frac{d}{dx}\left[x^2\frac{d}{dx}b_n(\alpha x)\right] + [(\alpha x)^2 - n(n+1)]b_n(\alpha x) = 0 \qquad (1\text{-}35)$$

and

$$\left[\frac{d^2}{dx^2} + \alpha^2 - \frac{n(n+1)}{x^2}\right]\hat{B}_n(\alpha x) = 0$$

It is also possible to define the spherical Bessel function of the first kind in terms of a generating function[30] just as we did for the cylindrical Bessel function in Sec. 1.4. The generating function for the spherical Bessel function of the first kind of order n is

$$e^{\pm jx\cos\theta} = \sum_{n=0}^{\infty}(2n+1)(\pm j)^n j_n(x)P_n(\cos\theta) \qquad (1\text{-}36)$$

By invoking the orthogonality of the Legendre functions in Sec. 1.5, one obtains an integral representation for the $j_n(x)$:

$$j_n(x) = \frac{(\pm j)^{-n}}{2}\int_0^{\pi} e^{\pm jx\cos\theta}P_n(\cos\theta)\sin\theta\,d\theta \qquad (1\text{-}37)$$

The functions $\hat{B}_n(\alpha x)$ are a modified form of the spherical Bessel functions that were introduced by Debye and used considerably by Schelkunoff. Hereafter, we will call them Debye-Schelkunoff functions.

---

[30]Sommerfeld, *Partial Differential Equations in Physics*, pp. 143–44.

It is interesting to note that the spherical Bessel functions for integer n are simpler in form than the cylindrical Bessel functions. In particular, they can be written in closed forms, e.g.,

$$j_0(\alpha x) = \frac{\sin \alpha x}{\alpha x}, \qquad h_0^{(1)}(\alpha x) = \frac{e^{j\alpha x}}{j\alpha x}$$

$$n_0(\alpha x) = -\frac{\cos \alpha x}{\alpha x}, \qquad h_0^{(2)}(\alpha x) = h_0^{(1)*}(\alpha x)$$

$$\hat{J}_0(\alpha x) = \sin \alpha x, \qquad \hat{H}_0^{(1)}(\alpha x) = -je^{j\alpha x}$$

$$\hat{N}_0(\alpha x) = -\cos \alpha x, \qquad \hat{H}_0^{(2)}(\alpha x) = \hat{H}_0^{(1)*}(\alpha x)$$

The recurrence relations can be obtained from (1-23) and become

$$\frac{(2n+1)}{x} b_n(x) = b_{n-1}(x) + b_{n+1}(x), \qquad \frac{(2n+1)}{x} \hat{B}_n(x) = \hat{B}_{n-1}(x) + \hat{B}_{n+1}(x)$$

$$(2n+1)b_n'(x) = nb_{n-1}(x) - (n+1)b_{n+1}(x),$$

$$(2n+1)\hat{B}_n'(x) = (n+1)\hat{B}_{n-1}(x) - n\hat{B}_{n+1}(x)$$

and the orthogonality integral obtained from (1-24) becomes

$$\int_a^b x^2 b_n(\alpha x)b_m(\alpha x)\, dx = \delta_{mn} N, \qquad \int_a^b \hat{B}_n(\alpha x)\hat{B}_m(\alpha x)\, dx = \delta_{mn}\hat{N}$$

where

$$N = \left[\frac{x^3}{2}\{b_n^2(\alpha x) - b_{n-1}(\alpha x)\, b_{n+1}(\alpha x)\}\right]_a^b$$

and

$$\hat{N} = \left[\frac{x}{2}\{\hat{B}_n^2(\alpha x) - \hat{B}_{n-1}(\alpha x)\hat{B}_{n+1}(\alpha x)\}\right]_a^b$$

or

$$\hat{N} = \left[\frac{x}{2}\left\{\hat{B}_n'^2(\alpha x) + \left[1 - \frac{n(n+1)}{\alpha^2 x^2}\right]\hat{B}_n^2(\alpha x) - \frac{\hat{B}_n(\alpha x)\hat{B}_n'(\alpha x)}{\alpha x}\right\}\right]_a^b$$

The Wronskian of the spherical Bessel functions is obtained from (1-27) and is

$$j_n(x)n_n'(x) - j_n'(x)n_n(x) = \frac{1}{x^2}, \qquad j_n(x)h_n^{(2)'}(x) - j_n'(x)h_n^{(2)}(x) = -\frac{j}{x^2} \qquad (1\text{-}38)$$

The Wronskian of other spherical Bessel and Hankel functions can be obtained from (1-38) by using the relations $h_n^{(1)}(x) = j_n(x) + jn_n(x)$ and

$h_n^{(2)}(x) = h_n^{(1)*}(x)$, etc. The Wronskian of several combinations of the Schel-kunoff functions is

$$\hat{J}_n(x)\,\hat{N}'_n(x) - \hat{J}'_n(x)\hat{N}_n(x) = \hat{H}_n^{(2)}(x)\hat{N}'_n(x) - \hat{H}_n^{(2)'}(x)\hat{N}_n(x) = 1$$

$$\hat{J}_n(x)\,\hat{H}_n^{(2)'}(x) - \hat{J}'_n(x)\hat{H}_n^{(2)}(x) = -j$$

Several other relations that are often useful are the following:

$$j_n(-x) = (-1)^n j_n(x), \qquad h_n^{(1)}(-x) = (-1)^n h_n^{(2)}(x)$$

$$n_n(-x) = (-1)^{n+1} n_n(x), \qquad h_n^{(2)}(-x) = (-1)^n h_n^{(1)}(x)$$

$$\hat{J}_n(-x) = (-1)^{n+1} \hat{J}_n(x), \qquad \hat{H}_n^{(1)}(-x) = (-1)^{n+1} \hat{H}_n^{(2)}(x)$$

$$\hat{N}_n(-x) = (-1)^n \hat{N}_n(x), \qquad \hat{H}_n^{(2)}(-x) = (-1)^{n+1} \hat{H}_n^{(1)}(x)$$

Additional relations may be found in Appendix C.3.

When the parameter α in (1-35) vanishes, the equation reduces to

$$\left[\frac{d}{dx} \left(x^2 \frac{d}{dx} \right) - n(n+1) \right] f(x) = 0 \qquad (1\text{-}39)$$

with two linearly independent solutions

$$f(x) = x^n$$
$$f(x) = x^{-(n+1)} \qquad (1\text{-}40)$$

and the Wronskian

$$W[x^n, x^{-(n+1)}] = -(2n+1)x^{-2}$$

The equation (1-39) occurs later in Chap. 5 in connection with Laplace's equation in spherical coordinates, in which case the independent variable becomes r.

1.9 Mathieu Functions

The Mathieu functions satisfy the Mathieu equation[31]

$$y'' + (a + 16b \cos 2x)y = 0 \qquad (1\text{-}41)$$

where a and b are real constants. These functions appear, for instance, when one considers the wave equation in elliptic coordinates. In this case the independent variable x represents an angle and the dependent variable y becomes an angular function,[32] i.e., a Mathieu function. We will not consider

[31]Margenau and Murphy, *The Mathematics of Physics and Chemistry*, p. 78.

[32]Stratton, *Electromagnetic Theory*, p. 375.

the wave equation in elliptic coordinates in this text but will find that the Mathieu equation appears when the wave equation is separated in rectangular coordinates in a medium that is inhomogeneous in a particular way. In this connection, it is more convenient to rewrite Mathieu's equation in the form[33]

$$y'' + (a - 2h^2 \cos 2x)y = 0 \tag{1-42}$$

Although the above equation can be solved by the usual power series method (also called method of Frobenius[34,35] and method of integration in series[36]), the solution is not valid when $x = n\pi$, n an integer. Since we are interested in the more general solution, it is necessary to solve the equation in a different manner. In particular, the solutions that are of interest are often periodic in x, i.e., $y(x) = y(x + n\pi)$, where n is an integer and π is the period. However, one obtains this periodicity only when the constant a is allowed certain values. When the constant a is different from these allowed values, then the solution is no longer periodic, i.e., $y(x) = ky(x + n\pi)$ where k can be a complex constant. However, this is just a statement of Floquet's theorem,[37,38,39] which finds many applications in problems involving propagation in periodic structures. Thus, we assume a solution to (1-42) in the form

$$y(x) = Ae^{j\nu x}P(x) + Be^{-j\nu x}P(-x) \tag{1-43}$$

where the term ν is called the characteristic exponent and in general is complex and $P(\pm x)$ is a periodic function that is periodic in x with a periodicity of π. One sees from (1-43) that when ν is a real integer the functions $y(x)$ are periodic and that when ν is complex, pure imaginary, or noninteger the functions $y(x)$ are not periodic. When ν is a real integer, $P(x)$ and $P(-x)$ are not linearly independent, so that (1-43) is not the general solution. Since the function P is periodic in any case, it is customary to expand it in an exponential Fourier series, i.e.,

$$P(x) = \sum_{n=-\infty}^{\infty} a_n e^{j2nx} \tag{1-44}$$

[33]T. Tamir and H. Wang, "Characteristic Relations for Nonperiodic Solutions of Mathieu's Equation," *J. Res. NBS*, Vol. 69B, pp. 101–19, January–June 1965.

[34]Reddick and Miller, *Advanced Mathematics for Engineers*, 2nd ed., p. 177.

[35]J. Dettman, *Mathematical Methods in Physics and Engineering*, McGraw-Hill Book Company, Inc., New York, 1962, p. 165.

[36]Margenau and Murphy, *The Mathematics of Physics and Chemistry*, pp. 59–88.

[37]*Ibid.*, p. 80.

[38]D. Watkins, *Topics in Electromagnetic Theory*, John Wiley & Sons, Inc., New York, 1958, p. 2.

[39]J. Slater, *Microwave Electronics*, D. Van Nostrand Company, Inc., Princeton, N.J., 1950, p. 170.

We can now combine (1-43) and (1-44) and state that the general solution to (1-42) is of the form

$$y(x) = e^{\pm jvx} \sum_{n=-\infty}^{\infty} a_n^{\pm} e^{j2nx} \qquad (1\text{-}45)$$

where the $\pm$ signs relate to the two linearly independent solutions. When v is a real integer in (1-45), it is customary to consider the even and odd periodic solutions to (1-42). These solutions are then called Mathieu functions.[40] If we substitute (1-45) into (1-42) we obtain an expression for the coefficients

$$a_{n+1} + D_n a_n + a_{n-1} = 0, \qquad \text{where } D_n = \frac{(v + 2n)^2 - a}{h^2} \qquad (1\text{-}46)$$

We see immediately that the problem of finding the coefficients a_n and the characteristic exponent v in terms of a and h (or a in terms of v and h) is not so simple as that encountered before. However, the general problem was solved by Hill[41] in connection with his attack on the equation which now bears his name, i.e.,

$$y'' + \left(b_0 + 2 \sum_{n=1}^{\infty} b_n \cos 2nx \right) y = 0 \qquad (1\text{-}47)$$

If we let $n = 1$ in Hill's equation, (1-47), we obtain Mathieu's equation as a special case. The process of solving (1-46) will not be considered here although the concept is simple. Since n takes on all positive and negative integer values in (1-46), we obtain an infinite set of homogeneous equations for the a_n. The determinant of the coefficients of the a_n must vanish in order for a solution to exist. When this procedure is performed, one obtains an equation for v as a function of a and h (or a as a function of v and h). This relation is often presented in a graphical form[42,43] by a "stability chart" as shown in Fig. 1-2. When v is real the solutions in the form of (1-45) are termed "stable" since $y(x)$ is bounded for all values of x. On the other hand, when v becomes complex, the solutions are not bounded for large values of x, whence the term "unstable." In order to consider the general case of complex v, let us define v as

$$v = m + j\alpha$$

[40]E. Whittaker and G. Watson, *A Course in Modern Analysis*, 4th ed., Cambridge University Press, London, 1940, Chap. 19.

[41]*Ibid.*, Secs. 19.12, 19.41, 19.42.

[42]Tamir and Wang, "Characteristic Relations for Nonperiodic Solutions of Mathieu's Equation."

[43]T. Tamir, "Characteristic Exponents of Mathieu Functions," *Math. of Computations*, Vol. 16, p. 100, January 1962.

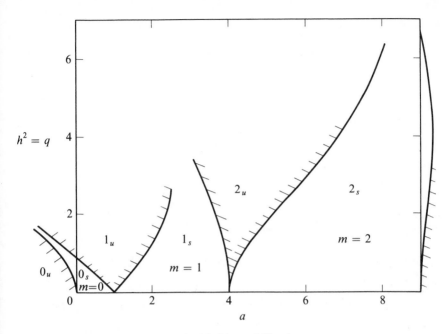

Fig. 1.2 Mathieu stability chart.

where α is real and $m = 0, 1, 2, \ldots$. When α vanishes, v is a real integer and we obtain a stable region on the chart for each value of m. The lines $v = m$ consist of two separate branches (except for $m = 0$) which intersect on the a-axis and are separated by a region of complex v. This is illustrated in Fig. 1-2 which shows the three stable regions corresponding to the values of $m = 0, 1, 2$. The unstable regions defined by $m = 0, 1, 2$ lie between the stable regions and just to the left of the one with the corresponding value of m. In Fig. 1-2, the unstable regions contain the crosshatched lines and are identified by the subscript u while the stable regions are identified by the subscript s, e.g., 2_u or 2_s.

PROBLEMS

1.1. Verify the steps leading to (1-4).

1.2. Derive (1-10) from (1-8) and (1-9).

1.3. Identify p, q, w, λ, and the orthogonality integral for the following:
 (a) Vibrating string of length L, both ends fixed.
 (b) Mathieu equation, $y'' + (a + 16b \cos 2x)y = 0$, a and b constants.

(c) Associated Legendre equation, $(1 - x^2)y'' - 2xy' + [n(n + 1) - m^2/(1 - x^2)]y = 0$.

(d) Bessel's equation, (1-18) or (2-42).

(e) Spherical Bessel's equation, (1-35).

1.4. Demonstrate that $J_n(x)$ in (1-17) satisfies (1-18).

1.5. Show that:

    (a) $J_n(x) = (-1)^n J_{-n}(x)$

    (b) $J_n(-x) = (-1)^n J_n(x)$

1.6. Show that

    (a) $e^{\pm jx \sin \phi} = \sum\limits_{n=-\infty}^{\infty} (\pm 1)^n J_n(x) e^{jn\phi}$

    (b) $e^{\pm jx \cos \phi} = \sum\limits_{n=-\infty}^{\infty} (\pm j)^n J_n(x) e^{jn\phi}$

1.7. Use orthogonality to derive two expressions for $J_n(x)$ from those in Prob. 1.6.

1.8. Combine the two recurrence relations in (1-23) to obtain the following two expressions:

$$\frac{d}{dx}[x^n B_n(x)] = x^n B_{n-1}(x)$$

$$\frac{d}{dx}[x^{-n} B_n(x)] = -x^{-n} B_{n+1}(x)$$, $\quad n \neq s$

1.9. Show that the expansion of the left side of (1-17) results in the following series representation:

$$J_n(x) = \left(\frac{x}{2}\right)^n \sum\limits_{m=0}^{\infty} \frac{(-1)^m}{m!(m + n)!} \left(\frac{x}{2}\right)^{2m}$$

This is the solution one obtains if (1-18) is solved by the method of Frobenius. *Note:* For noninteger values of n, replace $m!$ by $\Gamma(m + 1)$, etc.

1.10. Use the generating function for $J_n(x)$, (1-17), to obtain the recurrence relations in (1-23).

1.11. Verify the expressions for Φ_P in (1-30).

1.12. Show that (1-30) reduces to the common expression for the potential of a dipole when $Q_1 = -Q_2$ and $r \gg d$.

1.13. Use (1-28) to obtain the recurrence relations in (1-31).

1.14. Use (1-28) to obtain the following relation:

$$\frac{1}{R} = \frac{1}{r_0} \sum_{n=0}^{\infty} \left(\frac{r}{r_0}\right)^n P_n(\cos \xi), \qquad r < r_0$$

$$\frac{1}{R} = \frac{1}{r} \sum_{n=0}^{\infty} \left(\frac{r_0}{r}\right)^n P_n(\cos \xi), \qquad r > r_0$$

where $\mathbf{R} = \mathbf{r} - \mathbf{r}_0$ and ξ is the angle between $\mathbf{r}$ and $\mathbf{r}_0$, i.e., $R^2 = r^2 + r_0^2 - 2rr_0 \cos \xi$.

1.15. Show that the recurrence relation involving derivatives in (1-31) can be expressed as

$$(2n + 1)P_n(x) = P'_{n+1}(x) - P'_{n-1}(x)$$

2

Green's Functions

2.1 Introduction to Green's Functions and Dirac's Delta

The material that has been presented thus far is concerned with the viewpoint that arbitrary functions can be expressed in terms of sets of orthogonal functions. The boundary conditions determine the eigenvalues of the eigenfunctions (orthogonal functions) and the strengths of the sources determine the coefficients of the eigenfunctions. Another way of solving the same problem is to represent the solution directly in terms of its sources, or source-wise. In this case, a solution is first found for a unit source and then the general solution is written as a superposition of the effects of various sources at various locations. The strength of a general source at a particular point merely increases or decreases the effect of a unit source at that same point. The solution to the equation with a unit source is called a Green's function. The problem of finding the Green's function is the same problem that we discussed in the earlier sections. The only difference is that instead of finding the orthogonal function expansion for a general source, we find the orthogonal function expansion for a unit source. Obviously, this procedure involves the same amount of work and represents no advantage unless the same problem is to be solved many times for a variety of sources. Before we proceed further, let us consider the concept of a unit point source. A point mass or a point charge is a unit source although the terminology is not often used. Physically, it means that the product of density and volume is finite as the volume shrinks to that of a point. Mathematically, this means that the density becomes infinite as the volume becomes zero. Since their product is a constant, we may arbitrarily choose that constant to be one. Thus, we have a unit point source. The density function is called a delta function. Since it is not a well behaved mathematical function so much as a symbolic function, let us refer to it as Dirac's delta. (The other symbolic functions we have already used are Kronecker's delta and Neumann's number.) We shall define the Dirac delta as follows:

$$\int \delta(\mathbf{r}_i - \mathbf{r}_j)\, dV_i = \begin{cases} 1, & \mathbf{r}_j \text{ in } V_i \\ 0, & \mathbf{r}_j \text{ not in } V_i \end{cases}$$

where the zero result is included to take into account the fact that our density function or Dirac delta produces a contribution to the integral only when we include the source in our region of integration. We can extend the idea of a unit source to a general source to obtain

$$\int f(\mathbf{r})\delta(\mathbf{r} - \mathbf{r}_0)\, dV = \begin{cases} f(\mathbf{r}_0), & \mathbf{r}_0 \text{ in } V \\ 0, & \mathbf{r}_0 \text{ not in } V \end{cases}$$

The fact that $f(\mathbf{r})$ can be different at all points in V is irrelevant since $\delta(\mathbf{r} - \mathbf{r}_0)$ is zero everywhere except at $\mathbf{r} = \mathbf{r}_0$. Thus, one often speaks of the Dirac delta as possessing a "sifting or sampling" property. Further details concerning the Dirac delta and its representation in different coordinate systems are given in Appendix B.

Mathematically, we can introduce the Dirac delta in terms of the derivative of the unit step function (Heaviside function) shown in Fig. 2-1. Thus,

$$\delta(x - x_0) = \frac{d}{dx} u(x - x_0)$$

where $u(x - x_0)$ is the unit step defined as

$$u(x - x_0) = \begin{cases} 1, & x > x_0 \\ 0, & x < x_0 \end{cases}$$

If we integrate the Dirac delta we obtain

$$\int_a^b \delta(x - x_0)\, dx = \int_a^b \frac{d}{dx} u(x - x_0)\, dx = u(x - x_0)\Big]_a^b = \begin{cases} 1, & x_0 \text{ in } (a, b) \\ 0, & x_0 \text{ not in } (a, b) \end{cases}$$

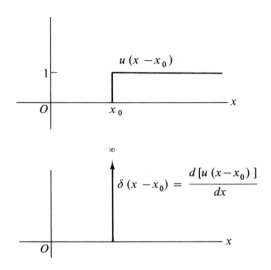

Fig. 2-1 Illustration of the unit step (Heaviside function) and its derivative, the Dirac delta.

since

$$u(x - x_0)\Big]_a^b = \operatorname*{Lim}_{\epsilon \to 0} u(x - x_0)\Big]_{x_0-\epsilon}^{x_0+\epsilon}$$

Let us apply these ideas to the differential equation of a string of length L excited by a unit source at $x = x_0$ where Γ is the displacement of the string

$$\left(\frac{d^2}{dx^2} + k^2\right)\Gamma = -\delta(x - x_0) \tag{2-1}$$

The homogeneous form of this equation is obtained from (1-8) by Fourier transforming the function $u(x, t)$ with respect to t or by assuming some harmonic time dependence and then canceling out the function of time. Both processes are essentially the same and give us the result above where $k = \omega\sqrt{m/T}$ and Γ is a function of x only. If we integrate the equation over the interval $(0, L)$ we obtain

$$\int_0^L (\Gamma'' + k^2\Gamma)\, dx = -\int_0^L \delta(x - x_0)\, dx = -u(x - x_0)\Big]_0^L$$

Since we do not have a prescription for evaluating the left-hand side of the equation as we do for the right-hand side, let us divide the interval $(0, L)$

into two parts[1] as follows:

$$\int_0^L = \lim_{\epsilon \to 0} \int_0^{x_0 - \epsilon} + \int_{x_0 + \epsilon}^L$$

which is the customary procedure for evaluating an improper integral. In this example, the integral is improper because the integrand becomes infinite at $x = x_0$ and the above limiting procedure defines Cauchy's principal value of the integral. Each integrand of the integrals over the intervals $(0, x_0 - \epsilon)$ and $(x_0 + \epsilon, L)$ is well behaved so we can now integrate to obtain information about Γ'. Unfortunately, the integrands of the integrals of both the left-hand side and the right-hand side of the above expression are zero in the two intervals $(0, x_0 - \epsilon)$ and $(x_0 + \epsilon, L)$. By definition the Dirac delta is discontinuous only in the vicinity of $x = x_0$ so there is no need to integrate over any region but the interval $(x_0 - \epsilon, x_0 + \epsilon)$. In other words, an inspection of our original differential equation gives us the following information: First, Γ must be continuous in the entire interval $(0, L)$ since Γ represents the displacement of an intact string. Second, $\Gamma'' = -k^2\Gamma$ everywhere in the interval $(0, L)$ *except* in the vicinity of the source at $x = x_0$. Since Γ is continuous everywhere in the entire interval $(0, L)$, then Γ'' is also continuous everywhere *except* in the vicinity of the source at $x = x_0$ where it becomes discontinuous as a Dirac delta. When we apply this information to our integral over the interval $(x_0 - \epsilon, x_0 + \epsilon)$, we obtain

$$\lim_{\epsilon \to 0} \Gamma' \Big]_{x_0 - \epsilon}^{x_0 + \epsilon} = -1$$

Thus, we find that in the vicinity of the source, Γ' is discontinuous by the strength of the source.

We are following Webster's notation[2] here by letting Γ denote the Green's function for k finite and K denote the Green's function for $k = 0$. Let us illustrate these ideas by considering as an example the equation

$$\frac{d^2K}{dx^2} = -\delta(x - x_0) \tag{2-2}$$

and the boundary conditions $K(0) = K(L) = 0$. Unfortunately, we do not know how to solve this equation on the basis of a course in ordinary differential equations. However, we know that the unit source causes K' to be

[1] A. Webster, *Partial Differential Equations of Mathematical Physics*, 2nd ed., edited by S. Plimpton, Hafner Publishing Company, Inc., New York, 1947, p. 109.

[2] *Ibid.*, pp. 109, 113.

discontinuous by minus one at the source. Thus, let us solve the equation

$$\frac{d^2K}{dx^2} = 0, \qquad x \neq x_0 \tag{2-3}$$

with the boundary conditions $K(0) = K(L) = 0$, and take into account the unit source by adding two new conditions, namely:
K continuous in $(0, L)$

$$\lim_{\epsilon \to 0} K' \Big]_{x_0 - \epsilon}^{x_0 + \epsilon} = -1 \tag{2-4}$$

The equation (2-3) is valid in the two regions on both sides of $x = x_0$, so we have two solutions

$$K_1 = A_1 x + B_1, \qquad x < x_0$$
$$K_2 = A_2 x + B_2, \qquad x_0 < x$$

The four (boundary) conditions give us four equations for the four unknown coefficients. Thus,

$$B_1 = 0$$
$$B_2 = x_0$$
$$A_1 = 1 - \frac{x_0}{L}$$
$$A_2 = \frac{-x_0}{L}$$

and

$$K_1 = \frac{x(L - x_0)}{L}$$
$$K_2 = \frac{x_0(L - x)}{L}$$

The complete solution is

$$K(x, x_0) = K_1(x, x_0)u(x_0 - x) + K_2(x, x_0)u(x - x_0) \tag{2-5}$$

We should also note that we are forced to divide the interval $(0, L)$ into two regions, one on each side of the source at $x = x_0$, in order to obtain a nontrivial solution for K. Later on we will find that it is always possible to express K in a Fourier-type series of normal modes, i.e., an orthogonal function expansion. One sees that this viewpoint is the same as our earlier viewpoint that any arbitrary function can be expanded in a set of orthogonal functions and that the Sturm-Liouville equation gives the general characteristics of the orthogonal functions. We are thus led to the conclusion that there are at least two alternate representations for the solution to (2-2). We will see later that this statement can be generalized to include the solutions of the Sturm-Liouville equation, (1-6).

An illustration of the solution to (2-2) as given by (2-5) is shown in Fig. 2-2. In this case, the unit source is at $x_0 = 2L/3$. One sees immediately that K is continuous throughout the interval $(0, L)$ and that K' and K'' are continuous throughout the same interval *except* in the vicinity of the source $(x = x_0)$.

In general we wish to solve (2-2) for an arbitrary source function, thus,

$$\frac{d^2y}{dx^2} = -f(x) \tag{2-6}$$

where $y(x)$ is the displacement of the string. The function we just discussed, $K(x, x_0)$, is the Green's function for this specific problem and satisfies the

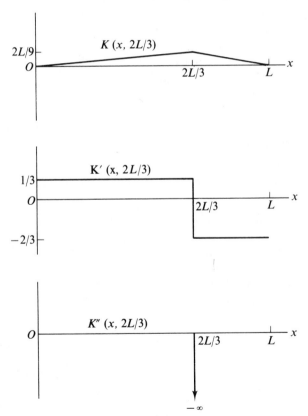

Fig. 2-2 Plots of K, K', and K'' for the equation $K'' = -\delta(x - 2L/3)$ and the boundary conditions $K(0) = K(L) = 0$.

same boundary conditions as y in (2-6), as well as either (2-2) or (2-3) and (2-4). Since K is a solution of (2-2) for a Dirac delta source of unit strength, it follows that the solution of (2-2) for a Dirac delta source of double strength would be $2K$. From this line of reasoning it follows that the solution of (2-2) for a Dirac delta source of strength $f(x_0)$ would be $f(x_0)K$. However, $f(x_0)$ is just the value of $f(x)$ in (2-6) at $x = x_0$ so the general solution of (2-6) is

$$y(x) = \int_0^L f(x_0)K(x, x_0)\, dx_0 + y_c \qquad (2\text{-}7)$$

where y_c is the usual complementary solution. The first term in (2-7) is the usual particular solution or particular integral.[3] The complementary solution to an ordinary differential equation results when there is no forcing or excitation function (right-hand side is zero) and the particular solution results from the presence of an excitation function. Although a nonhomogeneous differential equation, such as (2-6), always possesses a complementary solution, we will usually omit it as part of the general solution since it is not pertinent to our problems. For example, the particular solution to (2-1) is (2-11) and we have purposely omitted the complementary solution. This physical derivation of the solution to (2-6) indicates that $f(x)$ must have a finite value at every point in the interval $(0, L)$. Mathematically, one finds that $f(x)$ again must be a piecewise continuous function. We can also derive (2-7) from (2-2) and (2-6) by multiplying (2-2) by y and (2-6) by K, subtracting the two, and integrating both sides to obtain

$$\int_0^L (yK' - y'K)'\, dx = \int_0^L [-y\delta(x - x_0) + Kf]\, dx \qquad (2\text{-}8)$$

Since the left-hand side vanishes at both limits because of the boundary conditions, the right-hand side reduces to (2-7). In this case, the discontinuous nature of K' does not have to be taken into account explicitly since the Dirac delta on the right-hand side of (2-2) does this. On the other hand, we can use (2-3), (2-4), and (2-6) to arrive at (2-7). In this case, (2-4) takes into account the discontinuous nature of the source. This latter approach gives

$$\int_0^L (yK' - y'K)'\, dx = \int_0^L Kf\, dx \qquad (2\text{-}9)$$

Now, the left-hand side must be broken into two parts and (2-4) used to obtain (2-7). These two alternate approaches do not give (2-7) directly, but

---

[3]H. Margenau and G. Murphy, *The Mathematics of Physics and Chemistry*, D. Van Nostrand Company, Inc., Princeton, N.J., 1943, p. 53.

(2-7) with x replaced by x_0. This result is not the same unless

$$K(x, x_0) = K(x_0, x) \tag{2-10}$$

This relation expresses the symmetry of the Green's function and, physically, means that reciprocity is obeyed. The symmetry relation, (2-10), can be derived from (2-2) and (2-4) by considering two different unit sources at $x = x_1$ and $x = x_2$ and then using the approach indicated above in the derivation of (2-8) or (2-9). Symmetry is a very important property of Green's functions and applies to both K and Γ.

As another example, let us return again to the equation

$$\frac{d^2\Gamma}{dx^2} + k^2\Gamma = -\delta(x - x_0) \tag{2-1}$$

and the boundary conditions $\Gamma(0) = \Gamma(L) = 0$. As mentioned before, this equation describes the motion of a uniform string under the influence of a time periodic unit source. The boundary conditions on the Green's function, $\Gamma(x, x_0)$, are the same as those on $K(x, x_0)$ in (2-4). The solution to (2-1) is

$$\Gamma(x, x_0) = \Gamma_1(x, x_0)u(x_0 - x) + \Gamma_2(x, x_0)u(x - x_0) \tag{2-11}$$

where

$$\Gamma_1(x, x_0) = \frac{1}{k}\frac{\sin k(L - x_0)\sin kx}{\sin kL}, \qquad x < x_0$$

$$\Gamma_2(x, x_0) = \frac{1}{k}\frac{\sin k(L - x)\sin kx_0}{\sin kL}, \qquad x > x_0$$

Here, too, $\Gamma(x, x_0) = \Gamma(x_0, x)$, which expresses the symmetry of the Green's function as noted before in (2-10) for $K(x, x_0)$.

If we consider (2-11) for a general source function, $f(x)$, we find that the equation of motion becomes

$$\frac{d^2y}{dx^2} + k^2y = -f(x) \tag{2-12}$$

with boundary conditions $y(0) = y(L) = 0$. The solution to (2-12) now becomes, by analogy with (2-7),

$$y(x) = \int_0^L \Gamma(x, x_0) f(x_0)\, dx_0 \tag{2-13}$$

where $\Gamma(x, x_0)$ is given in (2-11). As mentioned before, we are omitting the complementary solutions as not being pertinent to our problems.

2.2   Green's Functions for Sturm-Liouville Equation

The preceding examples have indicated how we obtain Green's functions for several specific equations and their boundary conditions. Since we have already said that the Sturm-Liouville equation is the most general equation necessary to describe many physical systems, let us consider its Green's function. First, however, we consider the Green's function for $L(u)$ in (1-7). Thus, our inhomogeneous (nonhomogeneous) equation is

$$L(K) = -\delta(x - x_0) \qquad (2\text{-}14)$$

If we integrate both sides over an interval containing x_0, we find that

$$\underset{\epsilon \to 0}{\text{Lim}} \, K' \Big]_{x_0 - \epsilon}^{x_0 + \epsilon} = \frac{-1}{p(x_0)} \qquad (2\text{-}15)$$

As before, the solution of the equation

$$L(u) = -f(x) \qquad (2\text{-}16)$$

becomes

$$u(x) = \int_a^b K(x, x_0) f(x_0) \, dx_0 \qquad (2\text{-}17)$$

The expression (2-17) can be verified by showing that it satisfies (2-16) by performing the indicated operations, or (2-17) can be derived from (2-14) and (2-16) with a procedure analogous to that leading to the derivation of (2-8).

The Green's function for the general homogeneous Sturm-Liouville equation, (1-6), can be obtained from the differential equation

$$L(\Gamma) + \lambda w \Gamma = -\delta(x - x_0) \qquad (2\text{-}18)$$

with the two additional boundary conditions that Γ is continuous at $x = x_0$ and

$$\underset{\epsilon \to 0}{\text{Lim}} \, \Gamma' \Big]_{x_0 - \epsilon}^{x_0 + \epsilon} = \frac{-1}{p(x_0)} \qquad (2\text{-}19)$$

Thus, the solution of the general inhomogeneous Sturm-Liouville equation

$$L(v) + \lambda w v = -g(x) \qquad (2\text{-}20)$$

can be written in terms of the Green's function Γ as

$$v(x) = \int_a^b \Gamma(x, x_0) g(x_0) \, dx_0 \qquad (2\text{-}21)$$

As before, it can be shown that (2-21) is a solution of (2-20) either by performing the indicated operations on v or by obtaining (2-21) directly from (2-18), (2-19), and (2-20) by the proper manipulations. However, we can also find an expression for v in (2-21) in terms of $K(x, x_0)$ instead of $\Gamma(x, x_0)$. Thus, if we let $f(x) = \lambda wv + g(x)$ in (2-16), we find that (2-17) becomes

$$v(x) = \lambda \int_a^b K(x, x_0)w(x_0)v(x_0)\,dx_0 + \int_a^b K(x, x_0)g(x_0)\,dx_0 \qquad (2\text{-}22)$$

which must be a solution of (2-20) with its accompanying boundary conditions. The expression (2-22) can also be found directly by manipulating (2-14) and (2-20) and then using (2-15) after integrating both sides. Since $v(x)$ can be found in terms of either $K(x, x_0)$ or $\Gamma(x, x_0)$, it seems reasonable that there must be a relation between $K(x, x_0)$ and $\Gamma(x, x_0)$. Such a relation does exist and it can be obtained by manipulating (2-14) and (2-18) and using the conditions (2-15) and (2-19). However, we must not use the same unit source at $x = x_0$, but choose two unit sources at $x = x_1$ and $x = x_2$ for $K(x, x_1)$ and $\Gamma(x, x_2)$. Thus, we obtain

$$\Gamma(x_1, x_2) - K(x_2, x_1) = \lambda \int_a^b w(x)\Gamma(x, x_2)K(x, x_1)\,dx \qquad (2\text{-}23)$$

We are now in a position to outline a general procedure for evaluating $K(x, x_0)$ in the integrand of (2-17) and $\Gamma(x, x_0)$ in the integrand of (2-21). This general procedure does not depart particularly from the method used previously in the several examples but is more systematic. This method is not the only one for finding the Green's function and applies only when the region of interest is separated into two parts. Also, this procedure is valid when the region of interest becomes infinite. However, these points will be clarified by some of the examples and homework problems. We define the following items in the procedure for finding the Green's function, $K(x, x_0)$, that satisfies (2-14) or the Green's function, $\Gamma(x, x_0)$, that satisfies (2-18).

Properties of $K(x, x_0)$ and $\Gamma(x, x_0)$

(a) It is a continuous function of x.
(b) It satisfies the same boundary conditions as u in (2-16) or v in (2-20).
(c) Both the first and second derivatives are continuous in the interval (a, b) *except* at the point $x = x_0$, where the first derivative is discontinuous so that

$$\operatorname*{Lim}_{\epsilon \to 0} K' \Big]_{x_0-\epsilon}^{x_0+\epsilon} = \frac{-1}{p(x_0)}$$

or

$$\operatorname*{Lim}_{\epsilon \to 0} \Gamma' \Big]_{x_0-\epsilon}^{x_0+\epsilon} = \frac{-1}{p(x_0)}$$

(d) Except at $x = x_0$, $K(x, x_0)$ satisfies the differential equation

$$L(K) \equiv (pK')' - qK = 0$$

and $\Gamma(x, x_0)$ satisfies the differential equation

$$L(\Gamma) + \lambda w \Gamma = 0$$

where p, q, and w can be functions of x, and λ is positive real.

Procedure for finding $K(x, x_0)$ or $\Gamma(x, x_0)$

Find two linearly independent solutions of the differential equation $L(K) = 0$ and define these as K_a and K_b or find two linearly independent solutions of the differential equation $L(\Gamma) + \lambda w \Gamma = 0$ and define these as Γ_a and Γ_b. Since the rest of this procedure applies equally to K or Γ, let us consider the results for K with the understanding that they also apply to Γ. The general solution is then the sum of these two solutions multiplied by arbitrary constants. Now divide the interval (a, b) into two portions depending upon whether x is smaller or larger than x_0 and define

$$K_1(x) = (A - a)K_a(x) + (B - b)K_b(x), \qquad a \leq x \leq x_0$$
$$K_2(x) = (A + a)K_a(x) + (B + b)K_b(x), \qquad x_0 \leq x \leq b \tag{2-24}$$

Our Green's function $K(x, x_0)$ will be either K_1 or K_2 depending upon whether $x < x_0$ or $x > x_0$, respectively. However, K_1 and K_2 are defined in such a fashion that

$$K = AK_a + BK_b \mp (aK_a + bK_b) \tag{2-25}$$

where the upper sign is used when $x < x_0$ and the lower sign when $x > x_0$. If we apply the boundary (source) conditions (a) and (c) above, we obtain

$$a = \frac{1}{2p(x_0)} K_b(x_0)[K_a(x_0)K_b'(x_0) - K_a'(x_0)K_b(x_0)]^{-1}$$
$$b = \frac{-1}{2p(x_0)} K_a(x_0)[K_a(x_0)K_b'(x_0) - K_a'(x_0)K_b(x_0)]^{-1} \tag{2-26}$$

where the prime means differentiation with respect to x. If the argument of the functions K_a and K_b is x, then the quantity in the square brackets is just their Wronskian, i.e., $W[K_a, K_b]$, which was defined in (1-16). If the argument of K_a and K_b is αx, then the quantity in the square brackets is $\alpha W[K_a, K_b]$. On the other hand, if the strength of the delta source is $D(x)$ rather than one, then a and b in (2-26) must be multiplied by $D(x_0)$. It is also better terminology to call the properties in (a) and (c) source conditions rather than boundary conditions, which we shall do henceforth. The two constants A and B in (2-25) are found by applying the boundary condition (b).

Although this general procedure is straightforward, many of the steps are quite lengthy and tedious. It is frequently easier to solve for $A - a$ and $B - b$ in (2-24) and then write down K_2 immediately since

$$K_2(x, x_0) = K_1(x_0, x)$$

This relation follows directly from (2-10). However, it is usually better to solve directly for both K_1 and K_2 since their lack of symmetry indicates at once that an error has been made. In problems where a vanishes or b becomes infinite, the boundary condition is usually replaced by a statement concerning the finiteness of the function K_1 at the origin or the behavior of the function K_2 at infinity.

On the other hand, this general procedure has some advantages, too. For instance, the constants a and b remain the same for a given choice of K_a and K_b since they depend only on the source conditions. Thus, the Green's functions for a given differential equation with a variety of boundary conditions can be found by re-evaluating the constants A and B for each set of boundary conditions.

2.3 Green's Functions for Various Boundary Conditions

We mentioned earlier in Sec. 1.2 that general boundary conditions for problems involving the Sturm-Liouville equation can be specified as in (1-13). However, when we considered several Green's function solutions to equations that were of the Sturm-Liouville form, i.e., (2-2) and its solution (2-5) and (2-1) and its solution (2-11), we assumed in both cases that the functions vanished at the ends of the interval. Let us now consider the solutions to (2-2) and (2-1) when the boundary conditions are changed. Since we defined a general procedure for finding the Green's function in Sec. 2.2, let us use it for these particular examples. According to this procedure, we must first find two linearly independent solutions, K_a and K_b, to the given homogeneous ordinary differential equation and then determine the four constants a, b, A, B. If necessary, one can always check that the Wronskian of the two solutions does not vanish in order to verify their linear independence. On the other hand, one can proceed as outlined and determine the constants a and b. If they are finite, the two solutions are satisfactory since their Wronskian appears in the denominator of (2-26). One should also note that the constants a and b in (2-26) are fixed once K_a and K_b are given whereas the remaining constants A and B change each time the boundary conditions change.

First, we consider (2-2) and the boundary conditions $K(0) = K'(L) = 0$. If we define $K_a = x$ and $K_b = 1$, then according to (2-26) the constants a and b become the following:

$$a = -\frac{1}{2}$$

$$b = \frac{x_0}{2}$$

(2-27)

The constants A and B are determined by the boundary conditions; thus, according to (2-24) or (2-25), we obtain the two equations for A and B, $B - b = 0$, $A + a = 0$, whence we find that $B = x_0/2$ and $A = \frac{1}{2}$. The solution to (2-2) with the boundary conditions $K(0) = K'(L) = 0$ now becomes

$$K_1(x, x_0) = x, \qquad 0 \le x \le x_0$$

$$K_2(x, x_0) = x_0, \qquad x_0 \le x \le L$$

(2-28)

Second, we consider (2-2) again but with the boundary conditions $K'(0) = K'(L) = 0$. The constants a and b are still given by (2-27) while the constants A and B are determined from the two equations $A - a = 0$ and $A + a = 0$. The solutions to these equations are both trivial. In fact, the only solution to (2-2) is a trivial solution. This can also be verified by inspection of Fig. 2-2 since it is not possible for K' to vanish at both ends of the interval and yet be discontinuous at the source.

Third, let us examine the solution to (2-1) with the boundary conditions $\Gamma'(0) = \Gamma(L) = 0$. We define $\Gamma_a = \cos kx$ and $\Gamma_b = \sin kx$ so that the constants a and b become

$$a = (2k)^{-1} \sin kx_0$$

$$b = -(2k)^{-1} \cos kx_0$$

(2-29)

When we impose the boundary conditions, we get two equations for the two unknowns, A and B, i.e., $B - b = 0$ and $(B + b) \sin kL + (A + a) \cos kL = 0$, with the following solutions for A and B:

$$A = a + (k \cos kL)^{-1} \sin k(L - x_0)$$

$$B = b$$

(2-30)

The solution to (2-1) with the boundary conditions $\Gamma'(0) = \Gamma(L) = 0$ becomes

$$\Gamma_1(x, x_0) = (k \cos kL)^{-1} \sin k(L - x_0) \cos kx, \qquad 0 \le x \le x_0$$

$$\Gamma_2(x, x_0) = \Gamma_1(x_0, x), \qquad\qquad\qquad x_0 \le x \le L$$

(2-31)

As a fourth example, let us consider the solution to (2-1) when the boundary conditions are the same as those for the second example, i.e., $\Gamma'(0) = \Gamma'(L) = 0$. The two equations for A and B now become $B - b = 0$ and $(B + b) \cos kL - (A + a) \sin kL = 0$, with the solutions

$$A = -(k \sin kL)^{-1} \cos k(L - x_0) + a$$
$$B = b \tag{2-32}$$

Since Γ_a and Γ_b are the same as in the third example, the constants a and b are still given by (2-29). The solution to (2-1) with the boundary conditions that $\Gamma'(0) = \Gamma'(L) = 0$ becomes

$$\Gamma_1(x, x_0) = -(k \sin kL)^{-1} \cos k(L - x_0) \cos kx, \qquad 0 \leq x \leq x_0$$
$$\Gamma_2(x, x_0) = \Gamma_1(x_0, x), \qquad\qquad\qquad\qquad x_0 \leq x \leq L \tag{2-33}$$

2.4 Alternate Forms for Green's Functions

It often happens that the Green's function for a particular ordinary differential equation with its attendant boundary conditions will be expressed in several different forms. In general, these representations do not appear to be equivalent. Moreover, in many cases, it is more difficult to demonstrate in a direct manner that two alternate representations are equivalent than it is to solve the differential equation by two different methods and then state that the two solutions must be equivalent. If the problem being solved is of an electromagnetic nature, then the uniqueness theorem states that there is only one solution that satisfies the given equation and boundary conditions. In the following chapters we will often find it expedient to take advantage of the fact that alternate solutions of a problem are equivalent so that we may indirectly derive many useful identities.

Let us now consider several examples of how one may obtain alternate solutions to a given problem. Since we already have one solution to (2-1) as given by (2-11), let us find a second solution to (2-1). We start by assuming that Γ can be represented by a Fourier series, i.e., an orthogonal function expansion. We obtain the orthogonal functions by considering the homogeneous form of (2-1)

$$\left(\frac{d^2}{dx^2} + \lambda_m^2\right) u_m(x) = 0 \tag{2-34}$$

where

$$\lambda_m = \frac{m\pi}{L}, \qquad m = 1, 2, 3, \ldots \quad \text{and} \quad u_m(x) = \sin \lambda_m x$$

The orthogonal functions $u_m(x)$ are chosen to satisfy the boundary conditions $u_m(0) = u_m(L) = 0$. These are the same boundary conditions as those satisfied by Γ. Thus, our orthogonal function expansion becomes

$$\Gamma(x) = \sum_{m=1}^{\infty} A_m u_m(x) \tag{2-35}$$

If we substitute this expression into (2-1) and then use (2-34), we obtain

$$\sum_{m=1}^{\infty} (k^2 - \lambda_m^2) A_m u_m(x) = -\delta(x - x_0)$$

In view of the orthogonality of the $u_m(x)$ as expressed by (1-2), we multiply both sides of the above expression by $u_n(x)$, integrate over the interval $(0, L)$, and then replace n by m to obtain the coefficients

$$A_m = -(k^2 - \lambda_m^2)^{-1} N^{-1} u_m(x_0) \tag{2-36}$$

where $N = L/2$.

On the other hand, we can multiply both sides of (2-35) by $u_n(x)$, integrate over the interval $(0, L)$, and then replace n by m to obtain an alternate expression for the coefficients

$$A_m = N^{-1} \int_0^L \Gamma(y) u_m(y) \, dy \tag{2-37}$$

Thus, one alternate solution to (2-1) with the boundary conditions $\Gamma(0) = \Gamma(L) = 0$ is

$$\Gamma(x, x_0) = -N^{-1} \sum_{m=1}^{\infty} (k^2 - \lambda_m^2)^{-1} u_m(x_0) u_m(x) \tag{2-38}$$

where use was made of the expression for A_m in (2-36). If we substitute the values for λ_m and $u_m(x)$ from (2-34) into (2-38), we immediately obtain (2-39), i.e.,

$$\Gamma(x, x_0) = \frac{2}{L} \sum_{m=1}^{\infty} \left[\left(\frac{m\pi}{L} \right)^2 - k^2 \right]^{-1} \sin \frac{m\pi x}{L} \sin \frac{m\pi x_0}{L} \tag{2-39}$$

Since both solutions to (2-1) must be equivalent, we can equate (2-39) and (2-11) to obtain the identity

$$\Gamma(x, x_0) = \Gamma_1(x, x_0) u(x_0 - x) + \Gamma_2(x, x_0) u(x - x_0)$$

or

$$\frac{2}{L} \sum_{m=1}^{\infty} \left[\left(\frac{m\pi}{L} \right)^2 - k^2 \right]^{-1} \sin \frac{m\pi x}{L} \sin \frac{m\pi x_0}{L}$$
$$= (k \sin kL)^{-1} [\sin k(L - x_0) \sin kx \, u(x_0 - x)$$
$$+ \sin k(L - x) \sin kx_0 \, u(x - x_0)]$$

In order to prove that the left side of the above expression can be summed to yield the right side, we use an expression from Collin[4]

$$\sum_{n=1}^{\infty} \frac{\cos nx}{n^2 - \alpha^2} = \frac{1}{2\alpha^2} - \frac{\pi}{2\alpha} \frac{\cos (x - \pi)\alpha}{\sin \pi\alpha}, \qquad 0 \le x \le 2\pi$$

As a first step in the proof, we make the substitution $\alpha = kL/\pi$ and use a trigonometric identity to obtain

$$\Gamma(x, x_0) = \frac{2}{L} \sum_{m=1}^{\infty} \left[\left(\frac{m\pi}{L}\right)^2 - k^2\right]^{-1} \sin \frac{m\pi x}{L} \sin \frac{m\pi x_0}{L}$$

$$= \frac{L}{\pi^2} \sum_{m=1}^{\infty} (m^2 - \alpha^2)^{-1} \left[\cos \frac{m\pi}{L}(x - x_0) - \cos \frac{m\pi}{L}(x + x_0)\right]$$

Next, we make the substitutions

$$x_1 = \frac{\pi}{L}(x - x_0)$$

$$x_2 = \frac{\pi}{L}(x_0 - x), \qquad \text{where } 0 \le x_i \le 2\pi, \ i = 1, 2, 3$$

$$x_3 = \frac{\pi}{L}(x + x_0)$$

However, the above substitutions are not valid for all values of x in the interval $(0, L)$. Thus, we find that the substitution for x_1 is valid for $x_0 \le x \le L$, the substitution for x_2 is valid for $0 \le x \le x_0$, and the substitution for x_3 is valid for $0 \le x \le L$. If we solve for Γ_1 and Γ_2, we find that

$$\Gamma_1 = \frac{L}{\pi^2} \sum_{m=1}^{\infty} (m^2 - \alpha^2)^{-1}(\cos mx_2 - \cos mx_3), \qquad 0 \le x \le x_0$$

$$\Gamma_2 = \frac{L}{\pi^2} \sum_{m=1}^{\infty} (m^2 - \alpha^2)^{-1}(\cos mx_1 - \cos mx_3), \qquad x_0 \le x \le L$$

Now we reverse the order of the substitutions for α and the x_i to obtain our desired representations

$$\Gamma(x, x_0) = \Gamma_1(x, x_0)u(x_0 - x) + \Gamma_2(x, x_0)u(x - x_0) \qquad (2\text{-}11)$$

where

$$\Gamma_1(x, x_0) = (k \sin kL)^{-1} \sin k(L - x_0) \sin kx$$
$$\Gamma_2(x, x_0) = \Gamma_1(x_0, x)$$

---

[4]R. Collin, *Field Theory of Guided Waves*, McGraw-Hill Book Company, Inc., New York, 1960, p. 581.

One point that we should note in connection with the solutions in (2-11) and (2-39) is the fact that they both become indeterminate when $k = m\pi/L$, i.e., when the arbitrary constant k in the nonhomogeneous equation becomes an eigenvalue of the homogeneous equation. However, this condition just corresponds to resonance[5] of the system described by the equation (2-1). From a mathematical point of view, one can conclude that a solution to (2-1) exists and is unique only when k does not correspond to one of the eigenvalues. For this latter case, a solution may still exist but it will not be unique.

We can consider another viewpoint by proving that the right side of our expression is equivalent to the left side, i.e., (2-11) is equivalent to (2-39). To accomplish this we substitute the expression for A_m from (2-37) into (2-35) to obtain an alternate and equivalent expression for (2-38), namely,

$$\Gamma(x, x_0) = N^{-1} \sum_{m=1}^{\infty} \int_0^L \Gamma(y) u_m(y) \, dy \, u_m(x) \tag{2-40}$$

Now substitution of (2-11) for $\Gamma(y)$ in (2-40) yields the desired representation, (2-39). Since this involves a considerable amount of tedious work, the details are left as a problem at the end of the chapter. However, one should note that the problem is nothing other than the expansion of (2-11) in a set of orthogonal functions, i.e., a Fourier sine series.

As another example of alternate forms for Green's functions, let us consider the Green's function for Bessel's equation, i.e.,

$$\left[\frac{d}{dx}\left(x\frac{d}{dx}\right) + xk_m^2 - \frac{m^2}{x}\right]\Gamma(k_m x) = -\delta(x - x_0) \tag{2-41}$$

with the boundary conditions $\Gamma(x_1) = \Gamma(x_2) = 0$. One solution using the general procedure in Sec. 2.2 is given in Prob. 2.8 (for $k_m = 1$). We can find an alternate solution by expanding Γ in a series of orthogonal functions. The orthogonal functions are obtained from the homogeneous form of (2-41), i.e., Bessel's equation, (1-18), where we write it as

$$\left[\frac{d}{dx}\left(x\frac{d}{dx}\right) + x\beta_m^2 - \frac{m^2}{x}\right]B_m(\beta_m x) = 0 \tag{2-42}$$

so that it appears in the proper Sturm-Liouville form. Since the orthogonal functions must satisfy the same boundary conditions as those satisfied by Γ in (2-41), we choose a linear combination of $J_m(\beta_m x)$ and $N_m(\beta_m x)$ and impose the boundary conditions to obtain the eigenfunctions

$$B_m(\beta_{mn} x) = J_m(\beta_{mn} x) N_m(\beta_{mn} x_1) - N_m(\beta_{mn} x) J_m(\beta_{mn} x_1) \tag{2-43}$$

---

[5] J. Dettman, *Mathematical Methods in Physics and Engineering*, McGraw-Hill Book Company, Inc., New York, 1962, pp. 189, 230.

The eigenvalues, β_{mn}, are the nth roots of the transcendental equation

$$J_m(\beta_{mn}x_1)N_m(\beta_{mn}x_2) = J_m(\beta_{mn}x_2)N_m(\beta_{mn}x_1) \tag{2-44}$$

The orthogonal functions $B_m(\beta_{mn}x) = J_m(\beta_{mn}x)N_m(\beta_{mn}x_2) - N_m(\beta_{mn}x)J_m(\beta_{mn}x_2)$ also satisfy (2-42) and the boundary conditions. However, we shall not use them here since later we may wish to allow x_1 to vanish. In any case, we can now write down our desired series expansion for Γ as

$$\Gamma(x, x_0) = \sum_{n=1}^{\infty} A_{mn}B_m(\beta_{mn}x) \tag{2-45}$$

where the eigenfunctions $B_m(\beta_{mn}x)$ are given by (2-43) and the eigenvalues β_{mn} are determined from (2-44). If we substitute the expression for Γ from (2-45) into (2-41) and use (2-42) we obtain

$$\sum_{n=1}^{\infty} A_{mn}(k_m^2 - \beta_{mn}^2)xB_m(\beta_{mn}x) = -\delta(x - x_0)$$

Since the eigenfunctions $B_m(\beta_{mn}x)$ are orthogonal over the interval (x_1, x_2) for the same value of m but different values of n (with the weighting function x) according to (1-25), we multiply the above expression by $B_m(\beta_{mp}x)\,dx$ and integrate over the interval (x_1, x_2) to obtain the coefficients

$$A_{mn} = -(k_m^2 - \beta_{mn}^2)^{-1}N_{mn}^{-1}B_m(\beta_{mn}x_0) \tag{2-46}$$

where N_{mn} can be obtained from the first expression for N in Appendix C.1.6 and is

$$N_{mn} = \frac{x_2^2}{2}B_{m+1}^2(\beta_{mn}x_2) - \frac{x_1^2}{2}B_{m+1}^2(\beta_{mn}x_1)$$

To obtain an alternate expression for A_{mn}, we multiply (2-45) by $xB_m(\beta_{mp}x)\,dx$ and integrate over the interval (x_1, x_2). Now the orthogonality of the eigenfunctions as given by (1-25) yields the coefficients as

$$A_{mn} = N_{mn}^{-1} \int_{y_1}^{y_2} y\Gamma(y)B_m(\beta_{mn}y)dy \tag{2-47}$$

If we substitute the expression for A_{mn} from (2-46) into (2-45), we obtain one solution to (2-41) as

$$\Gamma(x, x_0) = -\sum_{n=1}^{\infty} N^{-1}(k_m^2 - \beta_{mn}^2)^{-1}B_m(\beta_{mn}x_0)B_m(\beta_{mn}x) \tag{2-48}$$

On the other hand, we can obtain an alternate expression for the solution to (2-41) by substituting the expression for A_{mn} from (2-47) into (2-45). Thus, the alternate solution becomes

$$\Gamma(x, x_0) = \sum_{n=1}^{\infty} N^{-1} \int_{y_1}^{y_2} y\Gamma(y)B_m(\beta_{mn}y)\, dy\, B_m(\beta_{mn}x) \qquad (2\text{-}49)$$

Now that we have considered several examples of how one may obtain alternate expressions for the Green's function for a particular ordinary differential equation with specific boundary conditions, let us be more general and discuss alternate solutions for the Green's function for the Sturm-Liouville equation, (2-18). As before, the eigenfunctions are determined from the homogeneous form of (2-18), i.e., (1-6), and must satisfy the same boundary conditions as those imposed on the Green's function, Γ. These boundary conditions can be of the general form indicated in (1-13). If we use the subscripts i on λ and Γ in (2-18), the subscripts j on λ and u in (1-6), and then replace λ by k in the latter, we can write an expansion for $\Gamma_i(x)$ as

$$\Gamma_i(x) = \sum_{j=1}^{\infty} A_{ij} u_j(\lambda_j x) \qquad (2\text{-}50)$$

Now substitution of (2-50) for Γ in (2-18) and the use of (1-6) gives us

$$\sum_{j=1}^{\infty} A_{ij}(\lambda_i - k_j)w(x)u_j(\lambda_j x) = -\delta(x - x_0)$$

Multiplication of the above expression by $u_k(\lambda_k x)\, dx$ and integration over the interval (a, b) give the coefficients A_{ij} as

$$A_{ij} = N_{ij}^{-1}(\lambda_i - k_j)^{-1}u_j(\lambda_j x_0) \qquad (2\text{-}51)$$

where use was made of the general orthogonality expression (1-15). The alternate expression for A_{ij} is obtained by multiplying (2-50) by $w(x)\, u_k(\lambda_k x)$ and integrating with respect to x over the interval (a, b). Use of the orthogonality expression (1-15) now yields the coefficients as

$$A_{ij} = N_{ij}^{-1} \int_a^b w(y)\Gamma_i(y)u_j(\lambda_j y)\, dy \qquad (2\text{-}52)$$

If we substitute (2-51) into (2-50), we obtain one expression for Γ as

$$\Gamma_i(x) = \sum_{j=1}^{\infty} N_{ij}^{-1}(\lambda_i - k_j)^{-1}u_j(\lambda_j x_0)u_j(\lambda_j x) \qquad (2\text{-}53)$$

The alternate expression is found by substituting (2-52) into (2-50), whence the expression for Γ becomes

$$\Gamma_i(x) = \sum_{j=1}^{\infty} N_{ij}^{-1}u_j(\lambda_j x)\int_a^b w(y)\Gamma_i(y)u_j(\lambda_j y)\, dy \qquad (2\text{-}54)$$

We should note that the procedure followed in this section for finding the Green's function differs from that given in Sec. 2.2. The difference lies in the fact that in this section three arbitrary constants are determined rather than four. The constants A and B defined in Sec. 2.2 are found here, too, and appear implicitly in the expression for the eigenfunctions used in the orthogonal function expansion of the Green's function. Since the eigenfunctions are continuous over the interval of the expansion, only one of the two constants a and b are needed. In this section we have denoted that constant (actually a series of constants) by A_m in (2-35), by A_{mn} in (2-45), and by A_{ij} in (2-50). This one constant is necessary in order to indicate the strength of the source. The number of arbitrary constants involved is always infinite for a unit source, because a unit source excites an infinite number of normal modes or eigenfunctions. This fact is made more evident by several of the homework problems where the Dirac delta is represented by a Fourier series of orthogonal functions.

Another item that is of interest concerns the fact that the expressions for Γ in (2-38), (2-48), and (2-53) are valid when k, k_m, and k_j, respectively, vanish. In this case, the Green's function Γ reduces to the Green's function K since (2-1) reduces to (2-2) and (2-18) reduces to (2-14). Note that in deriving (2-53) we replaced λ by k in (1-6) to avoid confusion with the λ in (2-18). Thus, the statement that (2-18) reduces to (2-14) applies when one remembers that the condition that k_j vanishes is equivalent to allowing λ to vanish in (2-18). The ordinary differential equation (2-41) reduces to

$$\left[\frac{d}{dx}\left(x\frac{d}{dx}\right) - \frac{m^2}{x}\right]K(x) = -\delta(x - x_0) \qquad (2\text{-}55)$$

This item is very useful since one always knows that the Green's function for a general equation is still valid when some of the terms in that equation vanish. More specifically, this means that the Green's function for the wave equation (Helmholtz equation) is valid when the wave equation reduces to Laplace's equation. For instance, the Green's function in (3-28) is the solution to the wave equation in unbounded space. When k vanishes, the wave equation reduces to Laplace's equation (actually Poisson's equation, since a source is present) and the three solutions in (3-28) reduce to a single solution. This single solution is still the Green's function for Laplace's equation in unbounded space. A one-dimensional differential equation that demonstrates this fact is (2-1) with its solution (2-11). If k vanishes in (2-1) the equation reduces to (2-2). If k vanishes in (2-11) the solution reduces to the solution in (2-5). This situation is quite different from that encountered with the solutions to the homogeneous form of the equations mentioned above. In this latter case, the solutions to the wave equation do not usually reduce to the solutions to Laplace's equation when one allows k to vanish. This situation is more obvious if one considers a one-dimensional differential equation such as the homogeneous form of (2-1). The solutions are $\sin kx$ and $\cos kx$ (for $k^2 > 0$) and they do not reduce to the solutions of the homogeneous form of (2-2) when k vanishes.

PROBLEMS

2.1. Show that $\Gamma(x_1, x_2) = \Gamma(x_2, x_1)$ where Γ satisfies Eq. (2-1). *Hint:* Choose two different sources at x_1 and x_2.

2.2. Solve Eq. (2-6) for $f(x) = \sin(\pi x/L)$ using conventional methods and compare with the solution obtained from Eq. (2-7).

2.3. Consider Eq. (2-1) and its solution as given by Eq. (2-11). Plot Γ, Γ', and Γ'' for $k = \pi/2$, $L = 1$, $x_0 = \frac{2}{3}$.

2.4. Find the solution to Eq. (2-1) when $k^2 = -\alpha^2$. Express the solution in a form similar to Eq. (2-11). Note that the solution can also be obtained by making the substitution $k = \pm j\alpha$ in (2-11).

2.5. Use the general procedure in Sec. 2.2 to verify the solution (2-5).

2.6. Use the general procedure in Sec. 2.2 to verify the solution (2-11).

2.7. Verify (2-33) and plot Γ, Γ', and Γ'' for the parameters given in Prob. 2.3.

2.8. Show that the Green's function for Bessel's equation, (1-18), with the boundary conditions $\Gamma(x_1) = \Gamma(x_2) = 0$, $x_1, x_2 \neq 0$, ∞ is

$$\Gamma_1(x, x_0) = \frac{\pi}{2}(J_{n1}N_{n2} - J_{n2}N_{n1})^{-1}(J_{n2}N_{n0} - J_{n0}N_{n2})(J_nN_{n1} - J_{n1}N_n),$$

$$x_1 \leq x \leq x_0$$

$$\Gamma_2(x, x_0) = \Gamma_1(x_0, x), \qquad x_0 \leq x \leq x_2$$

where $J_{n2} = J_n(x_2)$, $N_{n0} = N_n(x_0)$, $J_n = J_n(x)$, etc. Note that the solutions to Bessel's equation, (1-18), when $k_m = 1$ are unchanged when $k_m \neq 1$ so that (2-41) applies.

2.9. Repeat Prob. 2.8 for the modified Bessel's equation, (1-26).

2.10. Show that the Green's function for the associated Legendre equation, (1-32a), with the boundary conditions $\Gamma(x_1) = \Gamma(x_2) = 0$, $x_1, x_2 \neq |1|$ is

$$\Gamma_1(x, x_0) = [(1 - x_0^2)W(x_0)(P_{s1}^r Q_{s2}^r - P_{s2}^r Q_{s1}^r)]^{-1}$$

$$\times (P_{s2}^r Q_{s0}^r - P_{s0}^r Q_{s2}^r)(P_s^r Q_{s1}^r - P_{s1}^r Q_s^r), \qquad -1 < x_1 \leq x \leq x_0$$

$$\Gamma_2(x, x_0) = \Gamma_1(x_0, x), \qquad x_0 \leq x \leq x_2 < 1$$

where $P_{s2}^r = P_s^r(x_2)$, $Q_{s0}^r = Q_s^r(x_0)$, $P_s^r = P_s^r(x)$, etc., and $W(x_0)$ is the Wronskian of P_s^r and Q_s^r evaluated at $x = x_0$.

2.11. Show that the Green's function for Bessel's equation, (1-18), with the boundary conditions $\Gamma(x_2) = 0$, $x_2 \neq 0$, origin included, is

$$\Gamma_1(x, x_0) = \frac{\pi}{2} J_n J_{n2}^{-1}(J_{n0}N_{n2} - J_{n2}N_{n0}), \qquad 0 \leq x \leq x_0$$

$$\Gamma_2(x, x_0) = \Gamma_1(x_0, x), \qquad x_0 \leq x \leq x_2$$

Note that Γ can be obtained by using the procedure in Sec. 2.2 or by letting $x_1 = 0$ in Prob. 2.8.

2.12. Repeat Prob. 2.11 for the modified Bessel's equation, (1-26).

2.13. Repeat Prob. 2.10 for the boundary condition $\Gamma(x_1) = 0$, $x_2 = 1$.

2.14. Repeat Prob. 2.8 with the boundary condition $\Gamma(x_1) = 0$, $x_1 \neq 0$, $x_2 = \infty$. *Hint:* Use the result of Prob. 2.8 by replacing N_{n2} by J_{n2} and $H_{n2}^{(2)}$ and then letting $H_{n2}^{(2)} \rightarrow 0$ or use the general procedure in Sec. 2.2. The solution is

$$\Gamma_1(x, x_0) = -\frac{\pi}{2} \frac{H_{n0}^{(2)}}{H_{n1}^{(2)}}(J_n N_{n1} - J_{n1}N_n), \qquad 0 < x_1 \leq x \leq x_0$$

$$\Gamma_2(x, x_0) = \Gamma_1(x_0, x), \qquad x_0 \leq x \leq \infty$$

2.15. Extend the results of Prob. 2.14 so that the origin is included, i.e., $x_1 \rightarrow 0$. The result is

$$\Gamma_1(x, x_0) = \frac{\pi}{2j} H_{n0}^{(2)} J_n, \qquad 0 \leq x \leq x_0$$

$$\Gamma_2(x, x_0) = \Gamma_1(x_0, x), \qquad x_0 \leq x \leq \infty$$

This result is used by Papas and Levine[6] for the case where $n = 1$ and time dependence is of the form $e^{-j\omega t}$.

2.16. Show that the Green's function for Bessel's equation, (1-18), with the boundary conditions $\Gamma'(x_1) = \Gamma'(x_2) = 0$, $x_1, x_2 \neq 0$, is

$$\Gamma_1(x, x_0) = \frac{\pi}{2}(J_{n1}'N_{n2}' - J_{n2}'N_{n1}')^{-1}(J_{n2}'N_{n0} - J_{n0}N_{n2}')(J_n N_{n1}' - J_{n1}'N_n),$$

$$x_1 \leq x \leq x_0$$

$$\Gamma_2(x, x_0) = \Gamma_1(x_0, x), \qquad x_0 \leq x \leq x_2$$

2.17. Show that the Green's function for the spherical Bessel's equation, (1-35), $\alpha \rightarrow 1$, with the boundary conditions $\Gamma(x_1) = \Gamma(x_2) = 0$, x_1, $x_2 \neq 0, \infty$ is

$$\Gamma_1(x, x_0) = (j_{n1}n_{n2} - j_{n2}n_{n1})^{-1}(j_{n2}n_{n0} - j_{n0}n_{n2})(j_n n_{n1} - j_{n1}n_n),$$

$$x_1 \leq x \leq x_0$$

$$\Gamma_2(x, x_0) = \Gamma_1(x_0, x), \qquad x_0 \leq x \leq x_2$$

---

[6]C. Papas and H. Levine, "Theory of the Circular Diffraction Antenna," *J. Appl. Phys.*, Vol. 22, pp. 29–43, January 1951. (See p. 31.)

Note that the solutions to the spherical Bessel's equation, (1-35), when $\alpha \neq 1$ can be obtained from those given here for $\alpha = 1$ by multiplying the right-hand side by α.

2.18. Repeat Prob. 2.17 with the boundary condition $\Gamma(x_2) = 0$, $x_2 \neq 0$, origin included. The result is

$$\Gamma_1(x, x_0) = \frac{j_n}{j_{n2}}(j_{n0}n_{n2} - j_{n2}n_{n0}), \qquad 0 \leq x \leq x_0$$

$$\Gamma_2(x, x_0) = \Gamma_1(x_0, x), \qquad x_0 \leq x \leq x_2$$

2.19. Repeat Prob. 2.17 with the boundary condition $\Gamma(x_1) = 0$, $x_1 \neq 0$, $x_2 = \infty$. The result is

$$\Gamma_1(x, x_0) = -\frac{h_{n0}^{(2)}}{h_{n1}^{(2)}}(j_n n_{n1} - j_{n1} n_n), \qquad 0 < x_1 \leq x \leq x_0$$

$$\Gamma_2(x, x_0) = \Gamma_1(x_0, x), \qquad x_0 \leq x \leq \infty$$

2.20. Extend the results of Prob. 2.19 so that the origin is included. The result is

$$\Gamma_1(x, x_0) = -jh_{n0}^{(2)}j_n, \qquad 0 \leq x \leq x_0$$

$$\Gamma_2(x, x_0) = \Gamma_1(x_0, x), \qquad x_0 \leq x \leq \infty$$

2.21. Show that the Green's function for (1-39) with the boundary conditions $\Gamma(x_1) = \Gamma(x_2) = 0$, $x_1, x_2 \neq 0$, is

$$\Gamma_1(x, x_0) = [(2n + 1)(x_1^{2n+1} - x_2^{2n+1})]^{-1}[x_0^n - x_2^{2n+1}x_0^{-(n+1)}]$$
$$\times [x^n - x_1^{2n+1}x^{-(n+1)}], \qquad 0 < x_1 \leq x < x_0$$

$$\Gamma_2(x, x_0) = \Gamma_1(x_0, x), \qquad x_0 \leq x \leq x_2$$

2.22. Verify that (2-39) results from (2-40) when $\Gamma(y)$ is taken as $\Gamma(x, x_0)$ in (2-11).

2.23. Derive an alternate solution to Prob. 2.4 by using (2-40). In this case $\Gamma(y)$ is given by (2-11) with $k = j\alpha$. This alternate solution corresponds to (2-39).

2.24. Solve Prob. 2.23 in a direct manner by assuming a solution in the form of (2-35). Note that the final solution can be obtained from (2-38) by making the substitution $k^2 = -\alpha^2$.

2.25. Equate the solutions to Probs. 2.23 and 2.24 to obtain an identity similar to that obtained by equating (2-39) and (2-11). *Note:* The same result is obtained if the substitution $k = j\alpha$ is made in (2-39) and (2-11).

2.26. Find the solution to (2-1) for the boundary conditions $\Gamma(x_1) = \Gamma(x_2) = 0$, $x_1, x_2 \neq 0$ in a form similar to (2-35) and (2-36) except the $u_m(x)$ will consist of both cosines and sines. *Note:* This is the usual Fourier series viewpoint. Also, this solution is equally valid when x is replaced by the angle ϕ, e.g., see (5-61).

2.27. Verify that the eigenfunctions in (2-43) satisfy (2-42) and that the eigenvalues in (2-44) satisfy the boundary conditions imposed in connection with (2-41).

2.28. Find the Green's function for (2-41) with the accompanying boundary conditions using the general procedure in Sec. 2.2. The solution should reduce to that in Prob. 2.8 when $k_m = 1$.

2.29. Substitute the Green's function in Prob. 2.28 for $\Gamma(y)$ in (2-49) and demonstrate that (2-48) results.

2.30. Verify that when (2-48) is substituted into (2-49) the resulting expression becomes (2-48).

2.31. Verify that the expression for the norm N in (2-46) results from (1-25).

2.32. Show that the eigenfunctions in (2-43) become $J_m(\beta_{mn}x)$ and that the eigenvalues in (2-44) are determined from the equation $J_m(\beta_{mn}a) = 0$ when the interval (x_1, x_2) becomes the interval $(0, a)$.

2.33. Show that the Green's function in (2-48) becomes the following when the interval (x_1, x_2) becomes the interval $(0, a)$ and $k_m = 0$:

$$G(x, x_0) = 2 \sum_{n=1}^{\infty} [\beta_{mn}a\, J_{m+1}(\beta_{mn}a)]^{-2} J_m(\beta_{mn}x_0) J_m(\beta_{mn}x)$$

and the roots of $J_m(\beta_{mn}a) = 0$ define the eigenvalues β_{mn}.

2.34. Expand the Dirac delta in the orthogonal functions $u_m(x)$ given in (2-34). The result is

$$\delta(x - x_0) = N^{-1} \sum_{m=1}^{\infty} u_m(x)u_m(x_0)$$

2.35. Show that the result in Prob. 2.34 can be obtained quite easily from (2-40) by replacing $\Gamma(y)$ by δ.

2.36. Expand $\delta(x - x_0)/x$ in the orthogonal functions $B_m(\beta_{mn}x)$ given in (2.43). The result is

$$\frac{\delta(x - x_0)}{x} = \sum_{n=1}^{\infty} N_{mn}^{-1} B_m(\beta_{mn}x) B_m(\beta_{mn}x_0)$$

where N_{mn} is defined in (2-46).

2.37. Show that the result in Prob. 2.36 can be obtained quite easily from (2-49) by replacing Γ by δ/x inside the integral.

3

Transforms

3.1 Transform Methods

The preceding applications have been almost exclusively devoted to the expansion of a function in a set of orthogonal functions over a finite interval (a, b) or one of length L. We have seen that it is possible to obtain the Green's function for a region that is bounded in one direction and infinite in another direction provided the region is divided into two parts in the infinite direction. For instance, the Green's function for the uniform unit line source in an infinite rectangular pipe is given by (5-12), (5-13), and (5-4) when the interval $(0, b)$ is broken into two parts in the y-direction. Since the problem is solved in Sec. 5.2, we will write the solution for reference purposes as follows:

$$G(x, x_0; y, y_0) = \sum_{m=1}^{\infty} \frac{2}{m\pi} \frac{\sin (m\pi x/a) \sin (m\pi x_0/a)}{\sinh (m\pi b/a)} \left[\sinh\frac{m\pi y}{a} \sinh\frac{m\pi(b-y_0)}{a} \right.$$

$$\left. \times u(y_0 - y) + \sinh\frac{m\pi y_0}{a} \sinh\frac{m\pi(b-y)}{a} u(y - y_0) \right] \quad \text{(5-12), (5-13)}$$

$$= G_1 u(y_0 - y) + G_2 u(y - y_0) \quad \text{(5-4)}$$

If we let the width b in the y-direction become infinite, no difficulties are encountered with the series solution. This is demonstrated in Prob. 5.2.

However, if we let the height a in the x-direction become infinite, the series solution becomes indeterminate. This difficulty is caused by the eigenvalues $m\pi/a$ becoming zero because the interval a is approaching infinity. Since we know that the Green's function for the trough is as well defined as it is for the pipe, it follows that the difficulties are caused by the mathematics (or our improper use or interpretation of the mathematics). In this case, our difficulty can be resolved by replacing the sum by an integral, i.e., passing from a Fourier series to a Fourier integral.[1] Thus, in (5-12) and (5-13) we let

$$\alpha = \frac{m\pi}{a}$$

$$\alpha + \Delta\alpha = \frac{(m+1)\pi}{a}$$

and the infinite series with the eigenvalues $m\pi/a$ becomes an integral, e.g.,

$$\mathop{\text{Lim}}_{a\to\infty} \sum_{m=1}^{\infty} \frac{1}{m} F\left(\frac{m\pi}{a}\right) = \mathop{\text{Lim}}_{\Delta\alpha\to0} \sum_{m=1}^{\infty} \frac{\Delta\alpha}{\alpha} F(\alpha) = \int_0^{\infty} \frac{F(\alpha)\,d\alpha}{\alpha}$$

In this case, as a becomes infinite, (5-12) and (5-13) become

$$G_1(x, x_0; y, y_0) = \frac{2}{\pi} \int_0^{\infty} \frac{\sin\alpha x \sin\alpha x_0 \sinh\alpha y \sinh\alpha(b - y_0)}{\alpha \sinh\alpha b}\,d\alpha, y < y_0$$

$$G_2(x, x_0; y, y_0) = G_1(x_0, x; y_0, y), \qquad y > y_0$$

On the other hand, when the problem is unbounded in one direction, the solution can also be obtained directly by using a Fourier transform in the variable that corresponds to that direction. As an example, let us find the Green's function for the uniform line source in a grounded conducting trough of width a in the x-direction and open at infinity in the y-direction. This problem is illustrated in Fig. 3-1. In order to solve (5-1), i.e.,

$$\nabla_t^2 G = \left(\frac{\partial^2}{\partial x^2} + \frac{\partial^2}{\partial y^2}\right) G = -\delta(x - x_0)\delta(y - y_0) \qquad (5\text{-}1)$$

let us introduce the exponential Fourier transform pair

$$f(u) = \frac{1}{\sqrt{2\pi}} \int_{-\infty}^{\infty} \tilde{f}(\alpha) e^{j\alpha u}\,d\alpha$$

$$\tilde{f}(\alpha) = \frac{1}{\sqrt{2\pi}} \int_{-\infty}^{\infty} f(u) e^{-j\alpha u}\,du$$

$$(3\text{-}1)$$

---

[1] P. Morse and H. Feshbach, *Methods of Theoretical Physics*, McGraw-Hill Book Company, Inc., New York, 1953, p. 454. Also, see R. Churchill, *Fourier Series and Boundary Value Problems*, McGraw-Hill Book Company, Inc., New York, 1941, p. 88.

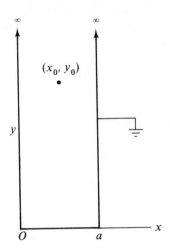

Fig. 3-1 Uniform unit dc line source in a grounded conducting trough.

the cosine Fourier transform pair

$$f(u) = \frac{2}{\sqrt{2\pi}} \int_0^\infty \tilde{f}(\alpha) \cos \alpha u \, d\alpha$$

$$\tilde{f}(\alpha) = \frac{2}{\sqrt{2\pi}} \int_0^\infty f(u) \cos \alpha u \, du \tag{3-2}$$

and the sine Fourier transform pair

$$f(u) = \frac{2}{\sqrt{2\pi}} \int_0^\infty \tilde{f}(\alpha) \sin \alpha u \, d\alpha$$

$$\tilde{f}(\alpha) = \frac{2}{\sqrt{2\pi}} \int_0^\infty f(u) \sin \alpha u \, du \tag{3-3}$$

The boundary condition on G at $y = 0$ forces us to choose the sine Fourier transform pair in (3-3) to represent the y-dependent portion of G. Thus, assume that G can be represented as

$$G = \frac{2}{\sqrt{2\pi}} \int_0^\infty \tilde{G} \sin \alpha y \, d\alpha \tag{3-4}$$

Now let us substitute (3-4) into (5-1) to obtain

$$\frac{2}{\sqrt{2\pi}} \int_0^\infty \left(\frac{d^2}{dx^2} - \alpha^2 \right) \tilde{G} \sin \alpha y \, d\alpha = -\delta(x - x_0)\delta(y - y_0) \tag{3-5}$$

Since we wish to replace (5-1) by an ordinary differential equation in x and G, let us find the sine Fourier transform of the Dirac delta, $\delta(y - y_0)$. From (3-3) we can write down at once that

$$\tilde{\delta}(\alpha - y_0) = \frac{2}{\sqrt{2\pi}} \int_0^\infty \delta(y - y_0) \sin \alpha y \, dy = \frac{2}{\sqrt{2\pi}} \sin \alpha y_0$$

whence we obtain the identity

$$\delta(y - y_0) = \frac{2}{\pi} \int_0^\infty \sin \alpha y_0 \sin \alpha y \, d\alpha \tag{3-6}$$

If we substitute (3-6) into (3-5), we find that

$$\left(\frac{d^2}{dx^2} - \alpha^2 \right) \tilde{G}(x, \alpha) = -\frac{2}{\sqrt{2\pi}} \sin \alpha y_0 \, \delta(x - x_0) \tag{3-7}$$

and indeed we have replaced the partial differential equation (5-1) in two independent variables x and y by an ordinary differential equation (3-7) in x. We can solve (3-7) by the standard method used to solve (2-1). Note that (3-7) with a unit source was solved in Prob. 2.4. The solution to (3-7) is

$$\tilde{G}_1(x, x_0; \alpha, y_0) = \frac{2}{\sqrt{2\pi}} \frac{\sin \alpha y_0 \sinh \alpha x \sinh \alpha(a - x_0)}{\alpha \sinh \alpha a}, \qquad x < x_0$$

$$\tilde{G}_2(x, x_0; \alpha, y_0) = \tilde{G}_1(x_0, x; y_0, \alpha), \qquad x > x_0$$

where $\tilde{G} = \tilde{G}_1 u(x_0 - x) + \tilde{G}_2 u(x - x_0)$. The solution to (5-1) is

$$G_1(x, x_0; y, y_0) = \frac{2}{\pi} \int_0^\infty \frac{\sin \alpha y \sin \alpha y_0 \sinh \alpha x \sinh \alpha(a - x_0) \, d\alpha}{\alpha \sinh \alpha a}, \qquad x < x_0$$

$$G_2(x, x_0; y, y_0) = G_1(x_0, x; y_0, y), \qquad x > x_0 \tag{3-8}$$

The product function solution[2] that is equivalent to (3-8) is given in Prob. 5.1.

Now that we have discussed an example of the use of the Fourier transform to solve an unbounded boundary value problem, let us briefly discuss the general theory behind the use of a Fourier transform in reducing the number of independent variables in a partial differential equation. The Fourier integral theorem states the following:

$$f(x) = \frac{1}{2\pi} \int_{-\infty}^\infty du \int_{-\infty}^\infty f(\alpha) e^{ju(x-\alpha)} \, d\alpha \tag{3-9}$$

---

[2]The equivalence of these two solutions is demonstrated in R. Collin, *Field Theory of Guided Waves*, McGraw-Hill Book Company, Inc., New York, 1960, Sec. 2.4.

where $f(x)$ is piecewise continuous and $\int_{-\infty}^{\infty} |f(x)|\, dx < \infty$. If (3-9) is rewritten in the more symmetrical form

$$f(x) = \frac{1}{\sqrt{2\pi}} \int_{-\infty}^{\infty} e^{jux}\, du \frac{1}{\sqrt{2\pi}} \int_{-\infty}^{\infty} f(\alpha)e^{-ju\alpha}\, d\alpha$$

we are led to the definition of an exponential Fourier transform pair

$$f(x) = \frac{1}{\sqrt{2\pi}} \int_{-\infty}^{\infty} \tilde{f}(u)e^{jux}\, du$$

$$\tilde{f}(u) = \frac{1}{\sqrt{2\pi}} \int_{-\infty}^{\infty} f(\alpha)e^{-ju\alpha}\, d\alpha \tag{3-10}$$

which is identical with (3-9). However, the variable u and the variable α in (3-10) are at our disposal. Thus, we can replace u in each term in (3-10) by α, and α in the second term in (3-10) by x to obtain the exponential Fourier transform pair in its usual form

$$f(x) = \frac{1}{\sqrt{2\pi}} \int_{-\infty}^{\infty} \tilde{f}(\alpha)e^{j\alpha x}\, d\alpha$$

$$\tilde{f}(\alpha) = \frac{1}{\sqrt{2\pi}} \int_{-\infty}^{\infty} f(x)e^{-j\alpha x}\, dx \tag{3-11}$$

Although (3-1) defines the exponential Fourier transform pair in the customary form, we find that when $\tilde{f}(\alpha)$ in (3-11) is substituted into the expression for $f(x)$ in (3-11) then (3-9) does not result. A similar problem[3] is encountered in passing from a Fourier series to a Fourier integral unless one pays close attention to which integral in (3-9) becomes improper first. One should also note that in (3-1), $f(u)$ is in general complex, that in (3-2) $f(u)$ is a real even function, and that in (3-3) $f(u)$ is a real odd function. It is common to call the second transform the direct transform. In other words, the direct transform takes the problem out of the physical coordinate system and into a transform space and the inverse transform returns the problem to physical space.

Let us now consider the problem encountered when we Fourier transform a partial differential equation such as the following:

$$(\nabla^2 + k^2)f(x, y, z) = g(x, y, z) \tag{3-12}$$

This equation is sufficiently general to demonstrate the utility of using a

---

[3]A. Sommerfeld, *Partial Differential Equations in Physics*, Academic Press, Inc., New York, 1949, Sec. 4.

single Fourier transform to reduce by one the number of independent variables. We can approach this problem by two alternate methods so let us consider both of them.

Method 1

Since the functions $f(x, y, z)$ and $g(x, y, z)$ are well behaved, i.e., Fourier transformable, we can define two Fourier transform pairs according to (3-1). Thus,

$$f(x, y, z) = \frac{1}{\sqrt{2\pi}} \int_{-\infty}^{\infty} \tilde{f}(x, y, \alpha_3) e^{j\alpha_3 z} \, d\alpha_3$$

$$\tilde{f}(x, y, \alpha_3) = \frac{1}{\sqrt{2\pi}} \int_{-\infty}^{\infty} f(x, y, z) e^{-j\alpha_3 z} \, dz$$

$$g(x, y, z) = \frac{1}{\sqrt{2\pi}} \int_{-\infty}^{\infty} \tilde{g}(x, y, \alpha_3) e^{j\alpha_3 z} \, d\alpha_3$$

$$\tilde{g}(x, y, \alpha_3) = \frac{1}{\sqrt{2\pi}} \int_{-\infty}^{\infty} g(x, y, z) e^{-j\alpha_3 z} \, dz$$

If we substitute the inverse transforms for $f(x, y, z)$ and $g(x, y, z)$ into our equation, we obtain

$$(\nabla^2 + k^2)\frac{1}{\sqrt{2\pi}} \int_{-\infty}^{\infty} \tilde{f}(x, y, \alpha_3) e^{j\alpha_3 z} \, d\alpha_3 = \frac{1}{\sqrt{2\pi}} \int_{-\infty}^{\infty} \tilde{g}(x, y, \alpha_3) e^{j\alpha_3 z} \, d\alpha_3$$

which becomes

$$\frac{1}{\sqrt{2\pi}} \int_{-\infty}^{\infty} (\nabla_t^2 + k^2 - \alpha_3^2)\tilde{f}(x, y, \alpha_3) e^{j\alpha_3 z} \, d\alpha_3 = \frac{1}{\sqrt{2\pi}} \int_{-\infty}^{\infty} \tilde{g}(x, y, \alpha_3) e^{j\alpha_3 z} \, d\alpha_3$$

whence

$$(\nabla_t^2 + k^2 - \alpha_3^2)\tilde{f}(x, y, \alpha_3) = \tilde{g}(x, y, \alpha_3) \tag{3-13}$$

where

$$\nabla_t^2 = \nabla^2 - \frac{\partial^2}{\partial z^2}$$

Thus we have replaced the inhomogeneous scalar Helmholtz equation (3-12) by another inhomogeneous scalar Helmholtz equation (3-13) but with one less independent variable. If we transform (3-13) again with respect to x and y we obtain the expression

$$(k^2 - \alpha^2)\tilde{f}(\alpha_1, \alpha_2, \alpha_3) = \tilde{g}(\alpha_1, \alpha_2, \alpha_3) \tag{3-14}$$

where $\alpha^2 = \alpha_1^2 + \alpha_2^2 + \alpha_3^2$,

$$\tilde{f}(\alpha_1, \alpha_2, \alpha_3) = (2\pi)^{-3/2} \iiint_{-\infty}^{\infty} f(x, y, z)e^{-j(\alpha_1 x + \alpha_2 y + \alpha_3 z)} \, dx \, dy \, dz$$

and

$$\tilde{g}(\alpha_1, \alpha_2, \alpha_3) = (2\pi)^{-3/2} \iiint_{-\infty}^{\infty} g(x, y, z)e^{-j(\alpha_1 x + \alpha_2 y + \alpha_3 z)} \, dx \, dy \, dz$$

In this case, we have introduced a triple exponential Fourier transform to replace (3-12) by (3-14). However, the triple transform is nothing other than the product of three single transforms. Thus, one can define the triple exponential cartesian Fourier transform pair as

$$\tilde{f}(\alpha_1, \alpha_2, \alpha_3) = (2\pi)^{-3/2} \iiint_{-\infty}^{\infty} f(x, y, z)e^{-j(\alpha_1 x + \alpha_2 y + \alpha_3 z)} \, dx \, dy \, dz$$
$$f(x, y, z) = (2\pi)^{-3/2} \iiint_{-\infty}^{\infty} \tilde{f}(\alpha_1, \alpha_2, \alpha_3)e^{j(\alpha_1 x + \alpha_2 y + \alpha_3 z)} \, d\alpha_1 \, d\alpha_2 \, d\alpha_3$$

$$(3\text{-}15)$$

and the double exponential cartesian Fourier transform pair as

$$\tilde{f}(\alpha_1, \alpha_2) = \frac{1}{2\pi} \iint_{-\infty}^{\infty} f(x, y)e^{-j(\alpha_1 x + \alpha_2 y)} \, dx \, dy$$
$$f(x, y) = \frac{1}{2\pi} \iint_{-\infty}^{\infty} \tilde{f}(\alpha_1, \alpha_2)e^{j(\alpha_1 x + \alpha_2 y)} \, d\alpha_1 \, d\alpha_2$$

$$(3\text{-}16)$$

Method 2

The other method for Fourier transforming (3-12) with respect to z is to multiply both sides by $(1/\sqrt{2\pi})e^{-j\alpha_3 z}$ and then integrate with respect to z from minus to plus infinity. Now we obtain

$$\frac{1}{\sqrt{2\pi}} \int_{-\infty}^{\infty} (\nabla^2 + k^2)f(x, y, z)e^{-j\alpha_3 z} \, dz = \frac{1}{\sqrt{2\pi}} \int_{-\infty}^{\infty} g(x, y, z)e^{-j\alpha_3 z} \, dz$$

The term on the right-hand side is the Fourier transform of $g(x, y, z)$, i.e., $\tilde{g}(x, y, \alpha_3)$. Now we integrate the term on the left-hand side with respect to z twice by parts to obtain

$$\frac{1}{\sqrt{2\pi}} \left[\frac{\partial f(x, y, z)}{\partial z}e^{-j\alpha_3 z} - f(x, y, z)\frac{\partial e^{-j\alpha_3 z}}{\partial z} \right]_{-\infty}^{\infty} + (\nabla_t^2 + k^2 - \alpha_3^2)\tilde{f}(x, y, \alpha_3)$$

$$(3\text{-}17)$$

Finally, we obtain the same partial differential equation in two independent variables

$$(\nabla_t^2 + k^2 - \alpha_3^2)\tilde{f}(x, y, \alpha_3) = \tilde{g}(x, y, \alpha_3) \qquad (3\text{-}13)$$

provided the first term in (3-17) vanishes at $|z| = \infty$ because either f and $\partial f/\partial z$ vanish there or because the term $e^{-j\alpha_3 z}$ vanishes at $|z| = \infty$. The mathematical meaning of these assumptions is that one often includes a "convergence factor" when evaluating a Fourier transform. The physical meaning is that finite sources do not produce an effect at infinity in the presence of a physical medium with loss.[4] If we rewrite the term in the square brackets in (3-17), we may obtain Sommerfeld's radiation condition[5] for a radiating plane wave. Thus, we write the expression as the product of two expressions

$$\left[\left\{ \frac{\partial f(x, y, z)}{\partial z} + j\alpha_3 f(x, y, z) \right\} e^{-j\alpha_3 z} \right]_{-\infty}^{\infty} \longrightarrow 0 \qquad (3\text{-}18)$$

The vanishing of the term in the curly brackets in (3-18) is the Sommerfeld radiation condition for an outgoing plane wave, i.e.,

$$\left[\frac{\partial f}{\partial z} + j\alpha_3 f \right]_{-\infty}^{\infty} = \underset{|z| \to \infty}{\text{Lim}} \left(\frac{\partial f}{\partial z} + j\alpha_3 f \right) = 0 \qquad (3\text{-}19)$$

In this case, time-varying finite sources in a medium without loss can produce fields at infinity but they must be waves traveling away from the source, i.e., outgoing waves. If the wave is an incoming plane wave, then the sign in front of α_3 has to be negative. For instance, we know that $e^{-j\alpha_3 z}$ represents a plane wave traveling in the $+z$ direction (for $e^{j\omega t}$ time dependence) and it is quite easy to verify that $f = e^{-j\alpha_3 z}$ satisfies (3-19) for all values of z. The wave $e^{j\alpha_3 z}$ satisfies (3-19) when the plus sign is replaced by a minus sign. In addition the waves $\cos \alpha_3 z$ and $\sin \alpha_3 z$ do not satisfy (3-19) since they are standing waves. Although the examples discussed here are quite simple, there are often situations where the function f is a complicated expression so that the radiation condition allows one to eliminate unwanted solutions to a given problem. The radiation condition also takes on different forms in space with more than one dimension. As a result, the general radiation condition can be written as

$$\underset{R \to \infty}{\text{Lim}} R^{(h-1)/2} \left(\frac{\partial f}{\partial R} + jkf \right) = 0 \qquad (3\text{-}20)$$

where the integer h indicates the number of dimensions of the physical space and R is an independent variable in the direction of power flow. In three-

---

[4]Collin, *Field Theory of Guided Waves*, pp. 207, 487.

[5]Sommerfeld, *Partial Differential Equations in Physics*, pp. 188–93.

dimensional space, $h = 3$ and $R = r$; in two-dimensional space, $h = 2$ and $R = \rho$. Our main conclusion from this discussion is that the radiation condition is used in exterior boundary value problems in place of the usual boundary condition.

Let us consider another example of the use of the Fourier transform to solve a boundary value problem in unbounded space (exterior problem). Let us solve the inhomogeneous Helmholtz equation

$$(\nabla^2 + k^2)G(x, y, z) = -\delta(\mathbf{r} - \mathbf{r}_0) = -\delta(x - x_0)\delta(y - y_0)\delta(z - z_0) \quad (3\text{-}21)$$

Physically, the solution to (3-21) could represent the potential at the point $P(x, y, z)$ of a radiating charge at $P_0(x_0, y_0, z_0)$. However, in electromagnetics it is more common for G to represent the electric potential of an electric dipole. Since the electric dipole is equivalent to an infinitesimal linear current element,[6] G can also represent a component of the vector magnetic potential. These concepts will be covered more thoroughly in Chap. 4. As usual, the vector $\mathbf{r}$ is from the origin to P and the vector $\mathbf{r}_0$ is from the origin to P_0. However, at present, it is not necessary to attach any physical meaning to G in (3-21) since applications will be considered later. Since (3-21) is the same as (3-12), except that $g(x, y, z)$ is replaced by a unit source, we can write down the Fourier transform of (3-21) immediately from (3-14) as

$$(k^2 - \alpha^2)\tilde{G}(\alpha_1, \alpha_2, \alpha_3) = -(2\pi)^{-3/2}e^{-j(\alpha_1 x_0 + \alpha_2 y_0 + \alpha_3 z_0)} \quad (3\text{-}22)$$

The solution to (3-22) as given by the inverse transform of (3-22) is

$$G(x, y, z) = (2\pi)^{-3/2} \int\!\!\!\int\!\!\!\int_{-\infty}^{\infty} \tilde{G}(\alpha_1, \alpha_2, \alpha_3)e^{j(\alpha_1 x + \alpha_2 y + \alpha_3 z)}\, d\alpha_1\, d\alpha_2\, d\alpha_3$$

$$= -(2\pi)^{-3} \int\!\!\!\int\!\!\!\int_{-\infty}^{\infty} \frac{e^{j[\alpha_1(x - x_0) + \alpha_2(y - y_0) + \alpha_3(z - z_0)]}}{k^2 - \alpha_1^2 - \alpha_2^2 - \alpha_3^2}\, d\alpha_1\, d\alpha_2\, d\alpha_3 \quad (3\text{-}23)$$

The difficult part of the problem of solving (3-21) is evaluating the triple integral in (3-23). Since the integral in (3-23) is over all α-space (and α-space is just as meaningful as r-space) in a cartesian form, let us change the variables of integration from the cartesian form to the spherical form. Thus, we can define the quantities

$$\boldsymbol{\alpha} = \alpha_1 \mathbf{a}_x + \alpha_2 \mathbf{a}_y + \alpha_3 \mathbf{a}_z$$

$$\mathbf{R} = \mathbf{r} - \mathbf{r}_0 = (x - x_0)\mathbf{a}_x + (y - y_0)\mathbf{a}_y + (z - z_0)\mathbf{a}_z \quad (3\text{-}24)$$

$$= \boldsymbol{\rho} - \boldsymbol{\rho}_0 + (z - z_0)\mathbf{a}_z = \boldsymbol{\rho}_1 + (z - z_0)\mathbf{a}_z$$

---

[6]R. Plonsey and R. Collin, *Principles and Applications of Electromagnetic Fields*, McGraw-Hill Book Company, Inc., New York, 1961, Sec. 11.1.

and note that $dV = d\alpha_1 \, d\alpha_2 \, d\alpha_3 = \alpha^2 \, d\alpha \sin \beta \, d\beta \, d\gamma$. The angles β and γ in the cartesian α-coordinate system are analogous to θ and ϕ, respectively, in the physical coordinate system defined by the unit vectors $\mathbf{a}_i$, $i = x, y, z$. We could have introduced these ideas earlier by noting that the triple exponential Fourier transform pair can be written as

$$\tilde{f}(\boldsymbol{\alpha}) = (2\pi)^{-3/2} \int_{-\infty}^{\infty} f(\mathbf{r}) e^{-j\boldsymbol{\alpha}\cdot\mathbf{r}} \, dV_r$$

$$f(\mathbf{r}) = (2\pi)^{-3/2} \int_{-\infty}^{\infty} \tilde{f}(\boldsymbol{\alpha}) e^{j\boldsymbol{\alpha}\cdot\mathbf{r}} \, dV_\alpha$$

(3-25)

where $dV_r = dx \, dy \, dz$ and $dV_\alpha = d\alpha_1 \, d\alpha_2 \, d\alpha_3$ in a cartesian system. When we write the transform pair in this form, it becomes rather obvious that α-space and r-space are indistinguishable. It is also clear that we can evaluate the integrals in any coordinate system we choose. If we substitute (3-24) into (3-23) we obtain

$$G(\mathbf{r}) = -(2\pi)^{-3} \int_0^{\infty} d\alpha \int_0^{\pi} d\beta \int_0^{2\pi} d\gamma \, \frac{e^{j\boldsymbol{\alpha}\cdot\mathbf{R}} \alpha^2 \sin \beta}{k^2 - \alpha^2}$$

(3-26)

Since the cartesian coordinate system in α-space is arbitrary, we find it expedient to orient it so that the positive α_3-axis in α-space is in the direction of $\mathbf{R}$. For this particular choice $\boldsymbol{\alpha}\cdot\mathbf{R} = \alpha R \cos \beta$, and we obtain

$$G(\mathbf{r}) = -(2\pi)^{-3} \int_0^{\infty} \frac{\alpha^2 \, d\alpha}{k^2 - \alpha^2} \left[\frac{-e^{j\alpha R \cos \beta}}{j\alpha R} \right]_0^{\pi} (2\pi)$$

$$= \frac{(2\pi)^{-2}}{jR} \int_0^{\infty} \frac{e^{-j\alpha R} - e^{j\alpha R}}{k^2 - \alpha^2} \alpha \, d\alpha = -\frac{(2\pi)^{-2}}{jR} \int_{-\infty}^{\infty} \frac{\alpha e^{j\alpha R}}{k^2 - \alpha^2} \, d\alpha$$

$$= \frac{(2\pi)^{-2}}{2jR} \left[\int_{-\infty}^{\infty} \frac{e^{j\alpha R}}{\alpha - k} \, d\alpha + \int_{-\infty}^{\infty} \frac{e^{j\alpha R}}{\alpha + k} \, d\alpha \right]$$

(3-27)

As explained by Tralli,[7] the last two integrals are meaningless since the first one becomes infinite when $\alpha = k$ and the second one becomes infinite when $\alpha = -k$. However, this dilemma is common with improper integrals and can be resolved by deleting the small interval where $\alpha = |k|$ from the integral and then using a limiting procedure to determine the final value of the integral as the interval shrinks to a point. Thus, one defines the principal value (Cauchy's principal value) of such an integral as

$$\lim_{\epsilon \to 0} \left[\int_{-\infty}^{k-\epsilon} \frac{e^{j\alpha R}}{\alpha - k} \, d\alpha + \int_{k+\epsilon}^{\infty} \frac{e^{j\alpha R}}{\alpha - k} \, d\alpha \right] = \int_{-\infty}^{\infty} \frac{e^{j\alpha R} \, d\alpha}{\alpha - k}$$

---

[7] N. Tralli, *Classical Electromagnetic Theory*, McGraw-Hill Book Company, Inc., New York, 1963, pp. 293–98.

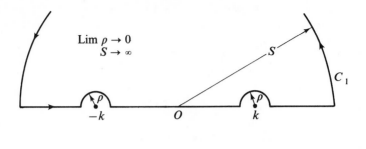

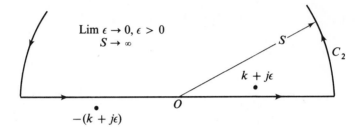

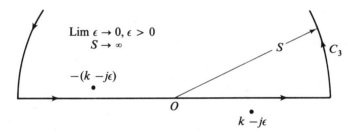

Fig. 3-2 Three contours for the evaluation of the inverse exponential Fourier transform.

Now, contour integration allows us to evaluate the integrals in (3-27) according to our choice for a contour. If we choose contour C_1 in Fig. 3-2 we obtain a value for the integral that corresponds to the principal value. In this case we have deformed the contour around the two poles and then straightened the contour through a limiting procedure. On the other hand, we could have moved the poles slightly off the real α-axis, which would avoid a deformation of the contour, and then used a limiting procedure to move the poles back to their original location.[8] This is illustrated in Fig. 3-2 with contours C_2

---

[8]G. Arfken, *Mathematical Methods for Physicists*, Academic Press, Inc., New York, 1966, Sec. 7.2.

and C_3. In contour C_3 the poles are moved into the second and fourth quadrants. Although this procedure for evaluating the integrals seems artificial, we will see later that each contour C_i yields a Green's function $G_i(\mathbf{r})$ that is physically meaningful. If we carry through the steps indicated above, we obtain three different Green's functions as solutions to (3-21). Thus,

$$G_i(\mathbf{r}) = \frac{\mathfrak{F}_i}{4\pi R}, \qquad i = 1, 2, 3 \qquad (3\text{-}28)$$

where

$$\mathfrak{F}_1 = \cos kR$$
$$\mathfrak{F}_2 = e^{jkR}$$
$$\mathfrak{F}_3 = e^{-jkR}$$

The solutions $G_i(\mathbf{r})$ in (3-28) represent a standing spherical wave and two traveling spherical waves, respectively. G_1 does not satisfy the general radiation condition (3-20) for $h = 3$ but G_3 does. Since G_2 represents an incoming traveling spherical wave, it satisfies the radiation condition with the sign of k changed. The details are left as Prob. 3.14. It is also possible to evaluate the integral in (3-23) in different forms so that alternate solutions are obtained. Let us examine the problem further in an effort to obtain alternate forms for these solutions. These alternate solutions will yield a large number of useful identities involving plane, cylindrical, and spherical waves. Since the present solutions (3-28) represent spherical waves, let us seek alternate solutions involving spherical Bessel functions. We do this by replacing the integral over β in (3-26) by the identity obtained in Sec. 1.8 for $n = 0$, i.e.,

$$j_0(kR) = \frac{1}{2} \int_0^\pi e^{jkR \cos \beta} \sin \beta \, d\beta \qquad (1\text{-}37a)$$

Now the expression for $G(\mathbf{r})$ becomes

$$G(\mathbf{r}) = \frac{1}{2\pi^2} \int_0^\infty \frac{j_0(\alpha R)\alpha^2 \, d\alpha}{\alpha^2 - k^2} \qquad (3\text{-}29)$$

This integral can be extended to include negative values of α by replacing $j_0(\alpha R)$ through the identities in Sec. 1.8 or Appendix C.3.9, i.e.,

$$j_0(\alpha R) = \frac{1}{2}[h_0^{(1)}(\alpha R) + h_0^{(2)}(\alpha R)] = \frac{1}{2}[h_0^{(1)}(\alpha R) + h_0^{(1)}(-\alpha R)]$$

whence we obtain

$$G(\mathbf{r}) = \frac{1}{8\pi^2}\left[\int_{-\infty}^\infty \frac{h_0^{(1)}(\alpha R)\alpha \, d\alpha}{\alpha - k} + \int_{-\infty}^\infty \frac{h_0^{(1)}(\alpha R)\alpha \, d\alpha}{\alpha + k} \right] \qquad (3\text{-}30)$$

When we evaluate the two integrals above using the three contours shown in Fig. 3-2, we find the following three Green's functions as solutions to (3-21):

$$G_i(\mathbf{r}) = \frac{k}{4\pi}\mathcal{G}_i, \qquad i = 1, 2, 3 \tag{3-31}$$

where

$$\mathcal{G}_1 = -n_0(kR)$$
$$\mathcal{G}_2 = jh_0^{(1)}(kR)$$
$$\mathcal{G}_3 = -jh_0^{(2)}(kR)$$

Since both (3-28) and (3-31) are solutions to the same problem, we may compare the Green's functions for identical contours and obtain expressions for the zero order spherical Bessel functions. Thus, we find directly that

$$n_0(kR) = -\frac{\cos kR}{kR}$$

$$h_0^{(1)}(kR) = \frac{e^{jkR}}{jkR} \tag{3-32}$$

$$h_0^{(2)}(kR) = -\frac{e^{-jkR}}{jkR}$$

and indirectly that

$$j_0(kR) = \frac{\sin kR}{kR}$$

The solutions (3-28) and (3-31) just obtained for (3-21) represent spherical waves as mentioned previously. On the other hand, if we Fourier transform out the z-dependence of (3-21), we obtain the two-dimensional equation

$$(\nabla_t^2 + k_1^2)\tilde{G}(x, y, \alpha_3) = -\frac{e^{-j\alpha_3 z_0}}{\sqrt{2\pi}}\delta(x - x_0)\delta(y - y_0) \tag{3-33}$$

where

$$k_1^2 = k^2 - \alpha_3^2$$

The solution to (3-33) could represent, for instance, the magnetic potential of a radiating, uniform, line source of strength $e^{-j\alpha_3 z_0}/\sqrt{2\pi}$ located at (x_0, y_0, z). Since (3-33) is a function only of the coordinates transverse to z, i.e., x and y or ρ and ϕ, we would expect the solution to (3-33) to possess cylindrical symmetry and to represent cylindrical standing and traveling waves. However, let us proceed with the solution of (3-33) and verify our expectations. Since we have the direct transform of (3-33) in (3-22), we may write the solution of (3-33) as

$$\tilde{G}(x, y, \alpha_3) = \frac{1}{2\pi} \int\int_{-\infty}^{\infty} \tilde{G}(\alpha_1, \alpha_2, \alpha_3) e^{j(\alpha_1 x + \alpha_2 y)} \, d\alpha_1 \, d\alpha_2 \qquad (3\text{-}34)$$

where

$$\tilde{G}(\alpha_1, \alpha_2, \alpha_3) = -(2\pi)^{-3/2}(k^2 - \alpha^2)^{-1} e^{-j(\alpha_1 x_0 + \alpha_2 y_0 + \alpha_3 z_0)} \qquad (3\text{-}22)$$

We now transform from cartesian to cylindrical coordinates in α-space by letting $\lambda = \alpha_1 \mathbf{a}_x + \alpha_2 \mathbf{a}_y = \alpha - \alpha_3 \mathbf{a}_z$ and $d\alpha_1 \, d\alpha_2 = \lambda \, d\lambda \, d\gamma$ where λ and γ are analogous to ρ and ϕ, respectively. With these substitutions and the use of (3-22), the expression in (3-34) becomes

$$\tilde{G}(x, y, \alpha_3) = -(2\pi)^{-5/2} \int\int_{-\infty}^{\infty} \frac{e^{j[\alpha_1(x-x_0) + \alpha_2(y-y_0)]}}{k_1^2 - \alpha_1^2 - \alpha_2^2} e^{-j\alpha_3 z_0} \, d\alpha_1 \, d\alpha_2$$

$$= -(2\pi)^{-5/2} \int_0^{\infty} \frac{e^{-j\alpha_3 z_0} \lambda \, d\lambda}{k_1^2 - \lambda^2} \int_0^{2\pi} e^{j\lambda\rho_1 \cos (\gamma-\delta)} \, d\gamma \qquad (3\text{-}35)$$

where

$$k_1^2 - \lambda^2 = k^2 - \alpha^2$$

$$|\mathbf{\rho}_1| = |\mathbf{\rho} - \mathbf{\rho}_0| = [\rho^2 + \rho_0^2 - 2\rho\rho_0 \cos (\phi - \phi_0)]^{1/2}$$

$$= [(x - x_0)^2 + (y - y_0)^2]^{1/2}$$

$$\tan \delta = \frac{(y - y_0)}{(x - x_0)}$$

$$\tan \gamma = \frac{\alpha_2}{\alpha_1}$$

The integral over γ can be evaluated by using the identity

$$J_0(\lambda\rho_1) = \frac{1}{2\pi} \int_0^{2\pi} e^{j\lambda\rho_1 \cos (\gamma-\delta)} \, d\gamma$$

which can be obtained from (1-20) by substituting $\lambda\rho_1$ for x and $\pi/2 - (\gamma - \delta)$ for ϕ. Now we obtain

$$\tilde{G}(x, y, \alpha_3) = -(2\pi)^{-3/2} e^{-j\alpha_3 z_0} \int_0^{\infty} \frac{J_0(\lambda\rho_1)}{k_1^2 - \lambda^2} \lambda \, d\lambda \qquad (3\text{-}36)$$

and use the identities from Sec. 1.4 or Appendices C.1.4 and C.1.9, i.e.,

$$J_0(\lambda\rho_1) = \frac{1}{2}[H_0^{(1)}(\lambda\rho_1) + H_0^{(2)}(\lambda\rho_1)] = \frac{1}{2}[H_0^{(1)}(\lambda\rho_1) - H_0^{(1)}(-\lambda\rho_1)]$$

to extend the above integral to include negative values of λ. This allows us to put the integral into the same form as that encountered in (3-27) and later in

(3-30). Thus, our final form is

$$\tilde{G}(x, y, \alpha_3) = \frac{1}{4}(2\pi)^{-3/2} e^{-j\alpha_3 z_0} \left[\int_{-\infty}^{\infty} \frac{H_0^{(1)}(\lambda\rho_1)\, d\lambda}{\lambda - k_1} + \int_{-\infty}^{\infty} \frac{H_0^{(1)}(\lambda\rho_1)\, d\lambda}{\lambda + k_1} \right]$$

$$(3\text{-}37)$$

Again we evaluate the two integrals for the three different contours specified in Fig. 3-2. This gives us the following three Green's functions as solutions to (3-33):

$$\tilde{G}_i(x, y, \alpha_3) = \frac{\pi}{2}(2\pi)^{-3/2} e^{-j\alpha_3 z_0}\mathfrak{IC}_i, \qquad i = 1, 2, 3 \qquad (3\text{-}38)$$

where

$$\mathfrak{IC}_1 = -N_0(k_1\rho_1)$$
$$\mathfrak{IC}_2 = jH_0^{(1)}(k_1\rho_1)$$
$$\mathfrak{IC}_3 = -jH_0^{(2)}(k_1\rho_1)$$

The Green's functions in (3-38) represent a standing cylindrical wave, an inwardly propagating cylindrical wave, and an outwardly propagating cylindrical wave, respectively. The outwardly propagating cylindrical wave satisfies the general Sommerfeld radiation condition (3-20) for $h = 2$. The details are left as Prob. 3.15.

Let us now proceed to find identities relating cylindrical waves to spherical waves. We can obtain a relation between spherical and cylindrical waves by using the fact that $\tilde{G}(x, y, \alpha_3)$ in (3-34) can be written as the direct exponential Fourier transform of $G(x, y, z)$, i.e.,

$$\tilde{G}(x, y, \alpha_3) = \frac{1}{\sqrt{2\pi}} \int_{-\infty}^{\infty} G(x, y, z) e^{-j\alpha_3 z}\, dz \qquad (3\text{-}39)$$

and

$$G(x, y, z) = \frac{1}{\sqrt{2\pi}} \int_{-\infty}^{\infty} \tilde{G}(x, y, \alpha_3) e^{j\alpha_3 z}\, d\alpha_3 \qquad (3\text{-}40)$$

Since $G(x, y, z)$ has already been evaluated for three different contours, we can substitute these expressions from (3-28) into (3-39) to derive the following three identities:

$$N_0(k_1\rho_1) = -\frac{1}{\pi} \int_{-\infty}^{\infty} \frac{\cos kR}{R} e^{-j\alpha_3(z-z_0)}\, dz$$

$$H_0^{(1)}(k_1\rho_1) = \frac{-j}{\pi} \int_{-\infty}^{\infty} \frac{e^{jkR}}{R} e^{-j\alpha_3(z-z_0)}\, dz$$

$$H_0^{(2)}(k_1\rho_1) = \frac{j}{\pi} \int_{-\infty}^{\infty} \frac{e^{-jkR}}{R} e^{-j\alpha_3(z-z_0)}\, dz$$

We can obtain a more useful form for these identities by allowing α_3 to vanish, thus

$$N_0(k\rho_1) = -\frac{1}{\pi} \int_{-\infty}^{\infty} \frac{\cos kR}{R} dz$$

$$H_0^{(1)}(k\rho_1) = \frac{-j}{\pi} \int_{-\infty}^{\infty} \frac{e^{jkR}}{R} dz \qquad\qquad (3\text{-}41)$$

$$H_0^{(2)}(k\rho_1) = \frac{j}{\pi} \int_{-\infty}^{\infty} \frac{e^{-jkR}}{R} dz$$

where R and ρ_1 are defined in (3-24). It is also possible to replace the integrands in (3-41) by the identities from (3-32), whence we obtain the general expression

$$B_0(k\rho_1) = \frac{k}{\pi} \int_{-\infty}^{\infty} b_0(kR) dz \qquad\qquad (3\text{-}42)$$

One could also obtain this expression directly by using (3-31) and (3-38) in (3-39). As before, $B_0(k\rho_1)$ is one of the zero order Bessel, Neumann, or Hankel cylinder functions and $b_0(kR)$ is the corresponding zero order spherical Bessel, Neumann, or Hankel function. It is interesting to note the physical meaning of the last integral in (3-41). We show in the next chapter in (4-28) that this integral is proportional to the z-component of the vector magnetic potential of an infinite, uniform, time-varying, z-directed, electric line source. However, the line source can be considered as an infinite number of infinitesimal electric current elements placed end to end. Since the z-component of the vector magnetic potential[9] of each of these elements, which are called electric dipoles, is given by G_3 in (3-28), it follows that the last integral in (3-41) represents the addition of each of these identical uniform spherical waves to form a uniform cylindrical wave. The same physical meaning can be attached to the last integral in (3-50). Thus, the uniform plane wave on the left is represented as a sum of the infinite number of uniform cylindrical waves on the right (let $\alpha_2 = 0$ for simplicity). This viewpoint is the same as that obtained by adding the sources, i.e., the sum of the uniform z-directed line sources is a uniform z-directed current sheet. This point is discussed further in connection with the derivation of (4-31) in Chap. 4.

There are also situations where it is desirable to have expressions that are inverse to those in (3-41) or (3-42), i.e., a spherical wave represented as a sum (integral) of cylindrical waves. In order to derive such identities, let us consider the transform pair (3-39) and (3-40) with attention directed toward the latter, since it represents a spherical wave in terms of a spectrum of cylindrical

---

[9]Plonsey and Collin, *Principles and Applications of Electromagnetic Fields*, p. 394.

waves. The validity of this last statement follows if one considers the representations for $G(x, y, z)$ in (3-28) and the representations for $\tilde{G}(x, y, \alpha_3)$ in (3-38). In problems dealing with point sources in planar regions it is often desirable to express the spherical waves from the sources in terms of cylindrical waves in order to facilitate the application of boundary conditions. A typical example is the Sommerfeld problem, i.e., a Hertz electric dipole above a plane earth, which will be treated in Chap. 7. Since the boundary conditions must be invoked at $z = 0$, it is convenient to modify the representation in (3-40) so that the variable of integration is not α_3. We shall derive this so-called Sommerfeld identity[10] later.

If we substitute (3-28) into the left side and (3-38) into the right side of (3-40) we obtain

$$\frac{\mathcal{F}_i}{R} = \frac{1}{2} \int_{-\infty}^{\infty} \mathcal{K}_i e^{j\alpha_3(z - z_0)} \, d\alpha_3 \tag{3-43}$$

Now, we use the relation

$$\alpha_3 = j\sqrt{k_1^2 - k^2}$$

to change the variable of integration from α_3 to k_1. The desired result is that

$$\frac{\mathcal{F}_i}{R} = \frac{j}{2} \int_{-\infty}^{\infty} \frac{\mathcal{K}_i e^{-\sqrt{k_1^2 - k^2}|z - z_0|}}{\sqrt{k_1^2 - k^2}} k_1 \, dk_1$$

The two cases that are of chief interest occur when $i = 2, 3$. Thus, the desired representations are

$$\frac{e^{\pm jkR}}{R} = \mp \frac{1}{2} \int_{-\infty}^{\infty} \frac{H_0^{(1,2)}(k_1\rho_1) e^{-\sqrt{k_1^2 - k^2}|z - z_0|}}{\sqrt{k_1^2 - k^2}} k_1 \, dk_1 \tag{3-44}$$

where $\rho_1^2 = R^2 - (z - z_0)^2$.

If one wishes a representation for the spherical Bessel functions in terms of cylindrical Bessel functions, then (3-31) should be substituted for $G(x, y, z)$ in the left side of (3-40). If this is done, one obtains the identity

$$\mathcal{G}_i = \frac{1}{2k} \int_{-\infty}^{\infty} \mathcal{K}_i e^{j\alpha_3(z - z_0)} \, d\alpha_3 \tag{3-45}$$

[10]Sommerfeld, *Partial Differential Equations in Physics*, p. 242. Also, see A. Baños, Jr., *Dipole Radiation in the Presence of a Conducting Half-Space*, Pergamon Press, Inc., New York, 1966, Sec. 2.12; and J. Van Bladel, *Electromagnetic Fields*, McGraw-Hill Book Company, Inc., New York, 1964, p. 398.

and the representations

$$n_0(kR) = \frac{1}{2k} \int_{-\infty}^{\infty} N_0(k_1\rho_1)e^{j\alpha_3(z-z_0)}\, d\alpha_3$$

$$h_0^{(1,2)}(kR) = \frac{1}{2k} \int_{-\infty}^{\infty} H_0^{(1,2)}(k_1\rho_1)e^{j\alpha_3(z-z_0)}\, d\alpha_3$$

On the other hand, if one substitutes the expressions for $\mathfrak{F}_i$ from (3-28) into the left side of (3-43) and the expressions for $\mathfrak{IC}_i$ from (3-38) into the right side of (3-43), it is possible to obtain expressions that are of the form of (3-44) but without the change of variable from α_3 to k_1. The desired representations[11] are

$$\frac{\cos kR}{R} = \frac{-1}{2} \int_{-\infty}^{\infty} N_0(k_1\rho_1)e^{j\alpha_3(z-z_0)}\, d\alpha_3$$

$$\frac{e^{\pm jkR}}{R} = \frac{\pm j}{2} \int_{-\infty}^{\infty} H_0^{(1,2)}(k_1\rho_1)e^{j\alpha_3(z-z_0)}\, d\alpha_3 \tag{3-46}$$

where $k_1^2 = k^2 - \alpha_3^2$ and R and ρ_1 are defined in (3-24), i.e., R and ρ_1 are the radial distances between the source and field points in spherical and cylindrical coordinates, respectively, and as vectors point toward the field point.

If we attempt to attach a physical meaning to the integrals in (3-45) and (3-46), we find that the problem is more difficult since we are not adding identical waves at different physical locations as we did in (3-41). In this case, we are adding waves at the same physical location but they are propagating in different directions. This viewpoint is discussed by Stratton[12] in connection with an expression similar to the last one in (3-35). This same viewpoint can also be used in connection with expressions such as (3-51) and (3-54).

We now wish to obtain identities involving plane waves and cylindrical waves. In order to do this, let us take the direct exponential Fourier transform of (3-33) with respect to y. This gives us the simple ordinary differential equation

$$\left(\frac{d^2}{dx^2} + k_2^2\right) \tilde{G}(x, \alpha_2, \alpha_3) = -D\delta(x - x_0) \tag{3-47}$$

where

$$k_2^2 = k_1^2 - \alpha_2^2$$

---

[11]Note that the expression in (3-46) with $i = 3$ is given by W. Weeks, *Electromagnetic Theory for Engineering Applications*, John Wiley & Sons, Inc., New York, 1964, p. 525.

[12]J. Stratton, *Electromagnetic Theory*, McGraw-Hill Book Company, Inc., New York, 1941, p. 370.

and

$$D = (2\pi)^{-1} e^{-j(\alpha_2 y_0 + \alpha_3 z_0)}$$

Since we know $\tilde{G}(\alpha_1, \alpha_2, \alpha_3)$, we can write down $\tilde{G}(x, \alpha_2, \alpha_3)$ as the inverse exponential Fourier transform with respect to α_1, i.e.,

$$\tilde{G}(x, \alpha_2, \alpha_3) = \frac{1}{\sqrt{2\pi}} \int_{-\infty}^{\infty} \tilde{G}(\alpha_1, \alpha_2, \alpha_3) e^{j\alpha_1 x} \, d\alpha_1$$

If we substitute $\tilde{G}(\alpha_1, \alpha_2, \alpha_3)$ from (3-22) into the above expression, we find that

$$\tilde{G}(x, \alpha_2, \alpha_3) = (2\pi)^{-2} e^{-j(\alpha_2 y_0 + \alpha_3 z_0)} \int_{-\infty}^{\infty} \frac{e^{j\alpha_1(x - x_0)}}{\alpha_1^2 - k_2^2} \, d\alpha_1$$

$$= \frac{e^{-j(\alpha_2 y_0 + \alpha_3 z_0)}}{2k_2(2\pi)^2} \left[\int_{-\infty}^{\infty} \frac{e^{j\alpha_1(x - x_0)}}{\alpha_1 - k_2} \, d\alpha_1 - \int_{-\infty}^{\infty} \frac{e^{j\alpha_1(x - x_0)}}{\alpha_1 + k_2} \, d\alpha_1 \right] \quad (3\text{-}48)$$

Since (3-48) is of the same form as (3-27), (3-30), and (3-39), we can once again evaluate the two integrals according to the three different contours indicated in Fig. 3-2. When this is accomplished, we obtain the following three Green's functions as solutions to (3-47):

$$\tilde{G}_i(x, \alpha_2, \alpha_3) = \frac{je^{-j(\alpha_2 y_0 + \alpha_3 z_0)}}{4\pi k_2} \mathcal{L}_i, \quad i = 1, 2, 3 \quad (3\text{-}49)$$

where

$$\mathcal{L}_1 = j \sin k_2(x - x_0)$$
$$\mathcal{L}_2 = e^{jk_2(x - x_0)}$$
$$\mathcal{L}_3 = -e^{-jk_2(x - x_0)}$$

and

$$k_2^2 = k^2 - \alpha_2^2 - \alpha_3^2$$

In order to obtain the desired identities between cylindrical and plane waves, let us consider the Fourier transform pair

$$\tilde{f}(x, \alpha_2, \alpha_3) = \frac{1}{\sqrt{2\pi}} \int_{-\infty}^{\infty} \tilde{f}(x, y, \alpha_3) e^{-j\alpha_2 y} \, dy$$

$$\tilde{f}(x, y, \alpha_3) = \frac{1}{\sqrt{2\pi}} \int_{-\infty}^{\infty} \tilde{f}(x, \alpha_2, \alpha_3) e^{j\alpha_2 y} \, d\alpha_2$$

If we substitute $\tilde{G}_i(x, y, \alpha_3)$ from (3-38) for $\tilde{f}(x, y, \alpha_3)$ and $\tilde{G}_i(x, \alpha_2, \alpha_3)$ from

(3-49) for $\tilde{f}(x, \alpha_2, \alpha_3)$ into the first expression above, we find that

$$\mathfrak{L}_i = \frac{k_2}{2j} \int_{-\infty}^{\infty} \mathfrak{K}_i e^{-j\alpha_2(y-y_0)} \, dy$$

whence we obtain the desired identities:

$$\frac{\sin k_2(x - x_0)}{k_2} = \frac{1}{2} \int_{-\infty}^{\infty} N_0(k_1 \rho_1) e^{-j\alpha_2(y-y_0)} \, dy$$

$$\frac{e^{jk_2(x-x_0)}}{k_2} = \frac{1}{2} \int_{-\infty}^{\infty} H_0^{(1)}(k_1 \rho_1) e^{-j\alpha_2(y-y_0)} \, dy \qquad (3\text{-}50)$$

$$\frac{e^{-jk_2(x-x_0)}}{k_2} = \frac{1}{2} \int_{-\infty}^{\infty} H_0^{(2)}(k_1 \rho_1) e^{-j\alpha_2(y-y_0)} \, dy$$

One should also note that, in the identities (3-41) and (3-50), the variable of integration can be a source coordinate or a field coordinate, i.e., z or z_0 in (3-41) and y or y_0 in (3-50).

The inverse of the above is the expression

$$\mathfrak{K}_i = \frac{j}{\pi} \int_{-\infty}^{\infty} \frac{\mathfrak{L}_i e^{j\alpha_2(y-y_0)}}{k_2} \, d\alpha_2$$

whence we obtain the identities:

$$N_0(k_1 \rho_1) = \frac{1}{\pi} \int_{-\infty}^{\infty} \frac{\sin k_2(x - x_0) e^{j\alpha_2(y-y_0)}}{k_2} \, d\alpha_2$$

$$H_0^{(1)}(k_1 \rho_1) = \frac{1}{\pi} \int_{-\infty}^{\infty} \frac{e^{jk_2(x-x_0)} e^{j\alpha_2(y-y_0)}}{k_2} \, d\alpha_2 \qquad (3\text{-}51)$$

$$H_0^{(2)}(k_1 \rho_1) = \frac{1}{\pi} \int_{-\infty}^{\infty} \frac{e^{-jk_2(x-x_0)} e^{j\alpha_2(y-y_0)}}{k_2} \, d\alpha_2$$

where

$$k_2^2 = k_1^2 - \alpha_2^2$$

Let us finally consider the derivation of the Sommerfeld identity or integral representation. The details are considered quite fully by Baños.[13] The desired representation is similar to (3-44) with a different variable of integration. However, we can derive the expression more directly (or alternately) by starting with the identity (3-40) and substituting $G(x, y, z)$ from (3-28) into the left side as before. Instead of substituting (3-38) for $\tilde{G}(x, y, \alpha_3)$

---

[13]Baños, *Dipole Radiation in the Presence of a Conducting Half-Space*, Sec. 2.12.

we use (3-36). This gives us the expression

$$\frac{\mathcal{F}_i}{4\pi R} = \frac{1}{(2\pi)^2} \int_0^\infty \lambda \, d\lambda \, J_0(\lambda \rho_1) \int_{-\infty}^\infty \frac{e^{j\alpha_3(z-z_0)}}{\alpha_3^2 - k_3^2} \, d\alpha_3$$

where

$$k_1^2 - \lambda^2 = k_3^2 - \alpha_3^2, \qquad k_1^2 = k^2 - \alpha_3^2, \qquad k_3^2 = k^2 - \lambda^2$$

and the integration over α_3 must be evaluated for the three contours in Fig. 3-2. This last integral is of the same form as that in (3-48) so that we can write down the result at once, i.e.,

$$\int_{-\infty}^\infty \frac{e^{j\alpha_3(z-z_0)}}{\alpha_3^2 - k_3^2} \, d\alpha_3 = \frac{1}{2k_3}\left[\int_{-\infty}^\infty \frac{e^{j\alpha_3(z-z_0)}}{\alpha_3 - k_3} \, d\alpha_3 - \int_{-\infty}^\infty \frac{e^{j\alpha_3(z-z_0)}}{\alpha_3 + k_3} \, d\alpha_3 \right]$$

$$= \frac{\pi j}{k_3}\mathcal{L}_i, \qquad i = 1, 2, 3$$

where $\mathcal{L}_i$ is given by (3-49) with k_2, x, and x_0 replaced by k_3, z, and z_0, respectively. This gives us the identity

$$\frac{\mathcal{F}_i}{R} = j \int_0^\infty \frac{\mathcal{L}_i J_0(\lambda \rho_1)\lambda \, d\lambda}{k_3}$$

and the representations[14]

$$\frac{\cos kR}{R} = -\int_0^\infty \frac{\lambda J_0(\lambda \rho_1) \sin k_3(z - z_0)}{k_3} d\lambda$$

$$\frac{e^{\pm jkR}}{R} = \pm j \int_0^\infty \frac{\lambda J_0(\lambda \rho_1) e^{\pm j k_3(z-z_0)}}{k_3} d\lambda \qquad (3\text{-}52)$$

The last expression can be put into the form of Sommerfeld's identity by the substitution

$$k_3 = \pm j\sqrt{\lambda^2 - k^2}$$

where the two different signs for the radical are required in order to insure the proper behavior of the integrand at $z = \infty$. The desired result[15] is

$$\frac{e^{\pm jkR}}{R} = \int_0^\infty \frac{\lambda J_0(\lambda \rho_1) e^{-(\lambda^2 - k^2)^{1/2}|z - z_0|}}{(\lambda^2 - k^2)^{1/2}} d\lambda \qquad (3\text{-}53)$$

---

[14]Stratton, *Electromagnetic Theory*, p. 391.

[15]*Ibid.*, p. 576.

The last identities we wish to derive are similar to (3-52) except that the right side will be given in terms of plane waves rather than cylindrical waves. Again, we start with the Fourier transform identity (3-40) but now replace $\tilde{G}(x, y, \alpha_3)$ by the expression given in (3-35). The integration with respect to α_3 can be evaluated immediately for the three contours in Fig. 3-2 so that we obtain the identity

$$\frac{\mathfrak{F}_i}{R} = \frac{j}{2\pi} \int\!\!\int_{-\infty}^{\infty} \frac{\mathfrak{L}_i e^{j[\alpha_1(x-x_0)+\alpha_2(y-y_0)]}}{k_3}\, d\alpha_1\, d\alpha_2$$

and the representations[16]

$$\frac{\cos kR}{R} = \frac{-1}{2\pi} \int\!\!\int_{-\infty}^{\infty} \frac{\sin k_3(z-z_0) e^{j[\alpha_1(x-x_0)+\alpha_2(y-y_0)]}}{k_3}\, d\alpha_1\, d\alpha_2$$

$$\frac{e^{\pm jkR}}{R} = \pm \frac{j}{2\pi} \int\!\!\int_{-\infty}^{\infty} \frac{e^{j[\alpha_1(x-x_0)+\alpha_2(y-y_0)\pm k_3(z-z_0)]}}{k_3}\, d\alpha_1\, d\alpha_2$$

(3-54)

3.2 Cylindrical Fourier Transforms

Let us consider the general exponential Fourier transform pair defined in (3-25) but with the coordinate system restricted to the cylindrical coordinate system. We saw earlier that when one restricted the coordinate system to the cartesian coordinate system in (3-25), the triple exponential cartesian Fourier transform pair in (3-15) resulted. We shall see that when one restricts (3-25) to the cylindrical coordinates, the Fourier-Bessel or Hankel transform pair[17] results.

We proceed by noting that in the cylindrical coordinate system $dV_r = \rho\, d\rho\, d\phi\, dz$, $dV_\alpha = \lambda\, d\lambda\, d\gamma\, d\alpha_3$, $\mathbf{r} = \rho\mathbf{a}_\rho + z\mathbf{a}_z$, and $\boldsymbol{\alpha} = \lambda\mathbf{a}_\lambda + \alpha_3\mathbf{a}_z$. The distance λ is analogous to ρ and the angle γ is analogous to ϕ as mentioned previously in the discussion leading to (3-35). If we substitute the above relations into (3-25), we obtain the following expressions:

$$\tilde{f}(\lambda, \alpha_3) = (2\pi)^{-3/2} \int_0^\infty \int_0^{2\pi} \int_{-\infty}^\infty f(\boldsymbol{\rho}, z) e^{-j[\lambda\rho \cos(\phi-\gamma)+\alpha_3 z]}\, \rho\, d\rho\, d\phi\, dz$$

$$f(\boldsymbol{\rho}, z) = (2\pi)^{-3/2} \int_0^\infty \int_0^{2\pi} \int_{-\infty}^\infty \tilde{f}(\lambda, \alpha_3) e^{j[\lambda\rho \cos(\phi-\gamma)+\alpha_3 z]}\, \lambda\, d\lambda\, d\gamma\, d\alpha_3$$

(3-55)

---

[16]D. Jones, *The Theory of Electromagnetism*, Pergamon Press, Inc., New York, 1964, p. 369.

[17]H. Margenau and G. Murphy, *The Mathematics of Physics and Chemistry*, D. Van Nostrand Company, Inc., Princeton, N.J., 1943, p. 250. Also, see Stratton, *Electromagnetic Theory*, p. 369.

Since the cylindrical coordinate system is different from the cartesian coordinate system only in the plane transverse to z, let us consider the above expressions as they apply to the transverse coordinates, ρ, ϕ, and λ, γ. Thus, we can reduce the expressions above from a triple to a double Fourier transform pair

$$\tilde{g}(\lambda) = (2\pi)^{-1} \int_0^\infty \int_0^{2\pi} g(\mathbf{\rho}) e^{-j\lambda\rho \cos (\phi - \gamma)} \rho \, d\rho \, d\phi$$

$$g(\mathbf{\rho}) = (2\pi)^{-1} \int_0^\infty \int_0^{2\pi} \tilde{g}(\lambda) e^{j\lambda\rho \cos (\phi - \gamma)} \lambda \, d\lambda \, d\gamma$$

by letting $f(\mathbf{\rho}, z) = g(\mathbf{\rho})h(z)$. We now assume that $g(\mathbf{\rho}) = \sum_{n=-\infty}^{\infty} g_n(\rho)e^{jn\phi}$. The first expression in (3-55) becomes

$$\tilde{g}(\lambda) = (2\pi)^{-1} \sum_{n=-\infty}^{\infty} \int_0^{2\pi} g_n(\rho) \rho \, d\rho \int_0^{2\pi} e^{-j[\lambda\rho \cos (\phi - \gamma) - n\phi]} \, d\phi$$

However, we have the identity obtained in Prob. 1.7, namely,

$$J_n(x) = \frac{(\mp j)^n}{2\pi} \int_0^{2\pi} e^{\pm j(x \cos \phi' - n\phi')} \, d\phi' \qquad (3\text{-}56)$$

We also note that $(\pm j)^{-n} = (\mp j)^n$. This identity was obtained by using the orthogonality of the functions $e^{\pm jn\phi}$ over an interval of length 2π. We now make the change of variable $\phi' = \phi - \gamma$ in the expression for $\tilde{g}(\lambda)$ above and use the identity for $J_n(x)$ in (3-56) (lower sign) to obtain

$$\tilde{g}(\lambda) = \sum_{n=-\infty}^{\infty} (-j)^n e^{jn\gamma} \int_0^\infty g_n(\rho) J_n(\lambda\rho)\rho \, d\rho = \sum_{n=-\infty}^{\infty} (-j)^n e^{jn\gamma} \tilde{g}_n(\lambda) \qquad (3\text{-}57)$$

If we substitute the expression for $\tilde{g}(\lambda)$ from (3-57) into the expression for $g(\mathbf{\rho})$ that results from (3-55), we obtain

$$g(\mathbf{\rho}) = (2\pi)^{-1} \sum_{n=-\infty}^{\infty} (-j)^n \int_0^\infty \tilde{g}_n(\lambda)\lambda \, d\lambda \int_0^{2\pi} e^{j[\lambda\rho \cos (\phi - \gamma) + n\gamma]} \, d\gamma$$

Once again we use the identity for $J_n(x)$ from (3-56) (upper sign) and change of variable $\phi' = \pi - (\phi - \gamma)$ in the expression above to obtain

$$g(\mathbf{\rho}) = \sum_{n=-\infty}^{\infty} e^{jn\phi} \int_0^\infty \tilde{g}_n(\lambda) J_n(\lambda\rho)\lambda \, d\lambda = \sum_{n=-\infty}^{\infty} e^{jn\phi} g_n(\rho) \qquad (3\text{-}58)$$

A comparison of (3-57) and (3-58) leads us to the desired Fourier-Bessel (Hankel) transform pair.

$$g_n(\rho) = \int_0^\infty \tilde{g}_n(\lambda) J_n(\lambda\rho)\lambda \, d\lambda$$

$$\tilde{g}_n(\lambda) = \int_0^\infty g_n(\rho) J_n(\lambda\rho)\rho \, d\rho$$

$$0 \le \rho < \infty, \lambda \ge 0, n \text{ an integer} \qquad (3\text{-}59)$$

In problems where $f(\rho)$ is defined in the region not including the origin, i.e., $0 < a \leq \rho < \infty$, it is possible to introduce a Weber[18] transform. Also, the cylindrical Hankel transform pair in (3-59) is valid for noninteger values of n even though the derivation assumed integer values of n. In the case of noninteger values of n, all positive values are permissible.[19]

Now we consider the problem encountered when one Fourier transforms the inhomogeneous scalar wave equation in cylindrical coordinates. This problem was previously discussed in Sec. 3.1 and led to the specific solution (3-36) when the excitation was a uniform line source. The approach we use now is more direct and general and should yield the same specific solution (3-36). We start with the inhomogeneous scalar wave equation as given by (3-12)

$$(\nabla^2 + k^2)f(\rho, \phi, z) = g(\rho, \phi, z) \tag{3-12}$$

and assume that $f(\rho, \phi, z)$ and $g(\rho, \phi, z)$ can be expanded as follows:

$$
\begin{aligned}
f(\rho, \phi, z) &= \sum_{n=-\infty}^{\infty} f_n(\rho, z)e^{jn\phi} \\
g(\rho, \phi, z) &= \sum_{m=-\infty}^{\infty} g_m(\rho, z)e^{jm\phi}
\end{aligned}
\tag{3-60}
$$

We now substitute these expressions into (3-12), multiply by $(1/\sqrt{2\pi})e^{-j\alpha_3 z}$, and integrate over z from minus to plus infinity to obtain

$$\sum_{n=-\infty}^{\infty}\left[\frac{1}{\rho}\frac{d}{d\rho}\left(\rho\frac{d}{d\rho}\right) + \left(k_1^2 - \frac{n^2}{\rho^2}\right)\right]\tilde{f}_n(\rho, \alpha_3)e^{jn\phi} = \sum_{m=-\infty}^{\infty}\tilde{g}_m(\rho, \alpha_3)e^{jm\phi}$$

where $k_1^2 = k^2 - \alpha_3^2$.

If we multiply both sides by $e^{jp\phi}$ and integrate over ϕ from zero to 2π, we get the inhomogeneous ordinary differential equation

$$\left[\frac{1}{\rho}\frac{d}{d\rho}\left(\rho\frac{d}{d\rho}\right) + \left(k_1^2 - \frac{n^2}{\rho^2}\right)\right]\tilde{f}_n(\rho, \alpha_3) = \tilde{g}_n(\rho, \alpha_3)$$

which is Bessel's differential equation (2-42) when $\tilde{g}_n(\rho, \alpha_3) = 0$. We multiply both sides of the above equation by $\rho J_n(\lambda\rho)$ and integrate over ρ from zero to infinity. Since the right-hand side is just the cylindrical Fourier transform defined in (3-59), we obtain

$$\int_0^{\infty}\left\{\left[\frac{1}{\rho}\frac{d}{d\rho}\left(\rho\frac{d}{d\rho}\right) + \left(k_1^2 - \frac{n^2}{\rho^2}\right)\right]\tilde{f}_n(\rho, \alpha_3)\right\}J_n(\lambda\rho)\rho\,d\rho = \tilde{g}_n(\lambda, \alpha_3)$$

---

[18]G. Duff and D. Naylor, *Differential Equations of Applied Mathematics*, John Wiley & Sons, Inc., New York, 1966, p. 337.

[19]*Ibid.*, pp. 333, 416. Also, see J. Dettman, *Mathematical Methods in Physics and Engineering*, McGraw-Hill Book Company, Inc., New York, 1962, p. 313.

In order to evaluate the left-hand side of the above equation, let us proceed as indicated in Method 2, Sec. 3.1. Thus, it is necessary to integrate with respect to ρ twice by parts to obtain

$$\left[\rho\left\{J_n(\lambda\rho)\frac{d\bar{f}_n(\rho,\alpha_3)}{d\rho} - \frac{dJ_n(\lambda\rho)}{d\rho}\bar{f}_n(\rho,\alpha_3)\right\}\right]_0^\infty + (k_1^2 - \lambda^2)\bar{f}_n(\lambda,\alpha_3) = \bar{g}_n(\lambda,\alpha_3)$$

$$(3\text{-}61)$$

Our desired result is that

$$\bar{f}_n(\lambda,\alpha_3) = (k_1^2 - \lambda^2)^{-1}\bar{g}_n(\lambda,\alpha_3) \tag{3-62}$$

provided the first term in (3-61) vanishes, i.e.,

$$\left[\rho\left\{J_n(\lambda\rho)\frac{d\bar{f}_n(\rho,\alpha_3)}{d\rho} - \frac{dJ_n(\lambda\rho)}{d\rho}\bar{f}_n(\rho,\alpha_3)\right\}\right]_0^\infty = 0 \tag{3-63}$$

Obviously this term vanishes at the lower limit as long as $\bar{f}_n(\rho,\alpha_3)$ and $(d/d\rho)\bar{f}_n(\rho,\alpha_3)$ are finite at the origin. In addition, both $J_n(\lambda\rho)$ and $(d/d\rho)J_n(\lambda\rho)$ must be finite at the origin so we must add the condition that $n \geq 0$ (see Appendix C.1.7). The fact that we have assumed n to be an integer here is not too relevant since that condition applies only when the ϕ-dependent function is expanded over the interval $(0, 2\pi)$. When the interval is of length $\phi_2 - \phi_1$, then a Fourier series expansion of the ϕ-dependent function is still possible as shown in (5-61). The details are considered in Prob. 2.26. In this case, n is not an integer but it is still a positive quantity. The behavior of the term at the upper limit is found by replacing $J_n(\lambda\rho)$ and $(d/d\rho)J_n(\lambda\rho)$ by their approximate values for large argument from Appendix C.1.8, i.e.,

$$\lim_{\lambda\rho\to\infty} J_n(\lambda\rho) = \sqrt{\frac{2}{\pi\lambda\rho}}\cos(\lambda\rho - \delta_1)$$

$$\lim_{\lambda\rho\to\infty} \frac{dJ_n(\lambda\rho)}{d\rho} = -\sqrt{\frac{2\lambda}{\pi\rho}}\sin(\lambda\rho - \delta_1)$$

where $\delta_1 = (\pi/4) + (n\pi/2)$. With these substitutions and the use of Euler's identities, the term becomes

$$\lim_{\rho\to\infty}\sqrt{\frac{\rho}{2\pi\lambda}}\left\{\left[\frac{d\bar{f}_n(\rho,\alpha_3)}{d\rho} - j\lambda\bar{f}_n(\rho,\alpha_3)\right]e^{j(\lambda\rho-\delta_1)}\right.$$

$$\left. + \left[\frac{d\bar{f}_n(\rho,\alpha_3)}{d\rho} + j\lambda\bar{f}_n(\rho,\alpha_3)\right]e^{-j(\lambda\rho-\delta_1)}\right\} = 0$$

The first and second expressions inside the curly brackets represent the asymptotic forms of inwardly and outwardly propagating cylindrical waves, respectively. If we assume that the medium has losses, then λ will always be complex and the terms $e^{\pm j\lambda\rho}$ will always vanish far from their sources. On the other hand, we can assume that the medium is loss-free and insist that the following conditions be met:

Radiation condition

$$\lim_{\rho\to\infty} \sqrt{\rho}\left[\frac{d\tilde{f}_n(\rho,\alpha_3)}{d\rho} + j\lambda\tilde{f}_n(\rho,\alpha_3)\right] = 0 \tag{3-64}$$

Absorption condition

$$\lim_{\rho\to\infty} \sqrt{\rho}\left[\frac{d\tilde{f}_n(\rho,\alpha_3)}{d\rho} - j\lambda\tilde{f}_n(\rho,\alpha_3)\right] = 0$$

However, these are just Sommerfeld's radiation and absorption conditions (3-20) as they apply to a cylindrical wave with time variation as $e^{j\omega t}$.

Let us now return to (3-62), take the inverse cylindrical Fourier transform with respect to λ, and then multiply by $e^{jn\phi}$ and sum over n to obtain

$$\tilde{f}(\rho,\phi,\alpha_3) = \sum_{n=-\infty}^{\infty} e^{jn\phi} \int_0^\infty \frac{\tilde{g}_n(\lambda,\alpha_3)J_n(\lambda\rho)\lambda\,d\lambda}{k_1^2 - \lambda^2} \tag{3-65}$$

If we let the excitation function $g(\rho,\phi,z)$ be a unit point source at (ρ_0,ϕ_0,z_0), then $f(\rho,\phi,z)$ becomes the Green's function for (3-12) and the expression for $\tilde{f}(\rho,\phi,\alpha_3)$ in (3-65) should reduce to (3-36). However, first we must find the cylindrical Fourier transform of the Dirac delta, i.e.,

$$g(\rho,\phi,z) = \sum_{n=-\infty}^{\infty} g_n(\rho,z)e^{jn\phi} = -\frac{\delta(\rho-\rho_0)\delta(\phi-\phi_0)\delta(z-z_0)}{\rho}$$

Let us multiply both sides by $(1/\sqrt{2\pi})e^{-j(n\phi+\alpha_3 z)}$ and integrate with respect to ϕ and z over the intervals $(0, 2\pi)$ and $(-\infty, \infty)$, respectively. Thus, we obtain

$$\tilde{g}_n(\rho,\alpha_3) = -(2\pi)^{-3/2}e^{-j(n\phi_0+\alpha_3 z_0)}\frac{\delta(\rho-\rho_0)}{\rho}\cdot$$

We now take the direct cylindrical Fourier transform according to (3-59) by multiplying the above expression by $\rho J_n(\lambda\rho)$ and integrating with respect to ρ over the interval $(0, \infty)$. This gives us the direct cylindrical Fourier transform of the Dirac delta as

$$\tilde{g}_n(\lambda,\alpha_3) = -(2\pi)^{-3/2}e^{-j(n\phi_0+\alpha_3 z_0)}J_n(\lambda\rho_0)$$

If we substitute this expression into (3-65) we find that

$$\tilde{f}(\rho, \phi, \alpha_3) \equiv \tilde{G}(\rho, \phi, \alpha_3)$$

$$= -(2\pi)^{-3/2} \sum_{n=-\infty}^{\infty} e^{-j\alpha_3 z_0} e^{jn(\phi-\phi_0)} \int_0^\infty \frac{J_n(\lambda\rho_0)J_n(\lambda\rho)\lambda \, d\lambda}{k_1^2 - \lambda^2} \quad (3\text{-}66)$$

This expression can be simplified considerably by using the addition theorem for Bessel functions of the first kind from Prob. 5.13, i.e.,

$$J_0(\lambda\rho_1) = \sum_{n=-\infty}^{\infty} \epsilon_n \cos n(\phi - \phi_0)J_n(\lambda\rho_0)J_n(\lambda\rho) \quad (3\text{-}67)$$

where $\rho_1 = \rho - \rho_0$ and $\epsilon_n = 1$, $n = 0$, and $\epsilon_n = 2$, $n > 0$. If we use Euler's identity to replace the cosine term in (3-67) by exponentials and use the identity $J_{-n}(x) = (-1)^n J_n(x)$, we can extend the sum to include both negative and positive values of n. Thus, we can write the addition theorem for Bessel functions of the first kind in the alternate (equivalent) form as

$$J_0(\lambda\rho_1) = \sum_{n=-\infty}^{\infty} e^{jn(\phi-\phi_0)}J_n(\lambda\rho_0)J_n(\lambda\rho) \quad (3\text{-}68)$$

If we substitute (3-68) into (3-66) we obtain our desired expression

$$\tilde{G}(\rho, \phi, \alpha_3) = -(2\pi)^{-3/2} e^{-j\alpha_3 z_0} \int_0^\infty \frac{J_0(\lambda\rho_1)\lambda \, d\lambda}{k_1^2 - \lambda^2} \quad (3\text{-}36)$$

which was derived previously in Sec. 3.1. The solution that satisfies the radiation condition, (3-64), is given by (3-38) with $i = 3$. Thus,

$$\tilde{G}(\rho, \phi, \alpha_3) = \frac{-j}{4\sqrt{2\pi}} e^{-j\alpha_3 z_0} H_0^{(2)}(k_1\rho_1)$$

and

$$G(\rho, \phi, z) = \frac{-j}{8\pi} \int_{-\infty}^\infty H_0^{(2)}(k_1\rho_1)e^{j\alpha_3(z-z_0)} \, d\alpha_3$$

However, this last expression is just (3-40) with $\tilde{G}(x, y, \alpha_3)$ obtained from (3-38) and with x and y replaced by ρ and ϕ. The final result is then (3-28) with $i = 3$, i.e.,

$$G(\rho, \phi, z) = G(x, y, z) = \frac{e^{-jkR}}{4\pi R}$$

which gives us another derivation of the expression in (3-46) for $i = 3$.

3.3 Spherical Fourier Transforms

We now return to the general exponential Fourier transform pair as given by (3-25) and derive the spherical Fourier or Hankel[20] transform pair. The procedure is completely analogous to that employed in the derivation of the cylindrical Fourier transforms in Sec. 3.2.

In the spherical coordinate system,

$$dV_r = r^2 \, dr \sin \theta \, d\theta \, d\phi$$

$$dV_\alpha = \alpha^2 \, d\alpha \sin \beta \, d\beta \, d\gamma$$

The distance α is analogous to r, the polar angle β is analogous to θ, and the azimuthal angle γ is analogous to ϕ as defined earlier in (3-24). If we substitute the above expressions into (3-25) we obtain the following:

$$\bar{f}(\boldsymbol{\alpha}) = (2\pi)^{-3/2} \int_0^\infty \int_0^\pi \int_0^{2\pi} f(\mathbf{r}) e^{-j\alpha r \cos \xi} r^2 \, dr \sin \theta \, d\theta \, d\phi$$

$$f(\mathbf{r}) = (2\pi)^{-3/2} \int_0^\infty \int_0^\pi \int_0^{2\pi} \bar{f}(\boldsymbol{\alpha}) e^{j\alpha r \cos \xi} \alpha^2 \, d\alpha \sin \beta \, d\beta \, d\gamma$$

(3-69)

since $\boldsymbol{\alpha} \cdot \mathbf{r} = \alpha r \cos \xi$, where $\cos \xi = \cos \theta \cos \beta + \sin \theta \sin \beta \cos (\phi - \gamma)$. Next we assume that

$$f(\mathbf{r}) = \sum_{n=0}^\infty \sum_{m=0}^n f_n(r) T^i_{mn}(\theta, \phi) \tag{3-70}$$

where the tesseral harmonics $T^i_{mn} (\theta, \phi)$ are defined in (1-33). If we substitute (3-70) into the first expression in (3-69), we find that the integrand contains a term $e^{-j\alpha r \cos \xi} T^i_{mn} (\theta, \phi)$. A similar situation occurs later involving the second expression in (3-69) with a resultant term $e^{j\alpha r \cos \xi} T^i_{mn} (\beta, \gamma)$. Consequently, in order to perform the integration with respect to θ, ϕ and β, γ above, it is necessary to derive identities involving the integrands, $e^{\pm j\alpha r \cos \xi} T^i_{mn}$.

Thus, let us consider the generating function for the spherical Bessel function of the first kind and order n from (1-36), i.e.,

$$e^{\pm j\alpha r \cos \xi} = \sum_{n=0}^\infty (\pm j)^n (2n + 1) j_n(\alpha r) P_n(\cos \xi) \tag{1-36}$$

where we used the substitutions $x = \alpha r$ and $\theta = \xi$. We now use the addition theorem for Legendre functions from Prob. 5.37, i.e.,

$$P_n(\cos \xi) = \sum_{m=0}^\infty \frac{\epsilon_m(n - m)!}{(n + m)!} P_n^m(\cos \theta) P_n^m(\cos \beta) \cos m(\phi - \gamma) \tag{3-71}$$

---

[20]Stratton, *Electromagnetic Theory*, p. 411.

where $\epsilon_m = 1$ for $m = 0$ and $\epsilon_m = 2$ for $m > 0$ to obtain the expression

$$e^{\pm j\alpha r \cos \xi} = \sum_{n=0}^{\infty} \sum_{m=0}^{n} (\pm j)^n (2n + 1) j_n(\alpha r) \frac{\epsilon_m(n - m)!}{(n + m)!} P_n^m(\cos \theta) P_n^m(\cos \beta)$$
$$\times \cos m(\phi - \gamma)$$
$$= \sum_{n=0}^{\infty} \sum_{m=0}^{n} (\pm j)^n (2n + 1) j_n(\alpha r) \frac{\epsilon_m(n - m)!}{(n + m)!} [T_{mn}^e(\theta, \phi) T_{mn}^e(\beta, \gamma)$$
$$+ T_{mn}^o(\theta, \phi) T_{mn}^o(\beta, \gamma)] \tag{3-72}$$

We can now obtain our desired identities by multiplying the second expression above by $\sin \theta \, T_{pq}^i(\theta, \phi)$ or $\sin \beta \, T_{pq}^i(\beta, \gamma)$ and integrating over the intervals $(0, \pi)$ in θ and β and $(0, 2\pi)$ in ϕ and γ. Use of the orthogonality condition (1-34) for the tesseral harmonics yields

$$j_n(\alpha r) T_{mn}^i(\beta, \gamma) = \frac{(\pm j)^{-n}}{4\pi} \int_0^{2\pi} \int_0^{\pi} e^{\pm j\alpha r \cos \xi} T_{mn}^i(\theta, \phi) \sin \theta \, d\theta \, d\phi$$

and $\tag{3-73}$

$$j_n(\alpha r) T_{mn}^i(\theta, \phi) = \frac{(\pm j)^{-n}}{4\pi} \int_0^{2\pi} \int_0^{\pi} e^{\pm j\alpha r \cos \xi} T_{mn}^i(\beta, \gamma) \sin \beta \, d\beta \, d\gamma$$

Substitution of (3-70) into the first integral in (3-69) and then the substitution of the first expression from (3-73) into the resultant expression gives

$$\tilde{f}(\alpha) = \sum_{n=0}^{\infty} \sum_{m=0}^{n} (-j)^n T_{mn}^i(\beta, \gamma) \sqrt{\frac{2}{\pi}} \int_0^{\infty} f_n(r) j_n(\alpha r) r^2 \, dr$$
$$= \sum_{n=0}^{\infty} \sum_{m=0}^{n} (-j)^n T_{mn}^i(\beta, \gamma) \tilde{f}_n(\alpha) \tag{3-74}$$

Note the similarity between (3-70) and the second expression in (3-74). This second expression for $\tilde{f}(\alpha)$ can now be inserted into the second integral in (3-69) to obtain

$$f(\mathbf{r}) = (2\pi)^{-3/2} \sum_{n=0}^{\infty} \sum_{m=0}^{n} (-j)^n \int_0^{2\pi} \int_0^{\pi} T_{mn}^i(\beta, \gamma) \tilde{f}_n(\alpha) e^{j\alpha r \cos \xi} \alpha^2 \, d\alpha \sin \beta \, d\beta \, d\gamma$$

We now use the second expression in (3-73) to simplify the above to the following:

$$f(\mathbf{r}) = \sum_{n=0}^{\infty} \sum_{m=0}^{n} T_{mn}^i(\theta, \phi) \sqrt{\frac{2}{\pi}} \int_0^{\infty} \tilde{f}_n(\alpha) j_n(\alpha r) \alpha^2 \, d\alpha$$
$$= \sum_{n=0}^{\infty} \sum_{m=0}^{n} T_{mn}^i(\theta, \phi) f_n(r) \tag{3-75}$$

If we compare the above expressions for $\bar{f}(\alpha)$ and $f(r)$, i.e., the first expressions in (3-74) and (3-75), respectively, we find that we have derived the spherical Fourier transform pair

$$\bar{f}_n(\alpha) = \sqrt{\frac{2}{\pi}} \int_0^\infty f_n(r) j_n(\alpha r) r^2 \, dr$$

$$f_n(r) = \sqrt{\frac{2}{\pi}} \int_0^\infty \bar{f}_n(\alpha) j_n(\alpha r) \alpha^2 \, d\alpha$$

(3-76)

It is also interesting to note that when $n = 0$, the spherical Fourier transform pair in (3-76) reduces to the sine Fourier transform pair[21] defined in (3-3) provided one defines $f_0(r)$ and $\bar{f}_0(\alpha)$ according to the Fourier integral theorem in (3-9).

Let us now consider the application of the spherical Fourier transform to the inhomogeneous scalar wave equation (3-12). If we assume that

$$f(r, \theta, \phi) = \sum_{n=0}^\infty \sum_{m=0}^n f_n(r) T_{mn}^i(\theta, \phi)$$

and

(3-77)

$$g(r, \theta, \phi) = \sum_{q=0}^\infty \sum_{p=0}^q g_q(r) T_{pq}^j(\theta, \phi)$$

then (3-12) becomes

$$\sum_{n=0}^\infty \sum_{m=0}^n \left[\frac{1}{r^2} \frac{d}{dr} \left(r^2 \frac{d}{dr} \right) - \frac{n(n+1)}{r^2} + k^2 \right] f_n(r) T_{mn}^i(\theta, \phi) = \sum_{q=0}^\infty \sum_{p=0}^q g_q(r) T_{pq}^j(\theta, \phi)$$

Now we multiply both sides by $\sin \theta \, T_{st}^k(\theta, \phi)$ and use the orthogonality of the tesseral harmonics as expressed by (1-34) to obtain the inhomogeneous ordinary differential equation

$$\left[\frac{1}{r^2} \frac{d}{dr} \left(r^2 \frac{d}{dr} \right) + k^2 - \frac{n(n+1)}{r^2} \right] f_n(r) = g_n(r)$$

(3-78)

The homogeneous form of this equation is the differential equation for the spherical Bessel functions, (1-35). Let us multiply both sides of the above equation by $\sqrt{2/\pi} r^2 j_n(\alpha r)$ and integrate over r from zero to infinity. Since the right side is the spherical Fourier transform defined in (3-76), we obtain

$$\sqrt{\frac{2}{\pi}} \int_0^\infty \left\{ \left[\frac{1}{r^2} \frac{d}{dr} \left(r^2 \frac{d}{dr} \right) + k^2 - \frac{n(n+1)}{r^2} \right] f_n(r) \right\} j_n(\alpha r) r^2 \, dr = \bar{g}_n(\alpha)$$

---

[21] *Ibid.*, p. 412.

If we evaluate the left-hand side of the above expression according to Method 2, Sec. 3.1, then we must integrate with respect to r twice by parts to obtain the final expression. Let us consider the details of the integration by parts. First, we make the substitution $r^2[df_n(r)/dr] = h(r)$, whence the first term becomes

$$\int_0^\infty \frac{dh(r)}{dr} j_n(\alpha r)\, dr = h(r) j_n(\alpha r)\Big]_0^\infty - \int_0^\infty h(r)\frac{dj_n(\alpha r)}{dr}\, dr$$

We evaluate the remaining integral by integrating once more by parts. Thus,

$$\int_0^\infty h(r)\frac{dj_n(\alpha r)}{dr}\, dr = \int_0^\infty r^2\frac{df_n(r)}{dr}\frac{dj_n(\alpha r)}{dr}\, dr$$

$$= f_n(r)r^2\frac{dj_n(\alpha r)}{dr}\Big]_0^\infty - \int_0^\infty f_n(r)\frac{d}{dr}\Big[r^2\frac{dj_n(\alpha r)}{dr}\Big]\, dr$$

After we collect these results, the desired expression becomes

$$\sqrt{\frac{2}{\pi}}\Big[r^2\frac{df_n(r)}{dr}j_n(\alpha r) - r^2 f_n(r)\frac{dj_n(\alpha r)}{dr}\Big]_0^\infty + (k^2 - \alpha^2)\tilde{f}_n(\alpha) = \tilde{g}_n(\alpha) \qquad (3\text{-}79)$$

From this we conclude that

$$\tilde{f}_n(\alpha) = (k^2 - \alpha^2)^{-1}\tilde{g}_n(\alpha) \qquad (3\text{-}80)$$

provided

$$\Big[r^2\Big\{j_n(\alpha r)\frac{df_n(r)}{dr} - \frac{dj_n(\alpha r)}{dr}f_n(r)\Big\}\Big]_0^\infty = 0 \qquad (3\text{-}81)$$

The second expression in (3-81) vanishes at the lower limit as long as $f_n(r)$ and $df_n(r)/dr$ are finite there (at the origin). In addition, all values of n are not permissible since $j_n(\alpha r)$ and $(d/dr)j_n(\alpha r)$ are well behaved for only positive values of n. However, both the integer and noninteger values are allowed. See Appendix C.3.6. In order to consider the behavior of the expression at the upper limit, let us substitute the approximate values of $j_n(\alpha r)$ and $(d/dr)j_n(\alpha r)$ for large values of r from Appendix C.3.7, i.e.,

$$\operatorname*{Lim}_{r\to\infty} j_n(\alpha r) = \frac{1}{\alpha r}\cos(\alpha r - \delta_2)$$

$$\operatorname*{Lim}_{r\to\infty} \frac{dj_n(\alpha r)}{dr} = -\frac{1}{r}\sin(\alpha r - \delta_2)$$

where $\delta_2 = (n + 1)\pi/2$. If we use the above substitutions and Euler's identities in the expression, we obtain

$$\lim_{r \to \infty} \frac{r}{2\alpha} \left\{ \left[\frac{df_n(r)}{dr} - j\alpha f_n(r) \right] e^{j(\alpha r - \delta_2)} + \left[\frac{df_n(r)}{dr} + j\alpha f_n(r) \right] e^{-j(\alpha r - \delta_2)} \right\} = 0$$

The first and second expressions inside the curly brackets represent the asymptotic (large argument) form of inwardly and outwardly propagating spherical waves, respectively. We can assume that the medium has losses so that α is complex in such a fashion that the terms $e^{\pm j\alpha r}$ vanish at points remote from the sources. On the other hand, we can assume that the medium is loss-free and impose the following conditions:

Radiation condition

$$\lim_{r \to \infty} r \left[\frac{df_n(r)}{dr} + j\alpha f_n(r) \right] = 0 \tag{3-82}$$

Absorption condition

$$\lim_{r \to \infty} r \left[\frac{df_n(r)}{dr} - j\alpha f_n(r) \right] = 0$$

Again, these are Sommerfeld's radiation and absorption conditions, (3-20), as they apply to a spherical wave and when the time dependence is of the form $e^{j\omega t}$.

Now that we have obtained the spherical Fourier transform pair in (3-76) and the spherical Fourier transform solution to the inhomogeneous scalar wave equation in (3-80), let us consider the solution of the inhomogeneous scalar wave equation in spherical coordinates. For convenience, we choose a unit point source for the excitation function, whence our solution should correspond to that previously obtained in (3-29). First, let us find the spherical Fourier transform of the Dirac delta, i.e.,

$$g(r, \theta, \phi) = \sum_{n=0}^{\infty} \sum_{m=0}^{n} g_n(r) T_{mn}^i(\theta, \phi) = -\frac{\delta(r - r_0)\delta(\theta - \theta_0)\delta(\phi - \phi_0)}{r^2 \sin \theta}$$

If we multiply both sides by $\sin \theta \, T_{pq}^j(\theta, \phi)$ and integrate over the intervals $(0, \pi)$ in θ and $(0, 2\pi)$ in ϕ, we obtain by virtue of the orthogonality of the tesseral harmonics in (1-34) the expression

$$g_n(r) = -\frac{\epsilon_m (2n + 1)(n - m)!}{4\pi(n + m)!} T_{mn}^i(\theta_0, \phi_0) \frac{\delta(r - r_0)}{r^2} \tag{3-83}$$

where $\epsilon_0 = 1$ and $\epsilon_m = 2$ for $m > 0$. We now take the spherical Bessel transform according to (3-76) and obtain finally that

$$\tilde{g}_n(\alpha) = -(2\pi)^{-3/2} \frac{\epsilon_m(2n+1)(n-m)!}{(n+m)!} T^i_{mn}(\theta_0, \phi_0) j_n(\alpha r_0) \qquad (3\text{-}84)$$

We substitute in turn, (3-84) into (3-80), (3-80) into (3-76), and (3-76) into (3-75) or (3-77) to obtain

$$f(\mathbf{r}) = -(2\pi^2)^{-1} \sum_{n=0}^{\infty} \sum_{m=0}^{n} \frac{\epsilon_m(2n+1)(n-m)!}{(n+m)!} T^i_{mn}(\theta_0, \phi_0) T^i_{mn}(\theta, \phi)$$
$$\times \int_0^{\infty} \frac{j_n(\alpha r_0) j_n(\alpha r)\alpha^2 \, d\alpha}{k^2 - \alpha^2}$$

Since the above expression is valid for $i = o$ and $i = e$, let us add two such expressions together to get

$$f(\mathbf{r}) = -(2\pi^2)^{-1} \sum_{n=0}^{\infty} \sum_{m=0}^{n} \frac{\epsilon_m(2n+1)(n-m)!}{(n+m)!} [T^e_{mn}(\theta_0, \phi_0) T^e_{mn}(\theta, \phi)$$
$$+ T^o_{mn}(\theta_0, \phi_0) T^o_{mn}(\theta, \phi)] \int_0^{\infty} \frac{j_n(\alpha r_0) j_n(\alpha r)\alpha^2 \, d\alpha}{k^2 - \alpha^2} \qquad (3\text{-}85)$$

However, the quantity in the square brackets is just $P_n^m(\cos\theta)P_n^m(\cos\theta_0)$ $\cos m(\phi - \phi_0)$. Thus, by using the addition theorem for the Legendre functions (3-71), we find that

$$f(\mathbf{r}) = -(2\pi^2)^{-1} \sum_{n=0}^{\infty} (2n+1) P_n(\cos\xi) \int_0^{\infty} \frac{j_n(\alpha r_0) j_n(\alpha r)\alpha^2 \, d\alpha}{k^2 - \alpha^2}$$

As the last step in simplifying the above expression, we use the addition theorem for the spherical Bessel functions of the first kind that can be obtained from Prob. 5.41, i.e.,

$$j_0(\alpha R) = \sum_{n=0}^{\infty} (2n+1) j_n(\alpha r_0) j_n(\alpha r) P_n(\cos\xi) \qquad (3\text{-}86)$$

where ξ is the angle between $\mathbf{r}$ and $\mathbf{r}_0$. It also follows that $\cos\xi = \cos\theta\cos\theta_0 + \sin\theta\sin\theta_0 \cos(\phi - \phi_0)$ and $R^2 = r^2 + r_0^2 - 2rr_0 \cos\xi$. If we substitute (3-86) into the above expression we get

$$f(\mathbf{r}) \equiv G(\mathbf{r}) = -(2\pi^2)^{-1} \int_0^{\infty} \frac{j_0(\alpha R)\alpha^2 \, d\alpha}{k^2 - \alpha^2} \qquad (3\text{-}29)$$

whose solution is (3-31) or (3-28) for $i = 3$, i.e., it must satisfy the radiation condition.

PROBLEMS

3.1. A uniform electric current sheet flows in the $+z$-direction and is located at $x = x_0$. The equation for the only finite component of the vector magnetic potential is

$$(\nabla^2 + k^2)A_z(x) = -J_s\delta(x - x_0)$$

Use an exponential Fourier transform to find A_z. The solution is given in (4-31). The inverse transform can be evaluated by using (3-49) for $i = 3$.

3.2. Obtain the solution to (5-1) for the problem in Fig. 3-1 when $a \rightarrow \infty$. This is the problem of a dc line source in front of a right angle corner. The solution is

$$G_1 = \frac{2}{\pi} \int_0^\infty \alpha^{-1} \sin \alpha y \sin \alpha y_0 \sinh \alpha x \, e^{-\alpha x_0} \, d\alpha, \qquad x < x_0$$

3.3. Obtain another (alternate) solution to Prob. 3.2 by the method of images.

3.4. Solve (3-33) for a unit source when $\alpha_3 \rightarrow 0$. This is the problem of a radiating uniform line source in unbounded space and the results in (3-38) are applicable.

3.5. Obtain the solution in Prob. 3.4 in the limit as $k \rightarrow 0$. This is the problem of a dc line source in unbounded space. The result can be obtained by using the small argument values of the Bessel functions in Appendix C.1.7 and is

$$G_3(x, y) = -\frac{1}{2\pi}\ln \rho_1 + \text{constant}$$

This result is useful in Probs. 3.3 and 3.4.

3.6. Consider Prob. 3.2 when no conducting plane exists at $x = 0$. This is the problem of a dc line source above a grounded conducting plane and the solution is

$$G = \frac{1}{\pi} \int_0^\infty \alpha^{-1} \sin \alpha y \sin \alpha y_0 \, e^{-\alpha|x-x_0|} \, d\alpha$$

Hint: The solution can be obtained from that in Prob. 3.2 by omitting the image term.

3.7. Obtain the solution in Prob. 3.6 by using the result in Prob. 7.2 for $k_0 \rightarrow 0$. The integration with respect to α_1 can be evaluated by contour integration[22] using a semicircle in the upper half plane for $x > x_0$ and the image semicircle for $x < x_0$.

[22]For a similar example, see Dettman, *Mathematical Methods in Physics and Engineering*, p. 271.

3.8. Consider Prob. 7.2 and obtain the solution by the method of images, i.e., solve the equation for A_z in completely unbounded space and then superimpose the solution for the image. The result is

$$A_z = \frac{1}{4j}[I(\mathbf{\rho}_0)H_0^{(2)}(k_0\rho_1) - I(\mathbf{\rho}_0')H_0^{(2)}(k_0\rho_2)]$$

where

$$\mathbf{\rho}_1 = (x - x_0)\mathbf{a}_x + (y - y_0)\mathbf{a}_y$$
$$\mathbf{\rho}_2 = (x - x_0)\mathbf{a}_x + (y + y_0)\mathbf{a}_y$$
$$\mathbf{\rho}_1 = \mathbf{\rho} - \mathbf{\rho}_0$$
$$\mathbf{\rho}_2 = \mathbf{\rho} - \mathbf{\rho}_0'$$

The geometry of this problem is shown in Fig. 7-1. The notation $I(\mathbf{\rho}_0)$ means that I is located at $\mathbf{\rho} = \mathbf{\rho}_0$ rather than being a function of $\mathbf{\rho}_0$.

3.9. Verify that the solutions in (3-28) are correct for the three contours shown in Fig. 3-2.

3.10. Repeat Prob. 3.9 for the solutions in (3-31).

3.11. Verify the steps leading from (3-33) to (3-36).

3.12. Repeat Prob. 3.9 for the solutions in (3-38).

3.13. Carry through the steps[23] leading to (3-52).

3.14. Verify that G_3 in (3-31) satisfies the general Sommerfeld radiation condition (3-20) for $h = 3$ and $R = r$ but that G_2 does not unless the sign of k in (3-20) is changed. *Hint:* Use relations in Appendix C.3.4.

3.15. Repeat Prob. 3.14 for G_3 in (3-38), i.e., for $H_0^{(2)}(k\rho)$. *Hint:* Use the large argument expansions for the Hankel functions from Appendix C.1.8.

3.16. Consider the identity

$$\tilde{G}_3(x, \alpha_2, \alpha_3) = \frac{1}{(2\pi)^2}\int_{-\infty}^{\infty}\int_{-\infty}^{\infty} G_3(x, y, z)e^{-j(\alpha_2 y + \alpha_3 z)}dy\, dz$$

and use (3-49) and (3-28) to obtain the following relation that represents a plane wave as a superposition of spherical waves:

$$e^{-jk|x-x_0|} = \frac{jk}{2\pi}\int_{-\infty}^{\infty}\int_{-\infty}^{\infty}\frac{e^{-jkR}}{R}dy\, dz$$

where $R^2 = (x - x_0)^2 + (y - y_0)^2 + (z - z_0)^2$. *Hint:* Let $\alpha_2 = \alpha_3 = 0$ for simplicity.

3.17. Verify the steps leading to (3-61). *Hint:* Use Bessel's equation, (2-42), in the last step.

---

[23]Note that one of these results is given by Stratton, *Electromagnetic Theory*, p. 576, in connection with the Sommerfeld problem of a vertical dipole on a finitely conducting earth.

3.18. Verify the steps leading to (3-79). *Hint:* Use the spherical Bessel's equation, (1-35), in the last step.

3.19. Show that if one uses Method 2 in Sec. 3.1 to Fourier transform a one-dimensional equation of the form of (3-12) then the following results are obtained for the cosine and sine transforms:

$$(k^2 - \alpha^2)\tilde{f}(\alpha) = \tilde{g}(\alpha)$$

provided

$$\left[\frac{df(x)}{dx}\cos\alpha x + \alpha f(x)\sin\alpha x\right]_0^\infty = 0$$

for the cosine Fourier transform, and provided

$$\left[\frac{df(x)}{dx}\sin\alpha x - \alpha f(x)\cos\alpha x\right]_0^\infty = 0$$

for the sine Fourier transform. From these results[24] we conclude that the cosine Fourier transform can be used if $(d/dx)f(x)$ vanishes at $x = 0$ and the sine Fourier transform can be used if $f(x)$ vanishes at $x = 0$. In each case, $f(x)$ must satisfy a radiation condition for $x \to \infty$.

3.20. If one repeats Prob. 3.19 for the cylindrical Fourier transform in Sec. 3.2, the expression (3-63) becomes

$$\left[\sqrt{\frac{2\rho}{\pi\lambda}}\left\{\frac{df(\rho)}{d\rho}\cos(\lambda\rho - \delta_1) + \lambda f(\rho)\sin(\lambda\rho - \delta_1)\right\}\right]_0^\infty = 0$$

which is similar to the expression in Prob. 3.19 for the cosine Fourier transform. As mentioned previously in deriving (3-63), the expression (3-63) vanishes at the lower limit provided f and $df/d\rho$ are finite there and provided J_n and $dJ_n/d\rho$ are finite there. This last condition restricts the order of the Bessel functions to positive values as one can verify from the small argument approximations in Appendix C.1.7.

3.21. Continue Prob. 3.20 for the spherical Fourier transform in Sec. 3.3 and show that the expression (3-81) becomes

$$\left[\frac{r}{\alpha}\left\{\frac{df(r)}{dr}\cos(\alpha r - \delta_2) + \alpha f(r)\sin(\alpha r - \delta_2)\right\}\right]_0^\infty = 0$$

which is similar to the expression in Prob. 3.20. Again, the vanishing of (3-81) at the lower limit is insured by restricting f and df/dr to finite values at the origin and by restricting the order of the spherical Bessel functions to positive values. Note that spherical Bessel functions of order $-\frac{1}{2}$ are allowed, although this point is chiefly of academic interest. See Appendix C.3.6 for the small argument approximations for the spherical Bessel functions.

---

[24]Note that these restrictions are more severe than those mentioned by Churchill, *Fourier Series and Boundary Value Problems*, pp. 91–92.

3.22. Demonstrate that the expression (3-53) can be used to obtain the identity[25]

$$\frac{1}{R} = \int_0^\infty e^{-\lambda|z-z_0|} J_0(\lambda \rho_1) \, d\lambda$$

where

$$R^2 = \rho_1^2 + (z - z_0)^2$$
$$\rho_1^2 = (x - x_0)^2 + (y - y_0)^2$$

Hint: Let $k \longrightarrow 0$ in (3-53).

---

[25]This expression is given in W. Smythe, *Static and Dynamic Electricity*, 2nd ed., McGraw-Hill Book Company, Inc., New York, 1950, p. 179.

4

Electromagnetics

4.1 Equations of Electromagnetics

We shall assume that the reader is familiar with the general equations of electromagnetics as formulated by Maxwell and presented in texts on the level of Plonsey and Collin.[1] A more comprehensive treatment of the theory of electromagnetics is exceptionally well presented by Stratton.[2] Thus, we write Maxwell's equations in point form as

$$\nabla \cdot \mathbf{D} = \rho$$
$$\nabla \cdot \mathbf{B} = 0$$
$$\nabla \times \mathbf{E} = -\frac{\partial \mathbf{B}}{\partial t} \qquad (4\text{-}1)$$
$$\nabla \times \mathbf{H} = \mathbf{J} + \frac{\partial \mathbf{D}}{\partial t}$$

---

[1] R. Plonsey and R. Collin, *Principles and Applications of Electromagnetic Fields*, McGraw-Hill Book Company, Inc., New York, 1961.

[2] J. Stratton, *Electromagnetic Theory*, McGraw-Hill Book Company, Inc., New York, 1941.

where **E** and **H** are the electric and magnetic field intensities, respectively, **D** is the electric displacement, and **B** is the magnetic induction. The quantities ρ and **J** are the electric charge and current densities, respectively, and are the sources of the four field quantities. The second equation above is not an independent relation since it can be obtained from the divergence of the third equation. Also, the divergence of the fourth equation leads to a fifth equation

$$\nabla \cdot \mathbf{J} + \frac{\partial \rho}{\partial t} = 0$$

which expresses the conservation of charge and is called the continuity of current equation. From the above statements, we conclude that three of the five equations are independent, which leaves three equations and six unknowns. Since the continuity of current equation gives a relation between the two sources, we are finally left with two equations and five unknowns. However, in many cases of interest, the electric current density is known in a certain region and the problem consists of finding the resultant field quantities. In such cases, we are left with two equations and four unknowns.

In order to make the problem determinate, i.e., obtain two more equations, we make the usual assumption that at every point in the region of interest **D** is related to **E**, and **B** is related to **H**. These so-called constitutive relations depend upon the material enclosed within the region and can be written as

$$\mathbf{D} = \epsilon \mathbf{E} = \epsilon_0 \mathbf{E} + \mathbf{P}$$
$$\mathbf{B} = \mu \mathbf{H} = \mu_0 (\mathbf{H} + \mathbf{M}) \tag{4-2}$$

where ϵ and μ are the electric and magnetic inductive capacities of the medium, which is assumed linear. The quantities **P** and **M** are the polarization and magnetization densities, respectively. The electric and magnetic inductive capacities of free space are denoted by ϵ_0 and μ_0, respectively, and their values and dimensions depend upon the system of units. We shall use the MKSC system of units in this text. Materials which are nonlinear, such as ferromagnets and ferroelectrics, will not be considered here except in the small signal range where they may be considered linear. In general the vectors **D**, **E** and **B**, **H** are not parallel so they must be related by a tensor. The usual cases of interest can be described by a dyadic,[3] i.e., a tensor of rank two. Thus, each component of **D** is linearly related to **E** and each component of **B** is linearly related to **H**, e.g.,

$$D_i = \epsilon_{ij} E_j$$
$$B_i = \mu_{ij} H_j \tag{4-3}$$

---

[3]H. Margenau and G. Murphy, *The Mathematics of Physics and Chemistry*, D. Van Nostrand Company, Inc., Princeton, N.J., 1943, p. 158.

where i and j can be x, y, z, etc. A material such as that just described is termed anisotropic. Both plasmas and ferrites are examples of anisotropic materials that are of current interest at microwave frequencies and can be assumed linear for small values of the field quantities. Biaxial and uniaxial materials[4] are examples of media that are often of interest at optic frequencies.

Many common materials are homogeneous, i.e., μ and ϵ are the same from point to point. However, there is interest in materials that are inhomogeneous, whence μ and ϵ are functions of position within the material. Unfortunately, boundary value problems involving inhomogeneous or anisotropic materials are more difficult than those involving homogeneous or isotropic materials as we shall see in the next two sections and in Chap. 6.

In much of the work since World War II, it has been found expedient to introduce Maxwell's equations in a dual sense so that sources can be electric or magnetic in nature. Such an approach is especially useful in antenna work[5] although it is also advantageous in connection with interior problems.[6] This approach differs from that used in Maxwell's equations in (4-1) in that two fictitious sources are added. Thus, we postulate the existence of two magnetic sources, J_m and ρ_m, i.e., a magnetic current density and a magnetic charge density, even though magnetic charge and current apparently do not exist. However, these two fictitious sources allow us to write Maxwell's equations in a symmetrical form so that we can use the concept of duality, which is useful in solving problems in electromagnetics, just as it is in solving problems in electrical circuits. We shall proceed with our postulate of magnetic charge and current and worry later about how one actually realizes such quantities in practice. Now, we write Maxwell's equations for isotropic and homogeneous materials in the form

$$\nabla \cdot \mathbf{E} = \frac{\rho}{\epsilon}$$

$$\nabla \cdot \mathbf{H} = \frac{\rho_m}{\mu} \tag{4-4}$$

$$\nabla \times \mathbf{E} = -j\omega\mu\mathbf{H} - \mathbf{J}_m$$

$$\nabla \times \mathbf{H} = j\omega\epsilon\mathbf{E} + \mathbf{J}$$

where we have assumed a time dependence of the form $e^{j\omega t}$ and constitutive relations of the form of (4-2). All of the above fields, sources, and inductive

---

[4] W. Smythe, *Static and Dynamic Electricity*, 2nd ed., McGraw-Hill Book Company, Inc., New York, 1950, pp. 22, 446.

[5] S. Silver, *Microwave Antenna Theory and Design*, McGraw-Hill Book Company, Inc., New York, 1949, p. 65.

[6] R. Harrington, *Time-Harmonic Electromagnetic Fields*, McGraw-Hill Book Company, Inc., New York, 1961, p. 134.

capacities can be complex. Rather than assume a time dependence of the form of $e^{j\omega t}$ and then suppress it, one could alternatively Fourier transform (4-1) with respect to time, after including the magnetic sources and assuming constitutive relations of the form of (4-2), and obtain equations that possess the same spatial behavior as those in (4-4) and are identical if one replaces j by $-j$. In this case one obtains the alternate but not equivalent form of Maxwell's equations as

$$\nabla \cdot \tilde{\mathbf{E}} = \frac{\tilde{\rho}}{\epsilon}$$

$$\nabla \cdot \tilde{\mathbf{H}} = \frac{\tilde{\rho}_m}{\mu} \tag{4-5}$$

$$\nabla \times \tilde{\mathbf{E}} = j\omega\mu\tilde{\mathbf{H}} - \tilde{\mathbf{J}}_m$$

$$\nabla \times \tilde{\mathbf{H}} = -j\omega\epsilon\tilde{\mathbf{E}} + \tilde{\mathbf{J}}$$

where the above field and source quantities are functions of the spatial coordinates and ω. Although this procedure is more rigorous and general mathematically, it will be sufficient for our purposes to use the procedure leading to (4-4) since it is adequate for steady-state problems. It appears that the use of time dependence in the form of $e^{j\omega t}$ was introduced by Steinmetz, or perhaps Heaviside, for the analysis of electrical circuits, whereas the form $e^{-j\omega t}$ is more consistent with the Fourier transform viewpoint leading to (4-5).

The boundary conditions that must be satisfied in order to insure the continuity of the field quantities across the interface between two regions are the following:

$$\mathbf{n} \times (\mathbf{H}_1 - \mathbf{H}_2) = \mathbf{J}_s$$

$$\mathbf{n} \times (\mathbf{E}_1 - \mathbf{E}_2) = -\mathbf{J}_{ms}$$

$$\mathbf{n} \cdot (\mathbf{D}_1 - \mathbf{D}_2) = \rho_s \tag{4-6}$$

$$\mathbf{n} \cdot (\mathbf{B}_1 - \mathbf{B}_2) = \rho_{ms}$$

where the unit vector $\mathbf{n}$ points into Region 1 as shown in Fig. 4-1. The source quantities are surface densities as indicated by the subscript s. As one might expect, the boundary conditions involving the tangential and normal components of $\mathbf{H}$ and $\mathbf{D}$ are not independent since the electric source quantities are connected by the continuity of electric current equation. A similar statement applies to the tangential and normal components of $\mathbf{E}$ and $\mathbf{B}$ and the continuity of magnetic current equation. A more comprehensive discussion of boundary conditions is given by Collin.[7]

The solution of (4-4) for the electric and magnetic field intensities can be accomplished by first introducing auxiliary quantities and then separating the electric and magnetic sources. This particular procedure is not the only

---

[7]R. Collin, *Field Theory of Guided Waves*, McGraw-Hill Book Company, Inc., New York, 1960, Sec. 1.4.

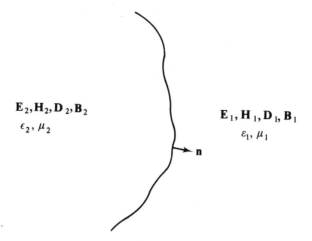

Fig. 4-1   Boundary between two regions.

method for solving (4-4) but it does have the advantage that the resultant equations for the auxiliary quantities are usually scalar. These auxiliary quantities are called electric and magnetic vector potentials and are of no particular importance in themselves. If we follow this procedure of separating the sources and introducing the vector potentials, we write Maxwell's equations for isotropic and homogeneous materials in the following form:

Electric sources

$$\nabla \cdot \mathbf{E} = \frac{\rho}{\epsilon}$$

$$\nabla \cdot \mathbf{H} = 0$$

$$\nabla \times \mathbf{E} = -j\omega\mu\mathbf{H} \tag{4-7}$$

$$\nabla \times \mathbf{H} = j\omega\mu\mathbf{E} + \mathbf{J}$$

$$\nabla \cdot \mathbf{J} + j\omega\rho = 0$$

$$\mathbf{H} = \nabla \times \mathbf{A}$$

Magnetic sources

$$\nabla \cdot \mathbf{E} = 0$$

$$\nabla \cdot \mathbf{H} = \frac{\rho_m}{\mu}$$

$$\nabla \times \mathbf{E} = -j\omega\mu\mathbf{H} - \mathbf{J}_m \tag{4-8}$$

$$\nabla \times \mathbf{H} = j\omega\epsilon\mathbf{E}$$

$$\nabla \cdot \mathbf{J}_m + j\omega\rho_m = 0$$

$$\mathbf{E} = -\nabla \times \mathbf{A}_m$$

The equations in (4-7) are equivalent to those in (4-8) if one invokes duality of the field and source quantities as prescribed in Table 4-1.[8] As a result, any solution of the equations in (4-7) is valid for the equations in (4-8) provided all of the field and source quantities are dualed. Also, it is necessary to dual the boundary conditions. The equation satisfied by the magnetic vector potential **A** in (4-7) can be obtained by first replacing **H** in the third equation by $\nabla \times \mathbf{A}$ to obtain an expression for **E**. This expression for **E** is then substituted into the fourth equation to obtain an expression involving **A**. Thus, one obtains from each step

$$\nabla \times (\mathbf{E} + j\omega\mu\mathbf{A}) = 0$$

whence

$$\mathbf{E} = -j\omega\mu\mathbf{A} - \nabla\Phi \tag{4-9}$$

TABLE 4-1 Duality of Maxwell's Equations for Electric and Magnetic Sources*

Electric sources	Magnetic sources
E	**H**
H	$-\mathbf{E}$
J	$\mathbf{J}_m$
A	$\mathbf{A}_m$
Φ	Φ_m
ρ	ρ_m
μ	ϵ
ϵ	μ
C	$\mathbf{C}_m$
μ_0	ϵ_0
ϵ_0	μ_0
K_m	K
K	K_m
$\mathcal{K}_m$	$\mathcal{K}$
$\mathcal{K}$	$\mathcal{K}_m$
Ψ	Ψ_m
k_0^2	k_0^2
B	$-\mathbf{D}$
D	**B**
S	**S**
Z	Y
Y	Z
p	m

*In part from R. Harrington, *Time-Harmonic Electromagnetic Fields*, McGraw-Hill Book Company, Inc., New York, 1961, p. 99.

[8]Harrington, *Time-Harmonic Electromagnetic Fields*, p. 99.

and

$$\nabla \times \nabla \times \mathbf{A} - k^2\mathbf{A} = \mathbf{J} - j\omega\epsilon\nabla\Phi \qquad (4\text{-}10)$$

where $k^2 = \omega^2\mu\epsilon$. We now take the divergence of the curl curl expression in (4-10) and replace $\nabla\cdot\mathbf{J}$ by its value from the continuity of electric current equation in (4-7) to obtain

$$\nabla^2\Phi + j\omega\mu\nabla\cdot\mathbf{A} = \frac{-\rho}{\epsilon} \qquad (4\text{-}11)$$

If we replace $\nabla \times \nabla \times \mathbf{A}$ in (4-10) by $\nabla\nabla\cdot\mathbf{A} - \nabla^2\mathbf{A}$, we get the equation

$$(\nabla^2 + k^2)\mathbf{A} = -\mathbf{J} + j\omega\epsilon\nabla\Phi + \nabla\nabla\cdot\mathbf{A} \qquad (4\text{-}12)$$

which is a vector wave equation for the magnetic vector potential. However, the equation in (4-11) is not of the form of a scalar wave equation unless $j\omega\mu\nabla\cdot\mathbf{A} = k^2\Phi$, which is the Lorentz condition, i.e., $\nabla\cdot\mathbf{A} = -j\omega\epsilon\Phi$. If we specify $\nabla\cdot\mathbf{A}$ in this fashion, then the expressions in (4-11) and (4-12) become

$$(\nabla^2 + k^2)\Phi = \frac{-\rho}{\epsilon} \qquad (4\text{-}13)$$

$$(\nabla^2 + k^2)\mathbf{A} = -\mathbf{J} \qquad (4\text{-}14)$$

From a mathematical viewpoint, the specification of the divergence of the magnetic vector potential is necessary since both the divergence and curl of a vector must be specified in order to uniquely determine the vector. On the other hand, one can show that the specification of the divergence of $\mathbf{A}$ according to the Lorentz condition merely insures that the continuity of electric current equation remains valid when applied to the vector magnetic potential $\mathbf{A}$ and the scalar potential Φ.[†] Both of these points are considered by Plonsey and Collin.[9] The specification that $\nabla\cdot\mathbf{A} = 0$ is called the Coulomb gauge and is discussed by Papas.[10]

We have now obtained the inhomogeneous scalar wave equation (4-13) for Φ and the inhomogeneous vector wave equation (4-14) for $\mathbf{A}$. However, the sources ρ and $\mathbf{J}$ are related through the continuity of current equation and the potentials Φ and $\mathbf{A}$ are related through the Lorentz condition. As a result, it is only necessary to solve one of the two wave equations. Since current sources are more commonly prescribed than charge sources, let us

---

[†]Note that Φ here and Ψ on p. 105 are scalar quantities although they appear boldface. The same problem may occur with other Greek symbols.

[9]Plonsey and Collin, *Principles and Applications of Electromagnetic Fields*, pp. 29, 323.

[10]C. Papas, *Theory of Electromagnetic Wave Propagation*, McGraw-Hill Book Company, Inc., New York, 1965, p. 11. Also, see Smythe, *Static and Dynamic Electricity*, 2nd ed., p. 459.

consider the problem of solving (4-14). Once the vector potential solution to (4-14) is found, the magnetic field can be obtained from the relation $\mathbf{H} = \nabla \times \mathbf{A}$ and the electric field can be obtained from the relation

$$\mathbf{E} = -j\omega\mu\mathbf{A} - \nabla\Phi = -j\omega\mu\mathbf{A} + (j\omega\epsilon)^{-1}\nabla\nabla\cdot\mathbf{A} \qquad (4\text{-}15)$$

A general solution for the vector potential can be obtained through the use of the Green's function in a manner similar to that used in Chap. 2. Thus, we consider the two inhomogeneous scalar wave equations

$$(\nabla^2 + k^2)A_i(\mathbf{r}) = -J_i(\mathbf{r}), \qquad i = x, y, z \qquad (4\text{-}16)$$

$$(\nabla^2 + k^2)G(\mathbf{r}, \mathbf{r}_0) = -\delta(\mathbf{r} - \mathbf{r}_0) \qquad (3\text{-}21)$$

where the first equation (or three equations in general) results from (4-14) provided a cartesian coordinate system is assumed. We proceed with the solution by multiplying (4-16) by G, (3-21) by $-A_i$, and adding to get

$$G\nabla^2 A_i - A_i\nabla^2 G = -J_i G + A_i\delta(\mathbf{r} - \mathbf{r}_0)$$

We next replace $\mathbf{r}$ by $\mathbf{r}_0$, $\mathbf{r}_0$ by $\mathbf{r}$, and solve for A_i, i.e.,

$$A_i(\mathbf{r}_0)\delta(\mathbf{r}_0 - \mathbf{r}) = G(\mathbf{r}_0, \mathbf{r})\nabla_0^2 A_i(\mathbf{r}_0) - A_i(\mathbf{r}_0)\nabla_0^2 G(\mathbf{r}_0, \mathbf{r}) + J_i(\mathbf{r}_0)G(\mathbf{r}_0, \mathbf{r})$$

where

$$\nabla_0^2 = \frac{\partial^2}{\partial x_0^2} + \frac{\partial^2}{\partial y_0^2} + \frac{\partial^2}{\partial z_0^2}$$

If this expression is integrated over the volume V_0, we obtain

$$A_i(\mathbf{r}) = \int G(\mathbf{r}_0, \mathbf{r})J_i(\mathbf{r}_0)\, dV_0 + \int [G(\mathbf{r}_0, \mathbf{r})\nabla_0^2 A_i(\mathbf{r}_0) - A_i(\mathbf{r}_0)\nabla_0^2 G(\mathbf{r}_0, \mathbf{r})]\, dV_0$$

The second integral can be converted to a surface integral by using Green's second identity[11] (see Appendix A), i.e.,

$$\int (\psi\nabla^2\phi - \phi\nabla^2\psi)\, dV = \int \left(\psi\frac{\partial\phi}{\partial n} - \phi\frac{\partial\psi}{\partial n}\right) dS = \int (\psi\nabla\phi - \phi\nabla\psi)\cdot d\mathbf{S}$$

whence we obtain the expression

$$A_i(\mathbf{r}) = \int G(\mathbf{r}, \mathbf{r}_0)J_i(\mathbf{r}_0)\, dV_0 + \int \left[G(\mathbf{r}, \mathbf{r}_0)\frac{\partial A_i(\mathbf{r}_0)}{\partial n_0} - A_i(\mathbf{r}_0)\frac{\partial G(\mathbf{r}, \mathbf{r}_0)}{\partial n_0}\right] dS_0$$

$$(4\text{-}17)$$

---

[11]Stratton, *Electromagnetic Theory*, p. 165.

where we have assumed that the Green's function G is symmetrical, i.e., $G(\mathbf{r}, \mathbf{r}_0) = G(\mathbf{r}_0, \mathbf{r})$. Since the above expression is valid for $i = x, y, z$, we can multiply each term by the unit vector $\mathbf{a}_i$ and sum over i to obtain the desired expression for the magnetic vector potential

$$\mathbf{A}(\mathbf{r}) = \int G(\mathbf{r}, \mathbf{r}_0)\mathbf{J}(\mathbf{r}_0)\, dV_0 + \int \left[G(\mathbf{r}, \mathbf{r}_0)\frac{\partial \mathbf{A}(\mathbf{r}_0)}{\partial n_0} - \mathbf{A}(\mathbf{r}_0)\frac{\partial G(\mathbf{r}, \mathbf{r}_0)}{\partial n_0} \right] dS_0$$

$$(4\text{-}18)$$

In the derivation of (4-18), we have assumed that V_0 is a volume enclosed by the surface $\mathbf{S}_0$ and that the unit normal points out of the volume. We have also assumed that both $\mathbf{r}$ and $\mathbf{r}_0$ are inside V_0.

The expression for the magnetic vector potential in (4-18) is the solution of the inhomogeneous vector wave equation (4-14), whereas the expression for each scalar component of the magnetic vector potential in (4-17) is the solution of the inhomogeneous scalar wave equation in (4-16). Since the equation for the scalar electric potential in (4-13) is of the same form as that for each scalar component of the magnetic vector potential in (4-16), it follows that the solution in (4-17) is valid with A_i replaced by Φ and J_i replaced by ρ/ϵ. Thus, the solution for the scalar electric potential in (4-13) is

$$\Phi(\mathbf{r}) = \frac{1}{\epsilon} \int G(\mathbf{r}, \mathbf{r}_0)\rho(\mathbf{r}_0)\, dV_0 + \int \left[G(\mathbf{r}, \mathbf{r}_0)\frac{\partial \Phi(\mathbf{r}_0)}{\partial n_0} - \Phi(\mathbf{r}_0)\frac{\partial G(\mathbf{r}, \mathbf{r}_0)}{\partial n_0} \right] dS_0$$

$$(4\text{-}19)$$

According to Stratton[12] we can interpret the first integral above as being the contribution to the electric potential from all of the sources inside V_0 and the second integral as being the contribution from all sources outside. Also, the first term of the integrand in the surface integral represents an equivalent surface charge of density $\epsilon(\partial \Phi/\partial n_0)$ and the second term represents an equivalent surface dipole layer of density $\epsilon\Phi$. The two integrals in (4-18) can be given a similar interpretation, i.e., the first one represents contributions from sources inside V_0 and the second one represents contributions from sources outside V_0. It also seems possible to consider the two terms in the integrand of the second integral as representing equivalent surface layers of currents and charges. Such an approach is employed by Stratton but will not be considered here since it is satisfactory in most practical problems to apply the boundary conditions in the form of (4-6).

The general solution for the magnetic vector potential in (4-18) allows one to determine completely the field quantities excited by the electric sources in (4-7). The similar solution for the scalar electric potential in (4-19) is depen-

---

[12]*Ibid.*, Sec. 3.17.

dent on the magnetic vector potential because of the Lorentz condition. However, this dependence vanishes when the frequency dependence of the sources and fields vanishes, i.e., when $\omega = 0$. As a result, the scalar electric potential is of interest chiefly in electrostatic problems.

The problem which is dual to the one considered above is formulated in (4-8). However, the duality relations in Table 4-1 allow us to write down immediately the general solutions for magnetic sources. The general expression for the vector electric potential which is dual to (4-18) can be written as

$$\mathbf{A}_m(\mathbf{r}) = \int G(\mathbf{r}, \mathbf{r}_0)\mathbf{J}_m(\mathbf{r}_0)\,dV_0 + \int \left[G(\mathbf{r}, \mathbf{r}_0)\frac{\partial \mathbf{A}_m(\mathbf{r}_0)}{\partial n_0} - \mathbf{A}_m(\mathbf{r}_0)\frac{\partial G(\mathbf{r}, \mathbf{r}_0)}{\partial n_0} \right] dS_0$$

(4-20)

This is the general solution to the inhomogeneous vector equation which is dual to (4-14), i.e.,

$$(\nabla^2 + k^2)\mathbf{A}_m = -\mathbf{J}_m \tag{4-21}$$

The general expression for the scalar magnetic potential which is dual to (4-19) can be written as

$$\Phi_m(\mathbf{r}) = \frac{1}{\mu}\int G(\mathbf{r}, \mathbf{r}_0)\rho_m(\mathbf{r}_0)\,dV_0 + \int \left[G(\mathbf{r}, \mathbf{r}_0)\frac{\partial \Phi_m(\mathbf{r}_0)}{\partial n_0} - \Phi_m(\mathbf{r}_0)\frac{\partial G(\mathbf{r}, \mathbf{r}_0)}{\partial n_0} \right] dS_0$$

(4-22)

As before, a knowledge of the vector electric potential in (4-20) allows one to determine completely the fields in (4-8) by using the relations

$$\mathbf{E} = -\nabla \times \mathbf{A}_m \quad \text{and} \quad \mathbf{H} = -j\omega\epsilon\mathbf{A}_m - \nabla\Phi_m = -j\omega\epsilon\mathbf{A}_m + (j\omega\mu)^{-1}\nabla\nabla\cdot\mathbf{A}_m$$

(4-23)

The scalar magnetic potential in (4-22) is of interest chiefly in magnetostatic problems.

It quite often happens that many problems of interest involve sources in regions of infinite extent, whence the surface integral vanishes in each of the expressions for the vector and scalar potentials in (4-18), (4-19), (4-20), and (4-22). In problems of this type, the usual boundary conditions on the fields as expressed in (4-6) are replaced by Sommerfeld's radiation condition, (3-20). This so-called radiation condition insures that the fields behave properly at infinity, i.e., represent waves traveling outwardly or away from the source. For regions of infinite extent, the Green's function solution of (3-21) that satisfies the radiation condition is given as $G_3(\mathbf{r})$ in (3-28) or (3-31). If we use the Green's function from (3-28), the expression for the vector magnetic potential in unbounded space becomes

$$\mathbf{A}(\mathbf{r}) = \frac{1}{4\pi} \int \frac{e^{-jkR}}{R} \mathbf{J}(\mathbf{r}_0) \, dV_0 \tag{4-24}$$

where $R = |\mathbf{r} - \mathbf{r}_0|$ and the expression for the scalar electric potential becomes

$$\Phi(\mathbf{r}) = \frac{1}{4\pi\epsilon} \int \frac{e^{-jkR}}{R} \rho(\mathbf{r}_0) \, dV_0 \tag{4-25}$$

The expressions for the duals become

$$\mathbf{A}_m(\mathbf{r}) = \frac{1}{4\pi} \int \frac{e^{-jkR}}{R} \mathbf{J}_m(\mathbf{r}_0) \, dV_0 \tag{4-26}$$

$$\Phi_m(\mathbf{r}) = \frac{1}{4\pi\mu} \int \frac{e^{-jkR}}{R} \rho_m(\mathbf{r}_0) \, dV_0 \tag{4-27}$$

Since the expression for the Green's function is valid for $k = 0$, whence the inhomogeneous wave equation reduces to the form of Poisson's equation, it follows that the expressions in (4-24), (4-25), (4-26), and (4-27) are valid for $k = 0$.

As an example of the preceding theory, let us find the vector magnetic potential of an infinite, uniform electric line source or current filament that flows in the direction of and parallel to the z-axis and exists in unbounded space. The general solution is given by (4-24). However, the current filament is of infinitesimal cross section so that

$$\lim_{A \to 0} \int_A dS_0 \int dz_0 \frac{e^{-jkR}}{R} \mathbf{J}(\boldsymbol{\rho}_0) = \int dz_0 \frac{e^{-jkR}}{R} I(\boldsymbol{\rho}_0) \mathbf{a}_{z_0}$$

The expression for the vector magnetic potential now reduces to the scalar expression

$$A_z(\boldsymbol{\rho}_1) = \frac{I(\boldsymbol{\rho}_0)}{4\pi} \int_{-\infty}^{\infty} \frac{e^{-jkR}}{R} \, dz_0 \tag{4-28}$$

where $\boldsymbol{\rho}_1 = \boldsymbol{\rho} - \boldsymbol{\rho}_0$, so that $R = [\rho_1^2 + (z - z_0)^2]^{1/2}$. The integral is easily evaluated by using the third identity in (3-41). Thus, the final expression for the z-component of the vector magnetic potential becomes

$$A_z(\boldsymbol{\rho}_1) = \frac{I(\boldsymbol{\rho}_0)}{4j} H_0^{(2)}(k\rho_1) \tag{4-29}$$

Another example consists of finding the vector magnetic potential of a uniform electric current sheet that flows in the direction of the positive z-axis. If we assume that the current sheet is located on the plane $x = x_0$, (4-24) reduces to the following scalar expression:

$$A_z(\mathbf{r}) = \frac{J_s(x_0)}{4\pi} \int_{-\infty}^{\infty} \int_{-\infty}^{\infty} \frac{e^{-jkR}}{R} \, dz_0 \, dy_0 \tag{4-30}$$

The integration with respect to z_0 can be performed by using again the third identity in (3-41). The integration with respect to y_0 can be performed by using the third identity (for $\alpha_2 = 0$) in (3-50). The final expression for the z-component of the vector magnetic potential becomes

$$A_z(x) = \frac{J_s}{2jk} e^{-jk(x-x_0)}, \qquad x > x_0$$

which represents a uniform plane wave traveling in the positive x-direction. The solution to (4-30) that is valid for $x < x_0$ does not follow directly from (4-24). It may be obtained by simply replacing $e^{-jk(x-x_0)}$ by $e^{jk(x-x_0)}$, i.e.,

$$A_z(x) = \frac{J_s}{2jk} e^{jk(x-x_0)}, \qquad x < x_0$$

These two expressions can be written in the form

$$A_z(x) = \frac{J_s}{2jk} e^{-jk|x-x_0|} \tag{4-31}$$

If the region of interest is of finite extent, the general expressions for the vector and scalar potentials in (4-18), (4-19), (4-20), and (4-22) must be used. However, if the region of interest is free of sources, the volume integral in each of the aforementioned equations vanishes and only the surface integral remains. Problems of this type often occur when considering two adjoining regions. In most cases, the electromagnetic field in one region is known and the field in the other region is unknown. Typical examples are problems involving coupling through apertures between waveguides and between a waveguide and free space, i.e., a radiating aperture. In such cases, the field in each region is known exactly in the absence of the aperture. With the aperture present, one often assumes that the field in one region is undisturbed whereas the field in the other region is excited solely by the field in the aperture. The field in the aperture is then approximated by the field that existed on the surface containing the aperture before the aperture was made. In other cases, the aperture field is known exactly but the final expressions for the electromagnetic fields must be approximated. Since the field in one region is usually known, it is often more convenient to specify the field in the other region and then use the general boundary conditions in (4-6) to complete their determination.

The electric and magnetic vector potentials that were introduced in (4-7) and (4-8), respectively, are not the only auxiliary vector potential functions commonly used. According to Stratton[13] it was Hertz in 1888 who first demonstrated the possibility of defining an electromagnetic field in terms of a single vector function. This single vector function is now referred to as the Hertz potential and is similar to the vector magnetic potential introduced in this section. In the absence of free charge and conduction current, the sources of the electromagnetic field are the electric and magnetic polarizations. If each source is considered separately, one may introduce a vector electric Hertz potential (or electric polarization potential) $\mathbf{\Pi}_e$ that satisfies a vector Helmholtz equation like (4-14) but with $\mathbf{J}$ replaced by $\mathbf{P}/\epsilon_0$ and k^2 replaced by k_0^2. In this case the vector electric Hertz potential is related to the vector magnetic potential as

$$\mathbf{\Pi}_e = (j\omega\epsilon_0)^{-1}\mathbf{A} \qquad (4\text{-}32)$$

where $\mathbf{D} = \epsilon_0\mathbf{E} + \mathbf{P}$. The vector magnetic Hertz potential (or magnetic polarization potential) $\mathbf{\Pi}_h$ is introduced in a dual fashion and satisfies the vector Helmholtz equation that is dual to (4-14), i.e., (4-21), but with $\mathbf{J}_m$ replaced by $\mathbf{M}$ and k^2 replaced by k_0^2. In this case the relation between the vector magnetic Hertz potential and the vector electric potential is

$$\mathbf{\Pi}_h = (j\omega\mu_0)^{-1}\mathbf{A}_m \qquad (4\text{-}33)$$

where $\mathbf{B} = \mu_0(\mathbf{H} + \mathbf{M})$. The notation used for the Hertz potentials above is that used by Collin[14] who also notes that the Hertz potentials are closely related to the vector potentials introduced in (4-7) and (4-8).

When free charge and conduction current are present, the single vector electric Hertz potential is sufficient to define the field. In this case the relation between the vector electric Hertz potential and the vector magnetic potential is given as

$$\mathbf{\Pi}_e = (j\omega\epsilon)^{-1}\mathbf{A} \qquad (4\text{-}34)$$

The inhomogeneous wave equation satisfied by $\mathbf{\Pi}_e$ is obtained from (4-14) by substituting for $\mathbf{A}$ from (4-34).

4.2 Inhomogeneous Media

The electric and magnetic inductive capacities in the preceding section were assumed to be complex scalars that were the same in all directions and did not vary from point to point, i.e., the material was assumed isotropic and

---

[13]*Ibid.*, pp. 28, 430.

[14]Collin, *Field Theory of Guided Waves*, Secs. 1.6, 1.7.

homogeneous, respectively. As a result, the Helmholtz equations (4-13) and (4-14) for the scalar electric and vector magnetic potentials are valid only for isotropic and homogeneous media. The same holds true for the Maxwell's equations in (4-7) and (4-8). Let us now rewrite Maxwell's equations so that they are valid for an inhomogeneous and isotropic material. In this case, we obtain the following:

Electric sources

$$\nabla \cdot \mathbf{D} = \rho$$
$$\nabla \cdot \mathbf{B} = 0$$
$$\nabla \times \mathbf{E} = -j\omega\mathbf{B}$$
$$\nabla \times \mathbf{H} = j\omega\mathbf{D} + \mathbf{J} \tag{4-35}$$
$$\nabla \cdot \mathbf{J} + j\omega\rho = 0$$
$$\mathbf{B} = \nabla \times \mathbf{C}$$

Magnetic sources

$$\nabla \cdot \mathbf{D} = 0$$
$$\nabla \cdot \mathbf{B} = \rho_m$$
$$\nabla \times \mathbf{E} = -j\omega\mathbf{B} - \mathbf{J}_m$$
$$\nabla \times \mathbf{H} = j\omega\mathbf{D} \tag{4-36}$$
$$\nabla \cdot \mathbf{J}_m + j\omega\rho_m = 0$$
$$\mathbf{D} = -\nabla \times \mathbf{C}_m$$

with the constitutive relations

$$\mathbf{D} = \epsilon_0 K(\mathbf{r})\mathbf{E} \tag{4-37}$$
$$\mathbf{B} = \mu_0 K_m(\mathbf{r})\mathbf{H} \tag{4-38}$$

where K and K_m are the specific inductive capacities of the material. The specific electric inductive capacity K is often called the dielectric constant or relative dielectric constant. The specific magnetic inductive capacity K_m is often called the permeability or relative permeability. The curl relations and the continuity of current relation in (4-35) and (4-36) are the same as those in (4-7) and (4-8). However, the divergence relations are different and **B** and **D** are defined in terms of the curl of two new vector potentials **C** and $\mathbf{C}_m$, respectively. On the other hand, this latter choice for vector potentials could have been used in (4-7) and (4-8), which would have avoided the need for introducing two different symbols, i.e., **A** and **C**. Since most problems of interest at the intermediate level involve isotropic and homogeneous materials, it is expedient to use the vector potentials $\mathbf{A}$ and $\mathbf{A}_m$ and Maxwell's equations in the form of (4-7) and (4-8). However, in problems at a more advanced level, it is preferable to use the vector potentials $\mathbf{C}$ and $\mathbf{C}_m$ and Maxwell's equations in the form of (4-35) and (4-36). In any case, the latter approach is more general.

Let us now proceed to find the equations satisfied by the potentials in (4-35). The procedure is analogous to that employed in the preceding section and leads to an equation for a magnetic vector potential and an equation for a scalar electric potential with a condition relating the two similar to the Lorentz condition. First, we replace $\mathbf{B}$ in the third equation in (4-35) by $\nabla \times \mathbf{C}$, which allows us to solve for $\mathbf{E}$. Next, we replace $\mathbf{H}$ in the fourth equation by $(\mu_0 K_m)^{-1} \nabla \times \mathbf{C}$ and $\mathbf{D}$ by $\epsilon_0 K \mathbf{E}$ to obtain the expressions

$$\mathbf{E} = -j\omega\mathbf{C} - \nabla\Psi \tag{4-39}$$

$$\nabla \times (K_m^{-1}\nabla \times \mathbf{C}) = \mu_0\mathbf{J} + k_0^2 K\mathbf{C} - j\omega\mu_0\epsilon_0 K\nabla\Psi \tag{4-40}$$

If we take the divergence of (4-40) and replace $\nabla \cdot \mathbf{J}$ by $-j\omega\rho$, we obtain the equation for the scalar electric potential, i.e.,

$$\nabla \cdot (K\nabla\Psi) + j\omega\nabla \cdot (K\mathbf{C}) = \frac{-\rho}{\epsilon_0} \tag{4-41}$$

However, both (4-40) and (4-41) contain a term involving the other potential. Also, each expression must reduce to (4-10) or (4-11) when the material becomes homogeneous. The expression in (4-41) can be put into the proper form by letting

$$j\omega\nabla \cdot (K\mathbf{C}) = k_0^2\Psi \tag{4-42}$$

which is the Lorentz condition for inhomogeneous media. If we use this condition in (4-40) and (4-41), the final equations[15] for the potentials become

$$K\nabla \cdot (K\mathbf{C}) - \nabla\mathbf{x}(K_m^{-1}\nabla \times \mathbf{C}) + k_0^2 K\mathbf{C} = -\mu_0\mathbf{J} \tag{4-43}$$

$$\nabla \cdot (K\nabla\Psi) + k_0^2\Psi = \frac{-\rho}{\epsilon_0} \tag{4-44}$$

and the expression for the electric field intensity becomes

$$\mathbf{E} = -j\omega[\mathbf{C} + k_0^{-2}\nabla\nabla \cdot (K\mathbf{C})] \tag{4-45}$$

It was mentioned in the preceding section that it is customary to introduce auxiliary potential functions for two reasons. First, the resulting inhomo-

---

[15] J. Van Bladel, *Electromagnetic Fields*, McGraw-Hill Book Company, Inc., New York, 1964, p. 238.

geneous vector equation for the potentials can usually be replaced by scalar equations in many practical problems. Second, the equations satisfied by the potentials are usually simpler in form than those satisfied by the electromagnetic fields provided that all material involved is isotropic and homogeneous. When anisotropic or inhomogeneous materials are involved in the problem, it is quite often advantageous to investigate the equations satisfied by the electromagnetic fields since the equations satisfied by the potentials may not be simpler. If we follow such a procedure here, we find that the inhomogeneous differential equations for the electric and magnetic field intensities are

$$\nabla \times (K_m^{-1} \nabla \times \mathbf{E}) - k_0^2 K \mathbf{E} = -j\omega\mu_0 \mathbf{J} \tag{4-46}$$

$$\nabla \times (K^{-1} \nabla \times \mathbf{H}) - k_0^2 K_m \mathbf{H} = \nabla \times (K^{-1}\mathbf{J}) \tag{4-47}$$

Indeed, the equations for the electric and magnetic field intensities in (4-46) and (4-47), respectively, are very similar to the equation for the vector magnetic potential in (4-43). In fact, the equation for the electric field intensity appears to be the simplest of the three in the general case. When the region of interest is free of sources, the equations for the electric and magnetic field intensities are of the same form and are still simpler than the equation for the vector potential.

4.3 Anisotropic Media

In media where the electric and magnetic inductive capacities are homogeneous but anisotropic, we still write Maxwell's equations in the form shown in (4-35) and (4-36). However, the constitutive relations in (4-37) and (4-38) now become

$$\mathbf{D} = \epsilon_0 \mathfrak{K} \cdot \mathbf{E} \tag{4-48}$$

$$\mathbf{B} = \mu_0 \mathfrak{K}_m \cdot \mathbf{H} \tag{4-49}$$

where $\mathfrak{K}$ and $\mathfrak{K}_m$ are tensors of rank two.[16] A general tensor is defined as a differential invariant that transforms according to a certain rule.[17] Many tensors have special names to indicate their rank, e.g., a tensor of rank two is a dyadic, a tensor of rank one is a vector, and a tensor of rank zero is an invariant scalar. An invariant scalar is independent of the orientation of the coordinate system and is also invariant to an orthogonal transformation of

---

[16]Stratton, *Electromagnetic Theory*, Sec. 1.20.

[17]H. Lass, *Vector and Tensor Analysis*, McGraw-Hill Book Company, Inc., New York, 1950, Sec. 126. Also, see M. Spiegel, *Schaum's Outline of Vector Analysis and an Introduction to Tensor Analysis*, McGraw-Hill Book Company, Inc., New York, 1959, Chap. 8.

coordinates. Examples of invariant scalars are such quantities as temperature, pressure, work, charge, and length. In problems involving time, it is often necessary to introduce another coordinate jct, where c is the velocity of light, whence the quantity ct has the dimension of length. Further details are adequately covered in Stratton or elsewhere.[18] Since the problems that are of interest to us at this level are concerned only with the steady-state solution, we can restrict ourselves to three-dimensional problems involving time dependence of the form $e^{j\omega t}$.

A second rank tensor, or dyadic, has nine tensor components, each one of which is a variant scalar. A variant scalar is a quantity that is not independent of an orthogonal coordinate transformation. Examples of variant scalars are the coordinates of a point or the components of a vector. We can express a dyadic in terms of variant scalars and unit vectors as follows:[19]

$$
\mathcal{C} = \begin{aligned}
&\mathbf{a}_1\mathbf{a}_1 C_{11} + \mathbf{a}_1\mathbf{a}_2 C_{12} + \mathbf{a}_1\mathbf{a}_3 C_{13} \\
+\,&\mathbf{a}_2\mathbf{a}_1 C_{21} + \mathbf{a}_2\mathbf{a}_2 C_{22} + \mathbf{a}_2\mathbf{a}_3 C_{23} \\
+\,&\mathbf{a}_3\mathbf{a}_1 C_{31} + \mathbf{a}_3\mathbf{a}_2 C_{32} + \mathbf{a}_3\mathbf{a}_3 C_{33}
\end{aligned} \tag{4-50}
$$

The quantities C_{ij} are variant scalars and will be different in different coordinate systems. On the other hand, it is possible to introduce the dyadic $\mathcal{C}$ as the sum of three vectors, each of which is associated with a particular unit vector. Thus, we define

$$\mathcal{C} = \mathbf{C}_1\mathbf{a}_1 + \mathbf{C}_2\mathbf{a}_2 + \mathbf{C}_3\mathbf{a}_3$$

where

$$
\begin{aligned}
\mathbf{C}_1 &= \mathbf{a}_1 C_{11} + \mathbf{a}_2 C_{21} + \mathbf{a}_3 C_{31} \\
\mathbf{C}_2 &= \mathbf{a}_1 C_{12} + \mathbf{a}_2 C_{22} + \mathbf{a}_3 C_{32} \\
\mathbf{C}_3 &= \mathbf{a}_1 C_{13} + \mathbf{a}_2 C_{23} + \mathbf{a}_3 C_{33}
\end{aligned} \tag{4-51}
$$

The vector $\mathbf{C}_1$ comes from the first column in (4-50) and the vectors $\mathbf{C}_2$ and $\mathbf{C}_3$ come from the second and third columns, respectively. This particular definition (or decomposition) of a dyadic in terms of the sum of three products of a vector and a unit vector allows one to express the dot and cross products of a dyadic with another dyadic, vector, or vector operator in a manner similar to that defined for vectors. However, one must use the rule that the dot or cross operation applies only to the two vectors which appear immediately adjacent to and on either side of the dot or cross. One should note that the decomposition scheme used in (4-51) is the only one that works

---

[18]R. Wangsness, *Introductory Topics in Theoretical Physics*, Part 1, John Wiley & Sons, Inc., New York, 1963. Also, see Stratton, *Electromagnetic Theory*, Sec. 1.19.

[19]Collin, *Field Theory of Guided Waves*, Sec. A.2.

using the procedure outlined above. For example, one could define

$$\mathbf{C} = \mathbf{a}_1 \mathbf{F}_1 + \mathbf{a}_2 \mathbf{F}_2 + \mathbf{a}_3 \mathbf{F}_3$$

where
$$\mathbf{F}_1 = \mathbf{a}_1 C_{11} + \mathbf{a}_2 C_{12} + \mathbf{a}_3 C_{13}$$
$$\mathbf{F}_2 = \mathbf{a}_1 C_{21} + \mathbf{a}_2 C_{22} + \mathbf{a}_3 C_{23}$$
$$\mathbf{F}_3 = \mathbf{a}_1 C_{31} + \mathbf{a}_2 C_{32} + \mathbf{a}_3 C_{33}$$

In this case, the vector $\mathbf{F}_1$ comes from the first row in (4-50), etc. However, the identity

$$\nabla \cdot \mathbf{C} = (\nabla \cdot \mathbf{C}_1)\mathbf{a}_1 + (\nabla \cdot \mathbf{C}_2)\mathbf{a}_2 + (\nabla \cdot \mathbf{C}_3)\mathbf{a}_3$$

is not valid with $\mathbf{C}_1$ replaced by $\mathbf{F}_1$, etc. On the other hand, this latter decomposition scheme can be made to work provided one introduces the proper procedure.[20]

Let us proceed to solve (4-35) in the same fashion as before except with the constitutive relations (4-48) and (4-49). Thus, we still obtain (4-39) for the electric field. However, the derivation of the equation similar to (4-40) will be more complicated since the solutions of (4-48) and (4-49) for **E** and **H** will be

$$\mathbf{E} = \epsilon_0^{-1} \mathcal{K}^{-1} \cdot \mathbf{D} \tag{4-52}$$

$$\mathbf{H} = \mu_0^{-1} \mathcal{K}_m^{-1} \cdot \mathbf{B} \tag{4-53}$$

where the dyadics $\mathcal{K}^{-1}$ and $\mathcal{K}_m^{-1}$ are the inverse of $\mathcal{K}$ and $\mathcal{K}_m$, respectively. However, the relations (4-52) and (4-53) express the same type of transformation as that defined in (4-48) and (4-49) so that the only real difficulty is in actually finding the inverse in terms of the given expressions for $\mathcal{K}$ or $\mathcal{K}_m$. Let us assume that $\mathcal{K}^{-1}$ and $\mathcal{K}_m^{-1}$ are known so that we may proceed. Thus, we replace **H** in the fourth equation of (4-35) by $\mu_0^{-1} \mathcal{K}_m^{-1} \cdot \nabla \times \mathbf{C}$ and **D** by $\epsilon_0 \mathcal{K} \cdot \mathbf{E}$ to obtain the expressions

$$\mathbf{E} = -j\omega \mathbf{C} - \nabla \Psi \tag{4-39}$$

$$\nabla \times (\mathcal{K}_m^{-1} \cdot \nabla \times \mathbf{C}) = \mu_0 \mathbf{J} + k_0^2 \mathcal{K} \cdot \mathbf{C} - j\omega \mu_0 \epsilon_0 \mathcal{K} \cdot \nabla \Psi \tag{4-54}$$

We note that (4-54) is the same as (4-40) provided that in the latter we replace K_m^{-1} by $\mathcal{K}_m^{-1} \cdot$ and K by $\mathcal{K} \cdot$. Since the continuity of current equation in (4-35) remains unchanged for anistropic media, it follows that we can write the expression for the scalar electric potential by making the same substitutions in (4-41). This gives us

$$\nabla \cdot (\mathcal{K} \cdot \nabla \Psi) + j\omega \nabla \cdot (\mathcal{K} \cdot \mathbf{C}) = \frac{-\rho}{\epsilon_0}$$

---

[20]Van Bladel, *Electromagnetic Fields*, Appendix 3.

which reduces to

$$\nabla\cdot(\mathfrak{K}\cdot\nabla\Psi) + k_0^2\Psi = \frac{-\rho}{\epsilon_0} \tag{4-55}$$

provided we assume the relation

$$j\omega\nabla\cdot(\mathfrak{K}\cdot\mathbf{C}) = k_0^2\Psi \tag{4-56}$$

which is the Lorentz condition for anisotropic material. If we take the gradient of (4-56) and substitute the expression for $\nabla\Psi$ into (4-54) we obtain

$$\mathfrak{K}\cdot[\nabla\nabla\cdot(\mathfrak{K}\cdot\mathbf{C})] - \nabla\times(\mathfrak{K}_m^{-1}\cdot\nabla\times\mathbf{C}) + k_0^2\mathfrak{K}\cdot\mathbf{C} = -\mu_0\mathbf{J} \tag{4-57}$$

which also could have been obtained from (4-43) by replacing K by $\mathfrak{K}\cdot$ and K_m^{-1} by $\mathfrak{K}_m^{-1}\cdot$.

Let us now derive the equations satisfied by the electric and magnetic fields to determine whether they're simpler than those satisfied by the potentials, i.e., (4-55) and (4-57). However, we again find that the expressions can be written immediately by using (4-46) and (4-47) after replacing K by $\mathfrak{K}\cdot$, K^{-1} by $\mathfrak{K}^{-1}\cdot$, K_m by $\mathfrak{K}_m\cdot$, and K_m^{-1} by $\mathfrak{K}_m^{-1}\cdot$. This procedure gives us the following expressions:[21]

$$\nabla\times(\mathfrak{K}_m^{-1}\cdot\nabla\times\mathbf{E}) - k_0^2\mathfrak{K}\cdot\mathbf{E} = -j\omega\mu_0\mathbf{J} \tag{4-58}$$

$$\nabla\times(\mathfrak{K}^{-1}\cdot\nabla\times\mathbf{H}) - k_0^2\mathfrak{K}_m\cdot\mathbf{H} = \nabla\times(\mathfrak{K}^{-1}\cdot\mathbf{J}) \tag{4-59}$$

If we use Table 4-1 we can write down the dual equations for the potentials and fields. This gives us the following relations:

$$\mathbf{H} = -j\omega\mathbf{C}_m - \nabla\Psi_m \tag{4-60}$$

$$j\omega\nabla\cdot(\mathfrak{K}_m\cdot\mathbf{C}_m) = k_0^2\Psi_m \tag{4-61}$$

$$\nabla\cdot(\mathfrak{K}_m\cdot\nabla\Psi_m) + k_0^2\Psi_m = \frac{-\rho_m}{\mu_0} \tag{4-62}$$

$$\mathfrak{K}_m\cdot[\nabla\nabla\cdot(\mathfrak{K}_m\cdot\mathbf{C}_m)] - \nabla\times(\mathfrak{K}^{-1}\cdot\nabla\times\mathbf{C}_m) + k_0^2\mathfrak{K}_m\cdot\mathbf{C}_m = -\epsilon_0\mathbf{J}_m \tag{4-63}$$

$$\nabla\times(\mathfrak{K}_m^{-1}\cdot\nabla\times\mathbf{E}) - k_0^2\mathfrak{K}\cdot\mathbf{E} = -\nabla\times(\mathfrak{K}_m^{-1}\cdot\mathbf{J}_m) \tag{4-64}$$

$$\nabla\times(\mathfrak{K}^{-1}\cdot\nabla\times\mathbf{H}) - k_0^2\mathfrak{K}_m\cdot\mathbf{H} = -j\omega\epsilon_0\mathbf{J}_m \tag{4-65}$$

The expression (4-60) is the dual of (4-39) and the expression (4-61) is the dual[22] of (4-56). We note that (4-56) is the Lorentz condition for electric

---

[21]Note that the source term in (4-59) differs in sign from the similar expression (8.34) in Van Bladel, *Electromagnetic Fields*, p. 234.

[22]We should note that the equations from (4-52) through (4-65) are valid if the material is both anisotropic and inhomogeneous. If the material is isotropic but inhomogeneous, then the dyadics reduce to scalars.

sources in anisotropic media whereas (4-61) is the Lorentz condition for magnetic sources in anisotropic media. Unfortunately, the terminology we are using to distinguish the various potentials is rather poor. For instance, the quantities A and C have been called magnetic vector potentials whereas A_m and C_m have been called electric vector potentials. On the other hand, the quantities Φ and Ψ have been called electric scalar potentials and the quantities Φ_m and Ψ_m have been called magnetic scalar potentials. The reason for doing this is because Φ is called the electric scalar potential, A is called the magnetic vector potential, and Ψ_m is called the magnetic scalar potential in introductory texts.[23] If we adhere to this usage, then the rest of the terminology follows. The concept of duality led us to call A_m the electric vector potential, Φ_m the magnetic scalar potential, and so on. When we introduced the Lorentz condition, which expresses the continuity of electric or magnetic current in terms of potentials, we found that the electric scalar potential was related to the magnetic vector potential for the case of electric sources and that the magnetic scalar potential was related to the electric vector potential for the case of magnetic sources. The reason that this confusion appears is because in electrostatics the scalar potential was called electric because the electric field could be obtained from it and not because the potential had an electric source. Concomitantly, the vector potential was called magnetic because the magnetic field could be obtained from it and not because its source was a magnetic current. Also, the magnetic scalar potential was introduced because one could derive the magnetic field from it. Its source turned out to be the nonphysical quantity of magnetic charge. The magnetic scalar potential was used in magnetostatic problems involving magnetic materials and the nonphysical magnetic charge density was found to be equivalent to the negative of the divergence of the magnetization in the material. One sees that the main disadvantage of the above nomenclature is the fact that the potentials are defined in terms of the derivable fields rather than in terms of the sources.

On the other hand, the Hertz potentials that were defined in (4-32) and (4-33) may be more appealing in this connection since they are defined by their sources. Thus, the electric vector Hertz potential has polarization as its source. In practice, the polarization and magnetization are usually externally controlled electric and magnetic dipole distributions. One other point that is worth noting is the fact that the scalar potentials are related to the vector potentials through a Lorentz condition and, thus, their use may be avoided entirely in solving for the electromagnetic fields. Such a technique is illustrated by the equivalence of (4-39) and (4-45). However, we have not

---

[23]Plonsey and Collin, *Principles and Applications of Electromagnetic Fields*, Chaps. 6, 7. Also, see R. Whitmer, *Electromagnetics*, 2nd ed., Prentice-Hall, Inc., Englewood Cliffs, N. J., 1963, Chap. 8.

pursued this point too far since the introduction of potentials in connection with problems concerned with anisotropic or inhomogeneous media is of doubtful value. As mentioned before, these auxiliary quantities were introduced because it was easier to solve for them than the field quantities. In most cases, their use allowed one to replace a vector problem by a scalar problem. On the other hand, it is well to note that in general they have no particular physical meaning, they are not observable quantities, and they are not independent of one another. An exception to this statement occurs with the electrostatic scalar potential.

One also finds that in considering axial wave propagation along cylindrical surfaces, a z-component of the electric vector Hertz potential leads to an electromagnetic field with a z-component of the electric field but no z-component of the magnetic field. Thus, the wave can be called an electric or E wave since it has an electric field in the direction of propagation. However, it can also be called a transverse magnetic or TM wave since the magnetic field is only in the transverse plane. Since the electric vector Hertz potential is related to the vector magnetic potential by (4-34), it also follows that the TM_z wave is derivable from the z-component of the vector magnetic potential. Similar statements apply for the magnetic or H wave which is also called a TE wave. A general electromagnetic field for the above problem can then be obtained by superposing the partial TM and TE fields.[24] However, the same statement is not necessarily true in any general coordinate system.

Let us now consider the problem of finding the inverse dyadic in terms of the given dyadic. This problem arose in connection with (4-52) and (4-53) and will be solved by using tensor notation. Thus, the problem is that of finding the reciprocal matrix and the solution can be found elsewhere.[25] If we use tensor notation[26] we can write (4-48) as

$$D_i = \epsilon_0 K_{ij} E_j \qquad (4\text{-}66)$$

and (4-52) as

$$E_j = \epsilon_0^{-1} K_{jk}^{-1} D_k \qquad (4\text{-}67)$$

---

[24]Stratton, *Electromagnetic Theory*, p. 351.

[25]Margenau and Murphy, *The Mathematics of Physics and Chemistry*, p. 295. Also, see G. Arfken, *Mathematical Methods for Physicists*, Academic Press, Inc., New York, 1966, Chap. 4.

[26]The expression (4-48) in matrix form would be written as

$$D_i = \epsilon_0 \sum_{j=1}^{3} K_{ij} E_j$$

or

$$\{D\} = \epsilon_0 K \{E\}$$

where $\{D\}$ and $\{E\}$ are 3-component column vectors and K is a 3 × 3 matrix.

If we substitute (4-67) into (4-66) we obtain

$$D_i = K_{ij} K_{jk}^{-1} D_k = \delta_{ik} D_k \qquad (4\text{-}68)$$

since by definition the product of a matrix and its inverse matrix is the unit matrix and a unit matrix can be represented by the Kronecker delta. Note that $K_{jk}^{-1} \neq \delta_{ik}/K_{ij}$ from (4-68) since the repeated indices on the right-hand side of (4-66) and (4-67) and (4-68) indicate a summation over all the values of j, k, and j and k, respectively. We have omitted the summation signs which is a fairly common practice in texts dealing with tensor analysis.[27] For instance, if we were to include the summation sign in (4-66), the expression would be written as

$$D_i = \epsilon_0 \sum_{j=1}^{n} K_{ij} E_j, \qquad i = 1, 2, \ldots, m$$

In all cases that are of interest to us $m = n = 3$, i.e., the matrix K_{ij} is a 3×3 square matrix. Also, the determinant of K_{ij} is nonzero, i.e., nonsingular. The quantities with a single subscript, e.g., D_i, E_j, are vectors (3-component column vectors) and the determinant of K_{ij} is evaluated through the Laplace development, i.e.,

$$|K| = \sum_{i=1}^{n} K_{ik} K^{ik} = \sum_{i=1}^{n} K_{ki} K^{ki}, \qquad k = 1, 2, \ldots, n$$

where the quantity on the left is the determinant of K_{ij} and the quantity K^{ik} is the cofactor of the element K_{ik}. The cofactor is defined as $(-1)^{i+k}$ times the complementary minor and the complementary minor is defined as the determinant obtained by omitting the row and column in which the element K_{ik} appears.

We are finally in a position to determine the inverse matrix in (4-68) which we now do by multiplying the second and third terms by K^{ij} and summing over i. This gives us

$$K_{ij} K^{ij} K_{jk}^{-1} D_k = \delta_{ik} K^{ij} D_k$$

By using the Laplace development for $|K|$ we can write the first term as

$$\sum_{i=1}^{n} K_{ij} K^{ij} K_{jk}^{-1} D_k = |K| K_{jk}^{-1} D_k$$

and the second term as

$$\sum_{i=1}^{n} \delta_{ik} K^{ij} D_k = K^{kj} D_k$$

---

[27]Lass, *Vector and Tensor Analysis*, p. 259.

If we equate the two expressions, we get the desired result

$$K_{jk}^{-1} = \frac{K^{kj}}{|K|}$$

where we note that K^{kj} is the adjoint matrix. The adjoint matrix is obtained by finding the cofactor of each element and then transposing the resulting matrix. Transposition indicates reflection about the diagonal, i.e., $A_{ij} \rightarrow A_{ji} \equiv \tilde{A}_{ij}$, where the $\sim$ sign above A indicates the transpose matrix.

As an example, let us consider the case of the 3 × 3 matrix below

$$[K_{ij}] = \begin{bmatrix} K_{11} & K_{12} & K_{13} \\ K_{21} & K_{22} & K_{23} \\ K_{31} & K_{32} & K_{33} \end{bmatrix}$$

and evaluate each term of the inverse matrix. The adjoint matrix, $[\tilde{K}_{ij}]$, is written by replacing each element above by its cofactor and then transposing the resulting matrix. The cofactor is $(-1)^{i+j}$ times the complementary minor. The complementary minor of K_{32} for instance is

$$\begin{vmatrix} K_{11} & K_{13} \\ K_{21} & K_{23} \end{vmatrix} = K_{11}K_{23} - K_{21}K_{13}$$

The cofactor of K_{32} is $(-1)^5$ times the complementary minor and is

$$K^{32} = K_{21}K_{13} - K_{11}K_{23}$$

The matrix formed from the cofactors is then transposed to give the adjoint matrix.

$$[\tilde{K}_{ij}] = [K_{ji}] = [K^{ji}] = \begin{bmatrix} K^{11} & K^{21} & K^{31} \\ K^{12} & K^{22} & K^{32} \\ K^{13} & K^{23} & K^{33} \end{bmatrix}$$

where

$$K^{11} = K_{22}K_{33} - K_{23}K_{32}$$
$$K^{12} = K_{23}K_{31} - K_{21}K_{33}$$
$$K^{13} = K_{21}K_{32} - K_{22}K_{31}$$
$$K^{21} = K_{13}K_{32} - K_{12}K_{33}$$
$$K^{22} = K_{11}K_{33} - K_{13}K_{31}$$
$$K^{23} = K_{12}K_{31} - K_{11}K_{32}$$
$$K^{31} = K_{12}K_{23} - K_{13}K_{22}$$
$$K^{32} = K_{13}K_{21} - K_{11}K_{23}$$
$$K^{33} = K_{11}K_{22} - K_{12}K_{21}$$

The inverse matrix is given by the adjoint matrix above divided by the determinant. If we define the inverse matrix by a different symbol L_{ij}, we obtain finally that

$$[K_{ij}]^{-1} \equiv L_{ij} \tag{4-69}$$

where

$$L_{11} = \frac{K^{11}}{\Delta}$$

$$L_{12} = \frac{K^{21}}{\Delta}$$

$$L_{13} = \frac{K^{31}}{\Delta}$$

etc., or symbolically,

$$L_{ij} = \frac{K^{ji}}{\Delta}$$

and

$$\Delta = |K_{ij}| = K_{11}K^{11} + K_{12}K^{12} + K_{13}K^{13}$$

Two physical materials that require the introduction of dyadic specific inductive capacities to specify their properties are magnetically biased plasmas and ferrites. In each case, we assume that the dyadic reduces to a scalar when the magnetostatic field vanishes. Also, the magnetostatic field is often impressed in the z-direction so that four of the nine elements are zero and the matrix representation is of the form

$$[G_{ij}] = \begin{bmatrix} g_1 & g_2 & 0 \\ -g_2 & g_1 & 0 \\ 0 & 0 & g_3 \end{bmatrix} \tag{4-70}$$

where $[G]$ can be either $[K]$ or $[K_m]$ depending upon whether the problem is concerned with a plasma or ferrite, respectively. For a plasma[28] we find that

$$g_1 = 1 - \omega_p^2(\omega^2 - \omega_g^2)^{-1}$$

$$g_2 = -j\omega_p^2\left(\frac{\omega_g}{\omega}\right)(\omega^2 - \omega_g^2)^{-1} \tag{4-71}$$

$$g_3 = 1 - \frac{\omega_p^2}{\omega^2}$$

---

[28]Papas, *Theory of Electromagnetic Wave Propagation*, p. 189. Also, see C.C. Johnson, *Field and Wave Electrodynamics*, McGraw-Hill Book Company, New York, 1965, Sec. 11.8.

where ω_p is the plasma frequency ($\omega_p^2 = ne^2/m\epsilon_0$), ω_g is the gyrofrequency ($\omega_g = |eB_0|/m$) of the electrons, n is the electron density, and e and m represent the magnitude of the charge and the mass of the electron, respectively. The angular frequency ω_g is also called the cyclotron frequency and is defined here as a positive quantity, i.e., the negative sign of the electronic charge is included in the expression for g_2.

For a microwave ferrite[29] one finds that

$$g_1 = 1 - \omega_g\omega_m(\omega^2 - \omega_g^2)^{-1}$$
$$g_2 = -j\omega\omega_m(\omega^2 - \omega_g^2)^{-1} \qquad (4\text{-}72)$$
$$g_3 = 1$$

where ω_m is the magnetization frequency ($\omega_m = \gamma M_0$), ω_g is the gyrofrequency ($\omega_g = \gamma H_0$), and γ is the magnetomechanical ratio[30] ($\gamma = ge\mu_0/2m$). Again, we have included the negative sign of the electronic charge in g_2. Since the quantity g (called the "g factor") in γ is approximately two for most microwave ferrites, the gyrofrequency for electrons in a plasma is the same as it is in a ferrite for a given internal magnetostatic field even though the physical mechanisms are completely different. We should also emphasize that the expressions in (4-71) and (4-72) do not take into account losses and that their derivation requires the use of a considerable number of approximations and simplifying assumptions. Further, the diagonal elements in (4-70) are real and the nondiagonal elements are imaginary and of opposite sign. Thus, (4-70) is a hermitian matrix, i.e., in general $G_{ij} = G_{ji}^*$.

The inverse matrix that corresponds to (4-70) is

$$[L_{ij}] = \begin{bmatrix} \ell_1 & -\ell_2 & 0 \\ \ell_2 & \ell_1 & 0 \\ 0 & 0 & \ell_3 \end{bmatrix} \qquad (4\text{-}73)$$

where

$$\ell_1 = g_1(g_1^2 + g_2^2)^{-1}$$
$$\ell_2 = g_2(g_1^2 + g_2^2)^{-1}$$
$$\ell_3 = \frac{1}{g_3}$$

We should note that since g_2 in (4-71) or (4-72) is a pure imaginary quantity, the quantity g_2^2 above is negative.

---

[29]Johnson, *Field and Wave Electrodynamics*, Sec. 12.3.

[30]The saturation magnetization M_0 is often given as $4\pi M_s$. The value for $\gamma/2\pi$ in cgs units is 2.8 MHz/oe when $g = 2$. Also, the term gyromagnetic ratio is often used rather than the proper term magnetomechanical ratio.

Another procedure that can be used in problems concerned with anisotropic media is to rotate or transform the coordinate system to one where the matrix is diagonal. This new coordinate system then defines the principal axes of the matrix and the constitutive relations reduce to

$$D_i = \epsilon_0 K_i E_i$$
$$B_i = \mu_0 K_{mi} H_i$$

(4-74)

where K_i and K_{mi} are scalars. When all three diagonal elements are different, the material is called biaxial and when only two of the elements are different the material is called uniaxial. The general problem of wave propagation in an anisotropic material thus reduces to the solution of three scalar Helmholtz equations with three different wavenumbers, provided one uses a cartesian coordinate system with axes that coincide with the principal axes. This subject will be considered further in Chap. 6.

PROBLEMS

4.1. Show that the vector wave equations for the field intensities in linear, homogeneous, isotropic media[31] are the following:

$$\nabla^2 E + k^2 E = j\omega\mu J + \epsilon^{-1}\nabla\rho = j\omega\mu J - (j\omega\epsilon)^{-1}\nabla\nabla\cdot J$$
$$\nabla^2 H + k^2 H = -\nabla\mathbf{x}J$$

Note: The current and charge densities do not enter into the above equations in a simple manner; thus, these equations are more difficult to solve than (4-13) and (4-14). However, the homogeneous forms are identical with (4-14).

4.2. Verify the expressions (4-46) and (4-47).

4.3. Show that in an isotropic medium that is inhomogeneous only in its electrical properties (4-46) and (4-47) reduce to ($K_m = 1$).

$$\nabla \times \nabla \times E - k_0^2 K E = -j\omega\mu_0 J$$

$$\nabla \times \nabla \times H - \frac{\nabla K}{K} \times \nabla \times H - k_0^2 KH = K\nabla \times (K^{-1}J)$$

$$= J \times \frac{\nabla K}{K} + \nabla \times J$$

4.4. Show that the equation[32] for the electric field intensity in Prob. 4.3 can

[31]Collin, *Field Theory of Guided Waves*, p. 21.

[32]This expression reduces to that given by Stratton, *Electromagnetic Theory*, p. 343, when the medium is source-free.

be written in the alternate form

$$\nabla^2 \mathbf{E} + k^2 \mathbf{E} = j\omega\mu_0 \mathbf{J} + \frac{\nabla \cdot \mathbf{J}}{j\omega\epsilon} \frac{\nabla K}{K} - \frac{\nabla\nabla \cdot \mathbf{J}}{j\omega\epsilon} - \nabla\left(\mathbf{E} \cdot \frac{\nabla K}{K}\right), \qquad k^2 = Kk_0^2$$

4.5. Consider the medium in Prob. 4.3 and show that the expression for the vector magnetic potential in (4-43) reduces to ($K_m = 1$).

$$\nabla^2 \mathbf{C} + k^2 \mathbf{C} = -\mu\mathbf{J} + \nabla\nabla \cdot \mathbf{C} - \left(\frac{k^2}{k_0^2}\right)\nabla\nabla \cdot (K\mathbf{C})$$

Compare this expression with that for the electric field intensity in Prob. 4.4.

4.6. Consider the cartesian form of the equations in Prob. 4.4 when the medium is source-free. Show that the three scalar equations involving the components of the electric field intensity are

$$(\nabla^2 + k^2)E_\alpha = -\frac{\partial Q}{\partial \alpha}$$

where $\alpha = x, y, z$

$$Q = K^{-1}\left(E_x \frac{\partial K}{\partial x} + E_y \frac{\partial K}{\partial y} + E_z \frac{\partial K}{\partial z}\right)$$

and the three scalar equations involving the components of the magnetic field intensity are

$$K(\nabla^2 + k^2)H_x = \frac{\partial K}{\partial y}\left(\frac{\partial H_x}{\partial y} - \frac{\partial H_y}{\partial x}\right) + \frac{\partial K}{\partial z}\left(\frac{\partial H_x}{\partial z} - \frac{\partial H_z}{\partial x}\right)$$

$$K(\nabla^2 + k^2)H_y = \frac{\partial K}{\partial z}\left(\frac{\partial H_y}{\partial z} - \frac{\partial H_z}{\partial y}\right) + \frac{\partial K}{\partial x}\left(\frac{\partial H_y}{\partial x} - \frac{\partial H_x}{\partial y}\right)$$

$$K(\nabla^2 + k^2)H_z = \frac{\partial K}{\partial x}\left(\frac{\partial H_z}{\partial x} - \frac{\partial H_x}{\partial z}\right) + \frac{\partial K}{\partial y}\left(\frac{\partial H_z}{\partial y} - \frac{\partial H_y}{\partial z}\right)$$

4.7. Assume a uniform plane wave solution to the equations in Prob. 4.6, e.g., $\mathbf{E} = E_x(z)\mathbf{a}_x$, $\mathbf{H} = H_y(z)\mathbf{a}_y$, and show that the following equations result:

$$\left(\frac{d^2}{dz^2} + Kk_0^2\right)E_x = 0$$

$$\left(\frac{d^2}{dz^2} - \frac{1}{K}\frac{dK}{dz}\frac{d}{dz} + Kk_0^2\right)H_y = 0$$

provided $K = K(z)$. Note that the latter equation can be put into the form of the Sturm-Liouville equation, (1-6), where $p = K^{-1}$, $q = -k_0^2$, $\lambda = 0$.

4.8. Assume a two-dimensional TM_z solution to the equations in Prob. 4.6, e.g., $\mathbf{H} = H_y(x, z)\mathbf{a}_y$, $\mathbf{E} = E_x(x, z)\mathbf{a}_x + E_z(x, z)\mathbf{a}_z$, and show that the following equations result:

$$\left(\frac{\partial^2}{\partial x^2} + \frac{\partial^2}{\partial z^2} + Kk_0^2\right)E_x = \frac{\partial}{\partial x}\left[K^{-1}\left(E_x\frac{\partial K}{\partial x} + E_z\frac{\partial K}{\partial z}\right)\right]$$

$$\left(\frac{\partial^2}{\partial x^2} + \frac{\partial^2}{\partial z^2} + Kk_0^2\right)E_z = \frac{\partial}{\partial z}\left[K^{-1}\left(E_x\frac{\partial K}{\partial x} + E_z\frac{\partial K}{\partial z}\right)\right]$$

$$K\left(\frac{\partial^2}{\partial x^2} + \frac{\partial^2}{\partial z^2} + Kk_0^2\right)H_y = \frac{\partial K}{\partial z}\frac{\partial H_y}{\partial z} + \frac{\partial K}{\partial x}\frac{\partial H_y}{\partial x}$$

4.9. Repeat Prob. 4.8 with a two-dimensional TE_z solution, e.g., $\mathbf{E} = E_y(x, z)\mathbf{a}_y$, $\mathbf{H} = H_x(x, z)\mathbf{a}_x + H_z(x, z)\mathbf{a}_z$, and show that the following equations are obtained:

$$K\left(\frac{\partial^2}{\partial x^2} + \frac{\partial^2}{\partial z^2} + k^2\right)H_x = \frac{\partial K}{\partial z}\left(\frac{-\partial H_z}{\partial x} + \frac{\partial H_x}{\partial z}\right)$$

$$K\left(\frac{\partial^2}{\partial x^2} + \frac{\partial^2}{\partial z^2} + k^2\right)H_z = \frac{\partial K}{\partial x}\left(\frac{-\partial H_x}{\partial z} + \frac{\partial H_z}{\partial x}\right)$$

$$\left(\frac{\partial^2}{\partial x^2} + \frac{\partial^2}{\partial z^2} + k^2\right)E_y = 0$$

where $k^2 = k_0^2 K$.

4.10. Show that the expression for A_z in (4-29) reduces to the usual static expression when k vanishes, i.e.,

$$A_z = -\frac{I(\boldsymbol{\rho}_0)}{2\pi}\ln \rho_1 + \text{constant}$$

See Prob. 3.5.

4.11. Obtain the static expression for the vector magnetic potential for Prob. 4.10 by using (4-24) for $k = 0$.

4.12. Show that the fields obtained from (4-29) are the following:

$$H_\phi = \frac{Ik}{4j}H_1^{(2)}(k\rho_1)$$

$$E_z = -\frac{Ik^2}{4\omega\epsilon}H_0^{(2)}(k\rho_1)$$

4.13. Show that the fields in Prob. 4.12 behave as $\rho_1^{-1/2}\,e^{-jk\rho_1}$ for $k\rho_1 \gg 1$ and, thus, appear as outwardly traveling plane waves except for the $\rho_1^{-1/2}$ damping factor. Note that the time-average real power flow is independent of distance from the source so that the field in Prob. 4.12 represents an outwardly traveling, uniform cylindrical wave.

4.14. Evaluate the integral in (4.28) by using the identity (Stratton, *Electromagnetic Theory*, p. 389)

$$H_p^{(2)}(z) = j^{p+1}\frac{2}{\pi}\int_0^\infty e^{-jz\cosh\alpha}\cosh p\alpha\,d\alpha$$

Hint: Use the substitution $\rho_1\cosh\alpha = [\rho_1^2 + (z - z_0)^2]^{1/2}$.

4.15. Verify the steps leading to (4-31).

4.16. Show that the fields obtained from (4-31) for $x > x_0$ are the following:

$$H_y = \frac{J_s}{2} e^{-jk(x-x_0)}$$

$$E_z = -\frac{J_s}{2} \sqrt{\frac{\mu}{\epsilon}} e^{-jk(x-x_0)}$$

4.17. Find the z-components of the vector magnetic potential given in (4-31) by solving (4-16) with $J_i(\mathbf{r}) = J_z(\mathbf{r}) = J_s \delta(x - x_0)$. Note that the resulting equation for $A_z(x)$ is the same as (2-1) except that the source is not of unit magnitude and the boundary conditions are replaced by the radiation conditions of (3-20).

4.18. Repeat Prob. 4.17 by using the Fourier transform method from Sec. 3.1. Note that this is Prob. 3.1.

4.19. Solve (4-16) for the case of a uniform line source at (ρ_0, ϕ_0, z) using the conventional method, i.e., the product function expansion. The source can be represented as $J_z = (I/\rho)\delta(\rho - \rho_0)\delta(\phi - \phi_0)$. Equate the result to (4-29) to obtain the addition theorem for cylindrical Hankel functions,

$$H_0^2(k\rho_1) = \sum_{n=-\infty}^{\infty} [J_n(k\rho)H_n^{(2)}(k\rho_0)u(\rho_0 - \rho)$$
$$+ J_n(k\rho_0)H_n^{(2)}(k\rho)u(\rho - \rho_0)] e^{jn(\phi-\phi_0)}$$

where $\rho_1 = |\boldsymbol{\rho} - \boldsymbol{\rho}_0|$. *Note*: The addition theorem is equally valid with $H_0^{(2)}$ replaced by $H_0^{(1)}$. Also, the representation for J_z above gives the correct answer although it is not correct dimensionally. This is because the integration with respect to z is already included in the right-hand side.

4.20. Consider the real and imaginary parts of the addition theorem in Prob. 4.19 to obtain the addition theorem for Bessel functions of the first and second kinds. Note that the first expression is already given in (3-68).

4.21. Solve (4-16) for the uniform line source as in Prob. 4.19 using a cylindrical Fourier transform. The solution is (4.29).

4.22. A plane wave in free space with components $\mathbf{e}_1 = \mathbf{a}_x e^{-jk_0 z}$ and $\mathbf{h}_1 = Y_0 \mathbf{a}_y e^{-jk_0 z}$, where $Y_0 = \sqrt{\epsilon_0/\mu_0}$ and $k_0^2 = \omega^2 \mu_0 \epsilon_0$, is incident on a ferrite[33] interface located at $z = 0$. Find the reflected and transmitted fields in each region and the reflection and transmission coefficients

---

[33] See Johnson, *Field and Wave Electrodynamics*, Sec. 11.13, for a similar analysis in connection with a plasma.

when the ferrite is magnetized to saturation by a magnetostatic field $H = H_0 a_z$. The results are the following:

$$E_r = (R_x a_x + R_y a_y)e^{jk_0 z}, \qquad H_r = (R_y a_x - R_x a_y)Y_0 e^{jk_0 z}$$

$$E_t = T^+(a_x - ja_y)e^{-jk^+ z} + T^-(a_x + ja_y)e^{-jk^- z},$$

$$H_t = jY^+ T^+(a_x - ja_y)e^{-jk^+ z} - jY^- T^-(a_x + ja_y)e^{-jk^- z}$$

$$T^\pm = \left(1 + \frac{Y^\pm}{Y_0}\right)^{-1}$$

$$R_x = -1 + T^+ + T^-$$

$$R_y = -j(T^+ - T^-)$$

The other quantities are defined in (6-35), (4-70), and (4-72).

5

Problems in Bounded and Unbounded Space

5.1 Normal Modes in Cartesian and Cylindrical Coordinates—Source-Free Problems

The normal modes in cartesian and right circular cylindrical coordinates result from the solutions of Laplace's equation or Helmholtz's equation in the two different coordinate systems. If one is interested in a waveguide type of propagation in the z-direction, it is convenient to separate the Laplacian operator into two parts, both acting on the z-component of a vector potential or an electromagnetic field. The simpler part of the Laplacian is the second derivative with respect to z and the more complicated part is the transverse Laplacian, i.e.,

$$\nabla_t^2 = \nabla^2 - \frac{\partial^2}{\partial z^2}$$

As mentioned by Stratton,[1] this separation procedure can be performed in any cylindrical coordinate system in which Helmholtz's equation is separable. Also, the TM_z and TE_z fields obtained from A_z and A_{mz}, respectively (or Hertz potentials, if one prefers them), can represent any general elec-

---

[1] J. Stratton, *Electromagnetic Theory*, McGraw-Hill Book Company, Inc., New York, 1941, p. 35.

tromagnetic field[2] inside a uniform cylindrical waveguide with perfectly conducting walls. The medium inside the waveguide is assumed isotropic and homogeneous but not necessarily loss-free. The uniform cylindrical waveguide is constructed by taking a z-directed line element and moving it parallel to itself around a simply connected closed curve. Waveguides with rectangular and round cross sections are two practical structures that result when the curve is simply connected and the coaxial line is a structure that results when the curve is doubly connected. Also, the TEM_z mode does not exist when the curve is simply connected.[3] The boundary conditions on the electromagnetic fields at the surface of the perfectly conducting wall are satisfied if A_z vanishes for the TM_z modes and $\partial A_{mz}/\partial n$ vanishes for the TE_z modes. This follows since A_z is proportional to E_z, and A_{mz} is proportional to H_z when the medium is isotropic and homogeneous. One can then view the TM_z modes as the eigenfunctions of a Dirichlet problem and the TE_z modes as the eigenfunctions of a Neumann[4] problem. It is equally easy to solve for the transverse electric and magnetic field components in terms of E_z and H_z. In this case, the boundary condition for the TM_z modes is that E_z vanishes on the wall and the boundary condition for the TE_z modes is that $\partial H_z/\partial n$ vanishes on the wall. Some of the typical combinations of modes that satisfy Laplace's equation in rectangular coordinates are shown in Table 5-1A and those that satisfy Helmholtz's equation in rectangular coordinates are shown in Table 5-1B.

When the walls of the guide are not perfectly conducting, the TM_z and TE_z modes are coupled (usually) so that a more general approach is necessary. One such method is discussed and illustrated by Collin[5] in connection with a general cylindrical waveguide with lossy walls. The method seems to be an application of the Schmidt orthogonalization method[6] whereby a nonorthogonal set of functions is expanded in an orthogonal set of functions. In the present case, the orthogonal set of functions is that of the waveguide with perfectly conducting walls. The question of mode completeness is also discussed by Collin.[7] This question becomes more difficult to answer when one considers exterior boundary value problems.

---

[2]R. Collin, *Field Theory of Guided Waves*, McGraw-Hill Book Company, Inc., New York, 1960, Chap. 5.

[3]D. Jones, *The Theory of Electromagnetism*, Pergamon Press, Inc., New York, 1964, p. 246. Also, see Stratton, *Electromagnetic Theory*, p. 538.

[4]J. Van Bladel, *Electromagnetic Fields*, McGraw-Hill Book Company, Inc., New York, 1964, Sec. 13.1.

[5]Collin, *Field Theory of Guided Waves*, pp. 185–95.

[6]G. Arfken, *Mathematical Methods for Physicists*, Academic Press, Inc., New York, 1966, Sec. 9.3.

[7]Collin, *Field Theory of Guided Waves*, pp. 204–9.

TABLE 5-1A   Solutions to Laplace's Equation
Cartesian Coordinates

$$\nabla^2\Phi(x, y, z) = \left(\frac{\partial^2}{\partial x^2} + \frac{\partial^2}{\partial y^2} + \frac{\partial^2}{\partial z^2}\right)\Phi(x, y, z) = 0$$

If

$$\Phi(x, y, z) = f(x)g(y)h(z)$$

then

$$\left(\frac{d^2}{dx^2} + k_1^2\right)f(x) = 0$$

$$\left(\frac{d^2}{dy^2} + k_2^2\right)g(y) = 0$$

$$\left(\frac{d^2}{dz^2} + k_3^2\right)h(z) = 0$$

and

$$k_1^2 + k_2^2 + k_3^2 = 0, \qquad \Gamma_{ij} = (k_i^2 + k_j^2)^{1/2} = -jk_m, \qquad i, j, m = 1, 2, 3;\; i \neq j \neq m$$

Typical Combinations

f(x)			g(y)			h(z)		
$k_1^2>0$	$k_1^2=0$	$k_1^2<0<\Gamma_{23}^2$	$k_2^2>0$	$k_2^2=0$	$k_2^2<0<\Gamma_{13}^2$	$k_3^2>0$	$k_3^2=0$	$k_3^2<0<\Gamma_{12}^2$
$\cos k_1 x$ $\sin k_1 x$			$\cos k_2 y$ $\sin k_2 y$					$\cosh \Gamma_{12} z$ $\sinh \Gamma_{12} z$
$\cos k_1 x$ $\sin k_1 x$					$\cosh \Gamma_{13} y$ $\sinh \Gamma_{13} y$	$\cos k_3 z$ $\sin k_3 z$		
		$\cosh \Gamma_{23} x$ $\sinh \Gamma_{23} x$	$\cos k_2 y$ $\sin k_2 y$			$\cos k_3 z$ $\sin k_3 z$		
$\cos k_1 x$ $\sin k_1 x$					$\cosh k_1 y$ $\sinh k_1 y$		z 1	
		$\cosh k_2 x$ $\sinh k_2 x$	$\cos k_2 y$ $\sin k_2 y$				z 1	
	x 1			y 1			z 1	

An even more general approach to the waveguide propagation problem is to consider the general cylinder when it is located in an infinite medium. Stratton[8] considers the case of a right circular cylinder when the media are homogeneous and isotropic but not loss-free and also the case of an imperfectly conducting right circular cylinder embedded in a dielectric. Collin[9] considers two similar circular cylinder problems, one involving a surface waveguide in the form of a dielectric-coated wire and the other involving surface waves along a dielectric rod.

Since the general solution to a partial differential equation always contains too many specific solutions, it is necessary to apply the boundary conditions in order to legitimately throw away many of the unwanted solutions.

---

[8]Stratton, *Electromagnetic Theory*, Chap. 9.

[9]Collin, *Field Theory of Guided Waves*, Chap. 11.

TABLE 5-1B   Solutions to Helmholtz's Equation
Cartesian Coordinates

$$(\nabla^2 + k^2)\Phi(x, y, z) = 0$$

If

$$\Phi(x, y, z) = f(x)g(y)h(z)$$

then

$$\left(\frac{d^2}{dx^2} + k_1^2\right)f(x) = 0, \qquad \left(\frac{d^2}{dy^2} + k_2^2\right)g(y) = 0, \qquad \left(\frac{d^2}{dz^2} + k_3^2\right)h(z) = 0$$

and

$$k_1^2 + k_2^2 + k_3^2 = k^2 > 0, \qquad \Gamma_{ij} = (k_i^2 + k_j^2)^{1/2} = -jk_m, \qquad i, j, m = 1, 2, 3; \; i \neq j \neq m$$

Typical Combinations

$f(x)$			$g(y)$			$h(z)$		
$k_1^2>0$	$k_1^2=0$	$k_1^2<0$	$k_2^2>0$	$k_2^2=0$	$k_2^2<0$	$k_3^2>0$	$k_3^2=0$	$k_3^2<0$
$\cos k_1 x$ $\sin k_1 x$			$\cos k_2 y$ $\sin k_2 y$			$\cos (k^2-\Gamma_{12}^2)^{1/2}z$ $\sin (k^2-\Gamma_{12}^2)^{1/2}z$		
$\cos k_1 x$ $\sin k_1 x$			$\cos k_2 y$ $\sin k_2 y$					$e^{\pm(\Gamma_{12}^2-k^2)^{1/2}z}$
$\cos k_1 x$ $\sin k_1 x$			$\cos k_2 y$ $\sin k_2 y$			$\cos k_3 z$ $\sin k_3 z$		
	x 1		$\cos k_2 y$ $\sin k_2 y$			$e^{\pm(k^2-k_2^2)^{1/2}z}$		
	x 1		$\cos k_2 y$ $\sin k_2 y$					$e^{\pm(k_2^2-k^2)^{1/2}z}$
	x 1			y 1		$e^{\pm jkz}$		
$\cos k_1 x$ $\sin k_1 x$			$\cos k_2 y$ $\sin k_2 y$				z 1	

If the problem is unbounded in one direction, then the absence of that boundary condition causes the problem to be indeterminate. However, the physical nature of the problem indicates that the solution should be bounded or well behaved in that direction. Thus, solutions to Laplace's equation must behave as damped exponentials or at least decrease in magnitude in this direction, whereas solutions to Helmholtz's equation satisfy a radiation condition (3-20) in this direction. This latter case corresponds to the physical situation where finite sources can create finite fields at infinity which is in contrast to the solutions to Laplace's equation. One noticeable feature of the eigenfunctions or modes in Tables 5-1A or 5-1B is that the sum of the square of the eigenvalues (or separation constants) is either zero or k^2, respectively. The similar constraint on the eigenvalues for cylindrical coordinates in Tables 5-2A or 5-2B is less stringent. Another feature of the eigenvalues for Laplace's equation in Tables 5-1A and 5-2A is that at least one of them has to behave as a damped or growing exponential and the others have to

TABLE 5-2A   Solutions to Laplace's Equation
Cylindrical Coordinates

$$\nabla^2\Phi(\rho,\phi,z) = \left[\frac{1}{\rho}\frac{\partial}{\partial\rho}\left(\rho\frac{\partial}{\partial\rho}\right) + \frac{1}{\rho^2}\frac{\partial^2}{\partial\phi^2} + \frac{\partial^2}{\partial z^2}\right]\Phi(\rho,\phi,z) = 0$$

If

$$\Phi(\rho,\phi,z) = f(\rho)g(\phi)h(z)$$

then

$$\left[\frac{1}{\rho}\frac{d}{d\rho}\left(\rho\frac{d}{d\rho}\right) + \left(\Gamma^2 - \frac{n^2}{\rho^2}\right)\right]f(\rho) = 0$$

$$\left(\frac{d^2}{d\phi^2} + n^2\right)g(\phi) = 0, \qquad \left(\frac{d^2}{dz^2} - \Gamma^2\right)h(z) = 0, \qquad \Gamma = -jk_3$$

$B_n(\Gamma\rho)$ is any linear combination of $J_n(\Gamma\rho)$, $N_n(\Gamma\rho)$, $H_n^{(1)}(\Gamma\rho)$, $H_n^{(2)}(\Gamma\rho)$.

All Combinations

$f(\rho)$			$g(\phi)$		$h(z)$		
$\Gamma^2 > 0$	$\Gamma^2 = 0$	$\Gamma^2 < 0 < k_3^2$	$n^2 > 0$	$n^2 = 0$	$\Gamma^2 > 0$	$\Gamma^2 = 0$	$\Gamma^2 < 0 < k_3^2$
$B_n(\Gamma\rho)$			$\cos n\phi$ $\sin n\phi$		$\cosh \Gamma z$ $\sinh \Gamma z$		
$B_0(\Gamma\rho)$				ϕ 1	$\cosh \Gamma z$ $\sinh \Gamma z$		
		$I_n(k_3\rho)$ $K_n(k_3\rho)$	$\cos n\phi$ $\sin n\phi$				$\cos k_3 z$ $\sin k_3 z$
		$I_0(k_3\rho)$ $K_0(k_3\rho)$		ϕ 1			$\cos k_3 z$ $\sin k_3 z$
	$\rho^{\pm n}$		$\cos n\phi$ $\sin n\phi$		z 1		
	$\ln \rho$ 1			ϕ 1	z 1		

behave as oscillating functions. For instance in Table 5-1A, when $g(y)$ behaves as a hyperbolic function or exponential with real exponent, $f(x)$ or $h(z)$ behaves as a sinusoid or exponential with imaginary exponent. Many simple examples of the solutions of Laplace's equation in cartesian, cylindrical, and spherical coordinates are available in other texts.[10]

The examples that will be discussed in the following sections of this chapter will be quite general in order to illustrate the mathematical procedure for solving Laplace's equation and Helmholtz's equation in cartesian, right circular cylindrical, and spherical coordinates. The source will not be general but will be a unit, i.e., a Dirac delta, source since the solution becomes too complicated if one considers a general source. However, the general solution for a general source can always be expressed in a form that requires a knowl-

---

[10] R. Plonsey and R. Collin, *Principles and Applications of Electromagnetic Fields*, McGraw-Hill Book Company, Inc., New York, 1961, Chap. 4. Also, see M. Javid and P. Brown, *Field Analysis and Electromagnetics*, McGraw-Hill Book Company, Inc., New York, 1963, Chap. 9.

TABLE 5-2B   Solutions to Helmholtz's Equation
Cylindrical Coordinates

$$(\nabla^2 + k^2)\Phi(\rho, \phi, z) = 0$$

If

$$\Phi(\rho, \phi, z) = f(\rho)g(\phi)h(z)$$

then

$$\left[\frac{1}{\rho}\frac{d}{d\rho}\left(\rho\frac{d}{d\rho}\right) + \left(k_1^2 - \frac{n^2}{\rho^2}\right)\right]f((\rho)) = 0, \qquad \left(\frac{d^2}{d\phi^2} + n^2\right)g(\phi) = 0,$$

$$\left(\frac{d^2}{dz^2} + k_3^2\right)h(z) = 0, \qquad k_1^2 + k_3^2 = k^2 > 0, \qquad \Gamma = -jk_1$$

$B_n(k\rho)$ is any linear combination of $J_n(k\rho)$, $N_n(k\rho)$, $H_n^{(1)}(k\rho)$, $H_n^{(2)}(k\rho)$.

All Combinations

$f(\rho)$			$g(\phi)$		$h(z)$		
$k_1^2>0$	$k_1^2=0$	$k_1^2<0<\Gamma^2$	$n^2>0$	$n^2=0$	$k_3^2>0$	$k_3^2=0$	$k_3^2<0$
$B_n(k_1\rho)$			$\cos n\phi$ $\sin n\phi$		$\cos(k^2-k_1^2)^{1/2}z$ $\sin(k^2-k_1^2)^{1/2}z$		
$B_0(k_1\rho)$				ϕ 1	$\cos(k^2-k_1^2)^{1/2}z$ $\sin(k^2-k_1^2)^{1/2}z$		
$B_0(k\rho)$				ϕ 1		z 1	
	$\rho^{\pm n}$		$\cos n\phi$ $\sin n\phi$		$\cos kz$ $\sin kz$		
	$\ln\rho$ 1			ϕ 1	$\cos kz$ $\sin kz$		
		$I_n(\Gamma\rho)$ $K_n(\Gamma\rho)$	$\cos n\phi$ $\sin n\phi$		$\cos(k^2+\Gamma^2)^{1/2}z$ $\sin(k^2+\Gamma^2)^{1/2}z$		
		$I_0(\Gamma\rho)$ $K_0(\Gamma\rho)$		ϕ 1	$\cos(k^2+\Gamma^2)^{1/2}z$ $\sin(k^2+\Gamma^2)^{1/2}z$		
$B_n(k\rho)$			$\cos n\phi$ $\sin n\phi$			z 1	
$B_n(k_1\rho)$			$\cos n\phi$ $\sin n\phi$				$\cosh(k_1^2-k^2)^{1/2}z$ $\sinh(k_1^2-k^2)^{1/2}z$
$B_0(k_1\rho)$				ϕ 1			$\cosh(k_1^2-k^2)^{1/2}z$ $\sinh(k_1^2-k^2)^{1/2}z$

edge of the Green's function so that the problem of finding the Green's function is quite relevant. In addition, for many problems, the solution for a prescribed source is obtained in the same manner as that used to obtain the Green's function. In general, it is possible to solve, i.e., separate, Helmholtz's equation[11] in ten different coordinate systems, all of which are degenerate

---

[11]P. Morse and H. Feshbach, *Methods of Theoretical Physics*, McGraw-Hill Book Company, Inc., New York, 1953, p. 513. Also, see Arfken, *Mathematical Methods for Physicists*, Chap. 2, and Stratton, *Electromagnetic Theory*, p. 349.

forms of the eleventh, which is the general ellipsoidal coordinate system. Laplace's equation is separable in the same eleven coordinate systems plus many additional ones. In addition, the two-dimensional Laplacian is separable in any coordinate system that is a conformal transformation of the cartesian coordinate system. This point is mentioned by Morse and Feshbach[12] and also by Smythe.[13] As illustrated by Arfken,[14] many of the problems where unusual coordinate systems are used occur in areas other than electromagnetic theory so that we will consider only the three coordinate systems mentioned previously. However, Smythe[15] considers a large number of electromagnetic problems involving Laplace's equation in a wide variety of coordinate systems and a smaller number of problems and coordinate systems involving Helmholtz's equation.[16] Also, a very detailed discussion of coordinate systems and the separation problem is presented by Moon and Spencer,[17] who introduce the concept of R-separation in contrast to simple separation. With simple separation, one assumes the usual product solution whereas with R-separation one assumes a product solution divided by a function R which can depend upon the coordinates. This latter technique allows one to separate Laplace's equation in an additional eleven coordinate systems.

5.2 Green's Functions for Laplace's Equation in Cartesian and Cylindrical Coordinates

Let us consider the Green's functions for several electrostatic problems. In these cases, the Green's functions will be analogous to the scalar electric potential. A two-dimensional problem that illustrates many of the basic concepts discussed heretofore is shown in Fig. 5-1. This is the problem of finding the potential everywhere inside an infinite rectangular pipe caused by a uniform line source at $x = x_0$, $y = y_0$. This problem is described mathematically by

$$\nabla_t^2 G = -\delta(x - x_0)\delta(y - y_0) \tag{5-1}$$

---

[12]Stratton, *Electromagnetic Theory*, p. 499.

[13]W. Smythe, *Static and Dynamic Electricity*, 2nd ed., McGraw-Hill Book Company, Inc., New York, 1950, p. 72.

[14]*Ibid.*, Chap. 2.

[15]*Ibid.*, Chaps. 4, 5, 15.

[16]We should note that the homogeneous Helmholtz's equation is a specific form of a wave (or propagation) equation and depends only upon spatial variables. When the angular frequency vanishes, Helmholtz's equation reduces to Laplace's equation.

[17]P. Moon and D. Spencer, *Field Theory for Engineers*, D. Van Nostrand Company, Inc., Princeton, N. J., 1961, Chap. 12.

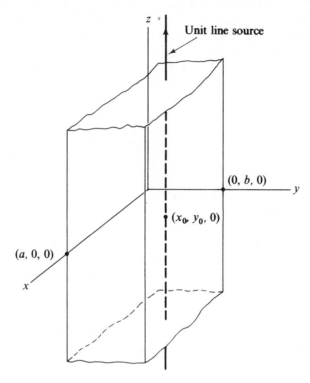

Fig. 5-1   Uniform dc line source inside a grounded rectan
gular conducting pipe.

where $\nabla_t^2 = \nabla^2 - (\partial^2/\partial z^2)$. The boundary condition on $G(x, x_0; y, y_0)$ is that
it vanishes on the four walls, i.e., the pipe is grounded. We begin the problem
by finding a solution of the homogeneous form of (5-1). Thus, we may write

$$G = (A \sin \alpha x + B \cos \alpha x)(C \sinh \alpha y + D \cosh \alpha y) \qquad (5\text{-}2)$$

or

$$G = (A_1 \sinh \beta x + B_1 \cosh \beta x)(C_1 \sin \beta y + D_1 \cos \beta y) \qquad (5\text{-}3)$$

Either formulation in (5-2) or (5-3) satisfies the homogeneous form of (5-1).
If we arbitrarily choose (5-2) and apply the boundary conditions that $G = 0$
at $x = 0$, a, we find that $B = 0$ and $\alpha = m\pi/a$, $m = 1, 2, \ldots$. When we
apply the other two boundary conditions that $G = 0$ at $y = 0$, b, we find that
$D = 0$ and $C = 0$. Since zero is a trivial solution of (5-1), we must conclude
that (5-2) cannot satisfy the homogeneous form of (5-1) and the four boundary conditions. However, we can still find a satisfactory solution to (5-1)
of the form of (5-2) by using the general method described in Sec. 2.2.

Unfortunately, that method applies only to an ordinary differential equation and we have a partial differential equation with two independent variables. Since the x-dependent portion of (5-2) satisfies the two boundary conditions at $x = 0$, a, let us divide the interval $(0, b)$ over y into two parts depending upon whether $y < y_0$ or $y > y_0$. Thus we may write

$$G = G_1 u(y_0 - y) + G_2 u(y - y_0) \tag{5-4}$$

where

$$G_1 = \sum_{m=1}^{\infty} G_x \left(C_1 \sinh \frac{m\pi}{a} y + D_1 \cosh \frac{m\pi}{a} y \right), \qquad 0 \le y \le y_0 \tag{5-5}$$

$$G_2 = \sum_{m=1}^{\infty} G_x \left(C_2 \sinh \frac{m\pi}{a} y + D_2 \cosh \frac{m\pi}{a} y \right), \qquad y_0 \le y \le b \tag{5-6}$$

and

$$G_x = \sin \frac{m\pi x}{a}$$

However, we now have four unknown coefficients and two boundary conditions at $y = 0$, b. Since (5-4) is a solution of (5-1), we know that the Green's function is continuous at $y = y_0$ while its y-derivative is discontinuous. This gives us two more boundary conditions at the source, which gives us the four we need. The boundary condition that $G = 0$ at $y = 0$ gives $D_1 = 0$ and the boundary condition that $G = 0$ at $y = b$ gives $D_2 = -C_2 \tanh (m\pi b/a)$. Now we obtain

$$G_1 = \sum_{m=1}^{\infty} G_x G_{1y} \tag{5-7}$$

$$G_2 = \sum_{m=1}^{\infty} G_x G_{2y} \tag{5-8}$$

where

$$G_{1y} =. C_m \sinh \frac{m\pi y}{a}$$

$$G_{2y} = D_m \sinh \frac{m\pi}{a} (b - y)$$

Let us now apply the two boundary conditions at the source (source conditions) to determine the constants C_m, D_m. The continuity of G at $x = x_0$, $y = y_0$ gives us

$$C_m \sinh \frac{m\pi y_0}{a} + D_m \sinh \frac{m\pi}{a} (y_0 - b) = 0 \tag{5-9}$$

Our last source condition concerns the discontinuity in the y-derivative of G at the source. We know from (2-19) that the derivative will be discontinuous by $-1/p(y_0)$ and from the discussion following (2-26) that the derivative will be discontinuous by $-D(y_0)/(p(y_0))$. In this latter case, $D(y)$ is the amplitude of the source for the one-dimensional equation and $p(y)$ is the usual $p(y)$ in the Sturm-Liouville operator (1-7). However, we do not yet know $D(y)$ and $p(y)$. To find them we must seek the ordinary differential equation satisfied by G_y. Since we know that (5-4) satisfies (5-1), let us substitute it into (5-1) and use (5-7) and (5-8) to obtain

$$\sum_{m=1}^{\infty} \left[-\left(\frac{m\pi}{a}\right)^2 + \frac{d^2}{dy^2} \right] G_x G_y = -\delta(x - x_0)\, \delta(y - y_0)$$

where

$$G_y = G_{1y} u(y_0 - y) + G_{2y} u(y - y_0)$$

If we multiply this expression by $\sin (n\pi x/a)\, dx$, integrate over the interval $(0, a)$, and then replace n by m, we obtain

$$\left[\frac{d^2}{dy^2} - \left(\frac{m\pi}{a}\right)^2 \right] G_y = -\frac{2}{a} \sin \frac{m\pi x_0}{a} \delta(y - y_0) \tag{5-10}$$

which is our desired ordinary differential equation for G_y. In this case, $p(y) = 1$ and $D(y) = (2/a) \sin (m\pi x_0/a)$ and our relation (2-19) yields

$$C_m \cosh \frac{m\pi y_0}{a} + D_m \cosh \frac{m\pi}{a}(b - y_0) = \frac{2}{m\pi} \sin \frac{m\pi x_0}{a} \tag{5-11}$$

We now have two equations, (5-9) and (5-11), in two unknowns, C_m and D_m, and their solution is

$$C_m = \frac{2}{m\pi} \frac{\sin (m\pi x_0/a) \sinh [(m\pi/a)(b - y_0)]}{\sinh (m\pi b/a)}$$

$$D_m = \frac{2}{m\pi} \frac{\sin (m\pi x_0/a) \sinh (m\pi y_0/a)}{\sinh (m\pi b/a)}$$

with the result that $G_{1,2}$ in (5-5) and (5-6) becomes

$$G_1(x, x_0; y, y_0) = \sum_{m=1}^{\infty} \frac{2}{m\pi}$$

$$\times \frac{\sin (m\pi x/a) \sin (m\pi x_0/a) \sinh (m\pi y/a) \sinh [(m\pi/a)(b - y_0)]}{\sinh (m\pi b/a)},$$

$$y < y_0 \tag{5-12}$$

and

$$G_2(x, x_0; y, y_0) = G_1(x_0, x; y_0, y), \qquad y > y_0 \tag{5-13}$$

The complete solution is given by (5-4).

As mentioned previously in solving (5-1), we arbitrarily chose the formulation (5-2) for G and found that the solution was given by (5-4) with G_1 and G_2 defined by (5-12) and (5-13), respectively. Let us now choose the formulation (5-3) for G and find the solution. The result is given in Prob. 5.1 and is the following:

$$G = G_1 u(x_0 - x) + G_2 u(x - x_0)$$

with

$$G_1(x, x_0; y, y_0) = \sum_{n=1}^{\infty} \frac{2}{n\pi}$$

$$\times \frac{\sin (n\pi y/b) \sin (n\pi y_0/b) \sinh (n\pi x/b) \sinh [(n\pi/b)(a - x_0)]}{\sinh (n\pi a/b)}, \qquad x < x_0$$

$$G_2(x, x_0; y, y_0) = G_1(x_0, x; y_0, y), \qquad x > x_0$$

Since the two different solutions for G must be identical, let us equate them and verify their equality. As the first step we obtain

$$\sum_{n=1}^{\infty} \frac{2}{n\pi} \frac{\sin (n\pi y/b) \sin (n\pi y_0/b)}{\sinh (n\pi a/b)} \left[\sinh \frac{n\pi x}{b} \sinh \frac{n\pi}{b}(a - x_0) u(x_0 - x) \right.$$

$$\left. + \sinh \frac{n\pi x_0}{b} \sinh \frac{n\pi}{b}(a - x) u(x - x_0) \right]$$

$$= \sum_{m=1}^{\infty} \frac{2}{m\pi} \frac{\sin (m\pi x/a) \sin (m\pi x_0/a)}{\sinh (m\pi b/a)} \left[\sinh \frac{m\pi y}{a} \sinh \frac{m\pi}{a}(b - y_0) u(y_0 - y) \right.$$

$$\left. + \sinh \frac{m\pi y_0}{a} \sinh \frac{m\pi}{a}(b - y) u(y - y_0) \right]$$

We can remove the summation signs by using the orthogonality of the sinusoids, i.e., we multiply both sides by $\sin (p\pi y/b) \, dy \sin (q\pi x/a) \, dx$ and integrate with respect to y over the interval $(0, b)$ and with respect to x over the interval $(0, a)$. Thus, we obtain the expression

$$\frac{\sin \alpha y_0}{\alpha \sinh \alpha a} \int_0^a [\sinh \alpha x \sinh \alpha (a - x_0) u(x_0 - x)$$

$$+ \sinh \alpha x_0 \sinh \alpha (a - x) u(x - x_0)] \sin \frac{q\pi x}{a} \, dx$$

$$= \frac{\sin \beta x_0}{\beta \sinh \beta b} \int_0^b [\sinh \beta y \sinh \beta (b - y_0) u(y_0 - y)$$

$$+ \sinh \beta y_0 \sinh \beta (b - y) u(y - y_0)] \sin \frac{p\pi y}{b} \, dy$$

where we used the substitution $\alpha = p\pi/b$ and $\beta = q\pi/a$ in all of the terms except the two shown above. We now use the identity that we can obtain from the expression following (2-39) (see Probs. 2.4 and 2.24), i.e.,

$$\frac{2}{L} \sum_{m=1}^{\infty} \left[\left(\frac{m\pi}{L}\right)^2 + \alpha^2 \right]^{-1} \sin \frac{m\pi x}{L} \sin \frac{m\pi x_0}{L}$$

$$= (\alpha \sinh \alpha L)^{-1} [\sinh \alpha (L - x_0) \sinh \alpha x \, u(x_0 - x)$$

$$+ \sinh \alpha (L - x) \sinh \alpha x_0 \, u(x - x_0)] \qquad (5\text{-}14)$$

to simplify the integrands above. Since the square bracket term in the identity is the same as the square bracket term in the integrand on the left above, provided we replace L by a, we can make this substitution in the integrand and then use the orthogonality of the sinusoids to obtain the simple expression

$$\sin \alpha y_0 \left[\left(\frac{q\pi}{a}\right)^2 + \alpha^2 \right]^{-1} \sin \frac{q\pi x_0}{a}$$

We can use the same substitution in the integrand on the right side above after replacing α by β, x by y, and L by b. Now, use of orthogonality of the sinusoids gives us a similar simplification. The final result is that

$$\sin \alpha y_0 \left[\left(\frac{q\pi}{a}\right)^2 + \alpha^2 \right]^{-1} \sin \frac{q\pi x_0}{a} = \sin \beta x_0 \left[\left(\frac{p\pi}{b}\right)^2 + \beta^2 \right]^{-1} \sin \frac{p\pi y_0}{b}$$

which is an identity as soon as we insert the values of α and β.

What we have just demonstrated in this problem is the equivalence between two alternate formulations for the Green's function. This is exactly the point that was introduced in Sec. 2.4, except that the present problem is two-dimensional rather than one-dimensional. Again, all we are doing is replacing a discontinuous function by its Fourier series representation, which is the meaning of the identity in (5-14).

We have now reached a position where we can make some comments about the procedure for separating a nonhomogeneous partial differential equation such as (5-1) into two ordinary differential equations. Let us use the two-dimensional problem that we have just worked as an example. When we chose the formulation (5-2) for the solution we found that G_x satisfied the homogeneous equation

$$\left(\frac{d^2}{dx^2} + \alpha^2\right) G_x = 0$$

The boundary conditions on G_x gave us the eigenfunctions $\sin \alpha x$ and the eigenvalues $\alpha = m\pi/a$. The Green's function G_y satisfied the nonhomogeneous equation

$$\left(\frac{d^2}{dy^2} - \alpha^2\right) G_y = -D(y)\,\delta(y - y_0) \qquad (5\text{-}10)$$

with

$$D(y) = \frac{2}{a} \sin \alpha x_0$$

On the other hand, when we used the formulation (5-3) we found that G_x satisfied the nonhomogeneous equation

$$\left(\frac{d^2}{dx^2} - \beta^2\right) G_x = -D(x)\,\delta(x - x_0)$$

with

$$D(x) = \frac{2}{b} \sin \beta y_0$$

and G_y satisfied the homogeneous equation

$$\left(\frac{d^2}{dy^2} + \beta^2\right) G_y = 0$$

with the eigenvalues $\beta = n\pi/b$. In both of the cases above the boundary conditions were particularly simple since G vanished on the four walls. This boundary condition is one of the three that constitute the homogeneous boundary conditions,[18] i.e.,

$$G = 0$$

$$\frac{dG}{dn} = 0$$

$$\frac{dG}{dn} + \gamma G = 0$$

The first two boundary conditions occur in electromagnetic problems involving perfect electric conductors whereas the third boundary condition occurs with imperfect electric conductors and perfect or imperfect materials. Let us consider some examples. For the voltage on a transmission line, the boundary conditions correspond, respectively, to the following terminations: short circuit, open circuit, general impedance. For the current on a transmission line, the boundary conditions correspond, respectively, to the following terminations: open circuit, short circuit, general admittance. For plane wave propagation normal to the boundary, the boundary conditions on the electric and magnetic fields correspond to those for the voltage and current, respectively, on a transmission line. For TM_z waves guided by perfect electric

---

[18] J. Dettman, *Mathematical Methods in Physics and Engineering*, McGraw-Hill Book Company, Inc., New York, 1962, p. 113.

conductors, the required boundary condition is that E_z or A_z vanish on the boundary. For TE_z waves guided by perfect electric conductors, the required boundary condition is that $\partial H_z/\partial n$ or $\partial A_{mz}/\partial n$ vanish on the boundary. For TM_z waves guided by a dielectric-slab guide, the required boundary condition is that $(\partial E_z/\partial n) + \delta_1 E_z = 0$ or $(\partial A_z/\partial n) + \gamma_1 A_z = 0$. For TE_z waves guided by a dielectric-slab guide, the required boundary condition is that $(\partial H_z/\partial n) + \delta_2 H_z = 0$ or $(\partial A_{mz}/\partial n) + \gamma_2 A_{mz} = 0$.

Thus, the problem that we are considering is one involving a nonhomogeneous partial differential equation with homogeneous boundary conditions. In both of the cases above we see that the partial differential equation can be separated into a homogeneous and a nonhomogeneous ordinary differential equation. If the partial differential equation were three-dimensional rather than two-dimensional, then two of the ordinary differential equations would be homogeneous and the third would be nonhomogeneous. We can also state that the homogeneous ordinary differential equations are the same as those obtained in the usual process of separating Laplace's or Helmholtz's equation. Further, the remaining nonhomogeneous ordinary differential equation will have the same homogeneous form as the remaining homogeneous ordinary differential equation separated from Laplace's or Helmholtz's equation. The only unknown will be the strength of the source for this nonhomogeneous ordinary differential equation. For example, in (5-10) the strength of the source was $D(y)$. Although your first inclination might be to state that the strength of the source is unity by definition, one must remember that this is true only when applied to the integral of the defining partial differential equation. For example, if we apply this criterion to (5-1), we must integrate both sides with respect to x and y over an area including x_0 and y_0. The right side becomes (minus) unity, which is the strength of the source. If we consider (5-15) and apply the same criterion, we find that again we get (minus) unity. In this case, the integration was a volume integration in cylindrical coordinates and the volume of integration included the source point. Since $dV = \rho\, d\rho\, d\phi\, dz$ in cylindrical coordinates, one sees at once why the right side of (5-15) contained the term $1/\rho$ whereas the right side of (5-1) required no additional factors. Further details are considered in Appendix B.

Another viewpoint is to realize that any space containing a delta source has to be divided into only two regions, i.e., the source has to be taken into account only once. If the space is two-dimensional, then there are two choices for the division: to the left and the right of the source or above and below the source. If the space is three-dimensional, then there are three choices for the division. The result of these choices is that a problem may have a considerable number of alternate solutions that appear to be different but must be the same.

Let us consider a three-dimensional problem involving cylindrical coordinates as shown in Fig. 5-2. This is the problem of finding the potential

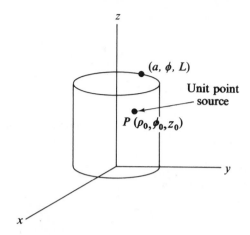

Fig. 5-2   Unit dc point source inside a grounded cylindrical conducting box.

inside a grounded, conducting, cylindrical box with a unit source inside it at the point $P(\rho_0, \phi_0, z_0)$. In this case, the equation for the Green's function is

$$\nabla^2 G = -\frac{\delta(\rho - \rho_0)\,\delta(\phi - \phi_0)\,\delta(z - z_0)}{\rho} \qquad (5\text{-}15)$$

and ∇^2 is expressed in cylindrical coordinates. By analogy with Prob. 5.3 concerning the Green's function for a rectangular box, we can divide the volume of the cylindrical box into two regions. Let us divide the volume into two regions depending upon whether $z > z_0$ or $z < z_0$. In this case, the problem differs from Prob. 5.3 only in the coordinates transverse to z, i.e., ρ and ϕ instead of x and y.

Thus, define

$$G = G_1 u(z_0 - z) + G_2 u(z - z_0)$$

$$G_1 = \sum_{m=0}^{\infty} \sum_{n=1}^{\infty} G_{m\phi} G_{n\rho} G_{1z}$$

$$G_2 = \sum_{m=0}^{\infty} \sum_{n=1}^{\infty} G_{m\phi} G_{n\rho} G_{2z}$$

$$G_{m\phi} = \cos m(\phi - \phi_0) \qquad (5\text{-}16)$$

$$G_{n\rho} = J_m(\beta_{mn}\rho)$$

$$G_{1z} = A_{mn} \sinh \beta_{mn} z$$

$$G_{2z} = B_{mn} \sinh \beta_{mn}(L - z)$$

and the eigenvalue β_{mn} is determined from the relation $J_m(\beta_{mn}a) = 0$. If we substitute (5-16) into (5-15), we obtain

$$\sum_{m=0}^{\infty} \sum_{n=1}^{\infty} \left[\frac{1}{\rho} \frac{\partial}{\partial \rho} \left(\rho \frac{\partial G_{n\rho}}{\partial \rho} \right) G_{m\phi} G_z + \frac{1}{\rho^2} \frac{\partial^2 G_{m\phi}}{\partial \phi^2} G_{n\rho} G_z + G_{m\phi} G_{n\rho} \frac{\partial^2 G_z}{\partial z^2} \right]$$

$$= - \frac{\delta(\rho - \rho_0)\, \delta(\phi - \phi_0)\, \delta(z - z_0)}{\rho}$$

where $G_z = G_{1z} u(z_0 - z) + G_{2z} u(z - z_0)$. Upon taking the derivative with respect to ϕ, we obtain

$$\sum_{m=0}^{\infty} \sum_{n=1}^{\infty} \left\{ \left[\frac{1}{\rho} \frac{\partial}{\partial \rho} \left(\rho \frac{\partial G_{n\rho}}{\partial \rho} \right) - \frac{m^2}{\rho^2} G_{n\rho} \right] G_{m\phi} G_z + G_{m\phi} G_{n\rho} \frac{\partial^2 G_z}{\partial z^2} \right\}$$

$$= - \frac{\delta(\rho - \rho_0)\, \delta(\phi - \phi_0)\, \delta(z - z_0)}{\rho}$$

However, by using Bessel's equation, we can simplify the above expression to

$$\sum_{m=0}^{\infty} \sum_{n=1}^{\infty} G_{n\rho} G_{m\phi} \left[-\beta_{mn}^2 G_z + \frac{d^2 G_z}{dz^2} \right] = - \frac{\delta(\rho - \rho_0)\, \delta(\phi - \phi_0)\, \delta(z - z_0)}{\rho}$$

By multiplying both sides by $J_q(\beta_{qn}\rho) \cos q(\phi - \phi_0)\rho \, d\rho \, d\phi$, integrating over the intervals $(0, a)$ and $(0, 2\pi)$, and then replacing q by m, we obtain the final result that

$$\left(\frac{d^2}{dz^2} - \beta_{mn}^2 \right) G_z = - \frac{\epsilon_m}{\pi a^2} \frac{J_m(\beta_{mn}\rho_0)}{J_{m+1}^2(\beta_{mn}a)} \delta(z - z_0) \qquad (5\text{-}17)$$

where, as usual $\epsilon_m = 1$ for $m = 0$, and $\epsilon_m = 2$ for $m > 0$. Since (5-17) is of the same form as (5-10), we may write the solution as

$$G_{1z} = \frac{D(z_0)}{\beta_{mn}} \frac{\sinh \beta_{mn}(L - z_0) \sinh \beta_{mn} z}{\sinh \beta_{mn} L}$$

$$G_{2z} = \frac{D(z_0)}{\beta_{mn}} \frac{\sinh \beta_{mn} z_0 \sinh \beta_{mn}(L - z)}{\sinh \beta_{mn} L}$$

where $D(z_0)$ is the coefficient of $-\delta(z - z_0)$ in (5-17). For this case,

$$D(z_0) = \frac{\epsilon_m}{\pi a^2} \frac{J_m(\beta_{mn}\rho_0)}{J_{m+1}^2(\beta_{mn}a)}$$

Our final result becomes

$$G_1(\rho, \rho_0; \phi, \phi_0; z, z_0) = \sum_{m=0}^{\infty} \sum_{n=1}^{\infty} \frac{\epsilon_m}{\pi a^2 \beta_{mn}} \frac{J_m(\beta_{mn}\rho_0)}{J_{m+1}^2(\beta_{mn}a)} \frac{J_m(\beta_{mn}\rho) \cos m(\phi - \phi_0)}{\sinh \beta_{mn} L}$$

$$\times \sinh \beta_{mn}(L - z_0) \sinh \beta_{mn} z, \qquad z < z_0 \qquad (5\text{-}18)$$

$$G_2(\rho, \rho_0; \phi, \phi_0; z, z_0) = G_1(\rho_0, \rho; \phi_0, \phi; z_0, z), \qquad z > z_0 \qquad (5\text{-}19)$$

Since the Green's function solution to (5-15) obtained in (5-18) and (5-19) is valid for all values of a and L, let us indicate how the Green's functions for other problems may be found from this one. For instance, if we allow L to become infinite, we can find the Green's function for a grounded, semi-infinite, cylindrical box with one end open. This problem can then be extended further to obtain the Green's function for a grounded, infinitely long cylinder. The solution to this last problem is

$$G_1(\rho, \rho_0; \phi, \phi_0; z, z_0) = \sum_{m=0}^{\infty} \sum_{n=1}^{\infty} \frac{\epsilon_m}{2\pi a^2 \beta_{mn}} \frac{J_m(\beta_{mn}\rho_0)J_m(\beta_{mn}\rho)}{J_{m+1}^2(\beta_{mn}a)}$$
$$\times \cos m(\phi - \phi_0) \exp(-\beta_{mn}|z - z_0|)$$
$$G_2(\rho, \rho_0; \phi, \phi_0; z, z_0) = G_1(\rho_0, \rho; \phi_0, \phi; z_0, z) \qquad (5\text{-}20)$$

As a final extension of this problem, we can let a become infinite so that we have the Green's function in cylindrical coordinates for a unit source in free space. However, now the interval $(0, a)$ becomes infinite so that our Fourier series representation for the radial dependence of the Green's function passes to a Fourier integral representation. When this is accomplished, the Green's function that is similar to that in (5-20) becomes

$$G = G_1 = G_2 = \sum_{m=0}^{\infty} \frac{\epsilon_m}{4\pi} \cos m(\phi - \phi_0) \int_0^{\infty} e^{-\alpha|z-z_0|} J_m(\alpha\rho_0) J_m(\alpha\rho) \, d\alpha$$
$$(5\text{-}21)$$

where we have replaced β_{mn} by α and used the values of $\beta_{mn}a$ from the asymptotic form of $J_m(\beta_{mn}a)$. However, we also know from electrostatics that an alternate representation for G in (5-21) is

$$G = \frac{1}{4\pi R} \qquad (5\text{-}22)$$

where

$$R^2 = \rho^2 + \rho_0^2 - 2\rho\rho_0 \cos(\phi - \phi_0) + (z - z_0)^2$$
$$= \rho_1^2 + (z - z_0)^2$$

5.3   Green's Functions for Helmholtz's Equation in Cartesian and Cylindrical Coordinates

The solution to the scalar Helmholtz equation in cartesian and cylindrical coordinates does not differ too much from the solution to Laplace's equation. This can be seen by inspecting the functions in Tables 5-1 and 5-2. The main difference is the constraint on the separation constants. Now, all of them can be real since k is nonzero. Some of the separation constants in the tables are actually eigenvalues. However, one must remember that the functions in the tables must satisfy a Sturm-Liouville equation in order to be called eigenfunctions, and the equations involving f, g, and h are of the Sturm-Liouville form for certain values of the separation constants but not others. For instance, in Table 5-1B, when k_1^2 is positive the equation involving $f(x)$ is of the Sturm-Liouville form so that its solutions are eigenfunctions and the separation constants are eigenvalues. This is not the case when k_1^2 is negative.

Let us consider the problem illustrated in Fig. 5-1 when the uniform line source varies harmonically with time. Physically, this problem could be the solution of (4-14) when the arbitrary current distribution becomes an infinitely long z-directed current filament as discussed in the derivation of (4-28). One can also think of this electric line source as composed of an infinite number of infinitesimal electric dipoles[19] laid end to end along the z-axis. This is exactly what the integration process in (4-28) means. This viewpoint was discussed earlier in Sec. 3.1 just after the derivation of (3-42). Alternatively, each infinitesimal electric dipole can be considered as two closely spaced point charges of opposite polarity[20] rather than as a short current filament. Since the structure in Fig. 5-1 constitutes a rectangular waveguide, it follows that the modes excited will be TM_z modes since the only finite component of the vector potential is A_z. The boundary condition is that A_z vanishes on the walls since they are assumed to be perfectly conducting. The Helmholtz equation in this case is only slightly different from (5-1) and is

$$(\nabla_t^2 + k^2)G = -\delta(x - x_0)\,\delta(y - y_0) \tag{5-23}$$

where we have replaced A_z by G since we are considering a unit source. The general solution to the homogeneous form of (5-23) can be obtained from Table 5-1B and is

$$G = (A \cos k_1 x + B \sin k_1 x)(C \cos k_2 y + D \sin k_2 y)$$

where $k_1^2 + k_2^2 = k^2 > 0$. The boundary conditions at $x = 0$ and $y = 0$ cause A and C to vanish and the boundary conditions at $x = a$ and $y = b$ force k_1 and k_2 to take on the values

$$k_1 = \frac{m\pi}{a}, \qquad m = 1, 2, 3, \ldots$$

$$k_2 = \frac{n\pi}{b}, \qquad n = 1, 2, 3, \ldots$$

[19]Jones, *The Theory of Electromagnetism*, p. 157.

[20]Plonsey and Collin, *Principles and Applications of Electromagnetic Fields*, p. 394.

Thus, we assume that the solution to (5-23) is

$$G = \sum_{m=1}^{\infty} \sum_{n=1}^{\infty} A_{mn} \sin \frac{m\pi x}{a} \sin \frac{n\pi y}{b} \tag{5-24}$$

We determine the constants A_{mn} by substituting (5-24) into (5-23) and using orthogonality of the sinusoids. The substitution of (5-24) into (5-23) gives us the relation

$$\sum_{m=1}^{\infty} \sum_{n=1}^{\infty} (k^2 - k_1^2 - k_2^2) A_{mn} \sin \frac{m\pi x}{a} \sin \frac{n\pi y}{b} = -\delta(x - x_0)\,\delta(y - y_0),$$

$$k^2 - k_1^2 - k_2^2 \neq 0$$

In this case, it is not necessary for the term in the parentheses to vanish since that constraint applies only to the homogeneous form of (5-23). (Physically, the constraint on the eigenvalues means that the system is in resonance so that zero excitation can cause a finite response.) Now, orthogonality of the sinusoids allows one to solve for A_{mn} so that the final expression for (5-23) becomes

$$G = \frac{4}{ab} \sum_{m=1}^{\infty} \sum_{n=1}^{\infty} \left[\left(\frac{m\pi}{a}\right)^2 + \left(\frac{n\pi}{b}\right)^2 - k^2 \right]^{-1} \sin \frac{m\pi x}{a} \sin \frac{m\pi x_0}{a} \sin \frac{n\pi y}{b} \sin \frac{n\pi y_0}{b}$$

$$\tag{5-25}$$

which is finite except at resonance and is of the same form as (2-38) except in two dimensions instead of one. Since the expression for G in (5-25) is valid for $k = 0$, it follows that (5-25) should reduce to (5-4) since the boundary conditions are the same for each problem. This viewpoint should lead to an alternate representation for the one-dimensional Green's function in y that corresponds to Prob. 2.23.

As another example of the solution of Helmholtz's equation in cartesian coordinates, let us return to Fig. 5.1 and consider a uniform finite electric line source located inside the pipe at (x_0, z_0) and directed parallel to the y-axis. This problem is considered in detail by Collin.[21] It is also considered by Van Bladel[22] for the case where the time dependence of the current source is described by a unit step. Since the current source is independent of y, it follows that the excited fields will be independent of y and the problem will be two-dimensional. Also, we know from (4-16) that the only finite compo-

---

[21]Collin, *Field Theory of Guided Waves*, p. 199.

[22]Van Bladel, *Electromagnetic Fields*, p. 429.

nent of the magnetic vector potential is A_y and that it satisfies the equation

$$\left(\frac{\partial^2}{\partial x^2} + \frac{\partial^2}{\partial z^2} + k^2\right) A_y = -I_0\,\delta(x - x_0)\,\delta(z - z_0) \qquad (5\text{-}26)$$

From (4-15) we know that the only finite component of electric field is

$$E_y = -j\omega\mu A_y$$

and the finite components of magnetic field are (Appendix A)

$$H_x = -\frac{\partial A_y}{\partial z}$$

$$H_z = \frac{\partial A_y}{\partial x}$$

The modes excited by the current source can be classified as TE_z modes, which is the customary classification. However, they can also be classified as LSM modes[23] or TM_y modes,[24] which is less customary but equally valid. The boundary conditions on A_y are that it vanish at $x = 0$, a and the physical conditions on A_y for large values of z are that the modes propagate away from the source. Inspection of Table 5-1B indicates that exponentials in z will satisfy the physical conditions and that a sinusoid in x will satisfy the boundary conditions. If we let the magnitude of the current element in (5-26) be unity instead of I_0, we can speak of A_y as a Green's function $G(x, z)$.

We could also let the Green's function be E_y since A_y and E_y only differ by a constant in this problem. However, let us proceed to solve (5-26) for A_y. We can immediately write down the solution as

$$A_y^> = \sum_{m=1}^{\infty} B_m \sin \frac{m\pi x}{a} e^{-k_3 z}, \qquad z > z_0 \qquad (5\text{-}27)$$

$$A_y^< = \sum_{m=1}^{\infty} C_m \sin \frac{m\pi x}{a} e^{k_3 z}, \qquad z < z_0 \qquad (5\text{-}28)$$

The relation between k_3, k, and the eigenvalues $m\pi/a$ can be determined by substituting (5-27) or (5-28) into (5-26) or by referring to Table 5-1B and is

$$k_3 = \left[\left(\frac{m\pi}{a}\right)^2 - k^2\right]^{1/2}$$

The coefficients B_m and C_m can be evaluated by using the boundary (or continuity) conditions involving the electric and magnetic fields at the source.

---

[23]Collin, *Field Theory of Guided Waves*, p. 225.

[24]R. Harrington, *Time-Harmonic Electromagnetic Fields*, McGraw-Hill Book Company, Inc., New York, 1961, Sec. 4-4.

From (4-6) it follows that at $z = z_0$

$$E_y^> = E_y^<$$

$$\mathbf{a}_z \mathbf{x}(\mathbf{H}^> - \mathbf{H}^<) = I_0\, \delta(x - x_0)\mathbf{a}_y$$

or

$$H_x^> - H_x^< = I_0\, \delta(x - x_0)$$

The equations involving A_y become at $z = z_0$

$$A_y^> = A_y^<$$

$$\frac{\partial A_y^>}{\partial z} - \frac{\partial A_y^<}{\partial z} = -I_0 \delta(x - x_0) \tag{5-29}$$

Substitution of (5-27) and (5-28) into (5-29) allows one to evaluate B_m and C_m as

$$B_m e^{-k_3 z_0} = C_m e^{k_3 z_0} = \frac{I_0}{ak_3} \sin \frac{m\pi x_0}{a}$$

and the solution to (5-27) and (5-28) becomes

$$A_y(x, z) = \frac{I_0}{a} \sum_{m=1}^{\infty} k_3^{-1} \sin \frac{m\pi x}{a} \sin \frac{m\pi x_0}{a} e^{-k_3|z - z_0|} \tag{5-30}$$

A slightly different approach to solving (5-26) is followed by Collin. A more basic procedure for solving (5-26) is to assume that

$$A_y = G_x G_z \tag{5-31}$$

and that G_x and G_z satisfy the ordinary differential equations

$$\left(\frac{d^2}{dx^2} + k_1^2\right)G_x = 0, \qquad k_1^2 > 0 \tag{5-32}$$

$$\left(\frac{d^2}{dz^2} - k_3^2\right)G_z = -D\, \delta(z - z_0), \qquad k_3^2 > 0 \tag{5-33}$$

Since G_x satisfies the boundary conditions at $x = 0, a$, the solution of (5-32) is

$$G_x = \sum_{m=1}^{\infty} \sin \frac{m\pi x}{a}$$

where $k_1 = (m\pi/a)$, $m = 1, 2, 3, \ldots$. If we insert this value of G_x into (5-31) and substitute the resultant expression for A_y into (5-26), we obtain the ordinary differential equation involving G_z, i.e.,

$$\sum_{m=1}^{\infty} \left[\frac{d^2}{dz^2} + k^2 - \left(\frac{m\pi}{a} \right)^2 \right] G_z \sin \frac{m\pi x}{a} = -I_0 \, \delta(x - x_0) \, \delta(z - z_0)$$

This equation can be put into the form of (5-33) by using the orthogonality of the sinusoids, i.e., multiplying both sides by $\sin (n\pi x/a)$ and integrating over the interval $(0, a)$. This gives us the result

$$\left(\frac{d^2}{dz^2} - k_3^2 \right) G_z = -\frac{2I_0}{a} \sin \frac{n\pi x_0}{a} \, \delta(z - z_0) \tag{5-34}$$

where again

$$k_3^2 = \left(\frac{n\pi}{a} \right)^2 - k^2$$

We can now identify D as

$$D = \frac{2}{a} I_0 \sin \frac{n\pi x_0}{a}$$

The solution to (5-34) can be obtained by the procedure outlined in Sec. 2.2. Thus, we define G_z as

$$G_z^< = (A - \alpha)G_{z\alpha}(z) + (B - \beta)G_{z\beta}(z), \qquad -\infty < z \leq z_0$$
$$G_z^> = (A + \alpha)G_{z\alpha}(z) + (B + \beta)G_{z\beta}(z), \qquad z_0 \leq z < \infty$$

where

$$G_{z\alpha} = e^{-k_3 z}$$
$$G_{z\beta} = e^{k_3 z}$$

The constants a and b in (2-26) (defined as α and β here) become

$$\alpha = (2ak_3)^{-1} I_0 e^{k_3 z_0} \sin \frac{n\pi x_0}{a}$$

$$\beta = -(2ak_3)^{-1} I_0 e^{-k_3 z_0} \sin \frac{n\pi x_0}{a}$$

The constants A and B are evaluated by using the physical conditions that $G_z^<$ must contain only the term proportional to $G_{z\beta}$ and $G_z^>$ must contain only the term proportional to $G_{z\alpha}$. We evaluate A and B in terms of α and β as

$$A = \alpha$$
$$B = -\beta$$

The final expressions for $G_z^<$ and $G_z^>$ are

$$G_z^< = (ak_3)^{-1} I_0 \sin\frac{n\pi x_0}{a} e^{-k_{3z_0}} e^{k_{3z}}$$

$$G_z^> = (ak_3)^{-1} I_0 \sin\frac{n\pi x_0}{a} e^{k_{3z_0}} e^{-k_{3z}}$$

(5-35)

One then finds that the final solution for A_y in (5-31) becomes the same as that already obtained in (5-30). Although this second approach is more tedious, it does illustrate the fact that the basic approach given in Chap. 2 is always valid. However, as one acquires more proficiency in solving boundary value problems, it is often possible to save many steps by judiciously guessing at parts of the solutions. This was the procedure we employed in writing the solutions for A_y in (5-27) and (5-28).

We now wish to consider a problem involving Helmholtz's equation in cylindrical coordinates. Let us solve the problem of a uniform magnetic current ring source located inside and concentric with a perfectly conducting right circular cylinder that is infinite in length as illustrated in Fig. 5-3. Once we have solved this problem, we can let the diameter of the magnetic ring source increase until it coincides with the conducting cylinder. The resultant problem is that of a cylindrical waveguide excited by a narrow circumferential gap[25] across which a uniform time harmonic voltage is applied. Since the magnetic current source is assumed uniform, i.e., it does not depend on ϕ,

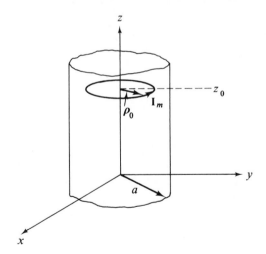

Fig. 5-3 Uniform time-varying magnetic current ring source inside a perfectly conducting infinitely long circular cylinder.

---

[25]Van Bladel, *Electromagnetic Fields*, p. 431.

the excited fields will be independent of ϕ and the problem will be two-dimensional. The relevant equation to be solved is not immediately obvious because the coordinate system is not cartesian. The first choice might be to use (4-21) and choose $A_{m\phi}$ since the current source is $J_{m\phi}$. However, this approach is not worthwhile in general since the vector wave equation separates only for the unit vectors $\mathbf{a}_x$, $\mathbf{a}_y$, and $\mathbf{a}_z$. On the other hand, it is always possible to derive the total field from two scalar potentials provided the vector potentials are expressed in cartesian coordinates or as the single radially directed component in spherical coordinates,[26] i.e., $\mathbf{A} = A_i\mathbf{a}_i$ and $\mathbf{A}_m = A_{mi}\mathbf{a}_i$, where i is x, y, z, or r. Even then, it is not always obvious as to how the source term should be introduced so that one can solve an inhomogeneous scalar equation that is of the form of (4-14). In the general case, it is usually easier to determine the field components excited by the specific source and then derive these components from a suitable scalar potential. The constants involving the source are evaluated by applying the boundary conditions (4-6) at the source. Both approaches will be used for the problem now under discussion to illustrate the general procedure.

If we examine the fields that are excited by the uniform magnetic current source, we see from (4-23) that the finite components will be E_ρ, E_z, and H_ϕ, which is a TM_z mode. One can also consider the physical nature of the problem to see that the slot excites E_z so that a TM_z mode is excited. Once we know that TM_z modes will be excited, we can use the fact that TM_z modes can be generated by A_z so that the relevant equation is

$$(\nabla^2 + k^2)A_z(\rho, z) = \frac{1}{\rho}\frac{\partial}{\partial\rho}\left(\rho\frac{\partial A_z}{\partial\rho}\right) + \frac{\partial^2 A_z}{\partial\rho^2} + k^2 A_z = 0 \qquad (5\text{-}36)$$

with the boundary condition

$$A_z = 0, \qquad \rho = a \qquad (5\text{-}37)$$

Actually, we are using the fact that the magnetic current loop[27] is equivalent to an electric dipole just as an electric current loop is equivalent to a magnetic dipole. The boundary condition on A_z results from the fact that it is proportional to E_z and E_z must vanish on the cylindrical wall. The physical condition on A_z is that it behaves as an outgoing wave for large values of z. Also, we see from Table 5-2B that the ρ-dependent solutions will be Bessel functions of the first kind and order zero and the z-dependent solutions will

---

[26]W. Weeks, *Electromagnetic Theory for Engineering Applications*, John Wiley & Sons, Inc., New York, 1964, p. 298. Also, see Morse and Feshbach, *Methods of Theoretical Physics*, p. 1764.

[27]Harrington, *Time-Harmonic Electromagnetic Fields*, p. 212.

be exponentials as in the preceding problem. Thus, we assume that A_z can be written as

$$A_{z1} = \sum_{n=1}^{\infty} a_n J_0(k_1\rho)e^{-k_3 z}, \qquad z > z_0 \tag{5-38}$$

$$A_{z2} = \sum_{n=1}^{\infty} b_n J_0(k_1\rho)e^{k_3 z}, \qquad z < z_0$$

where

$$k_1 = \frac{p_{0n}}{a}, \qquad n = 1, 2, 3, \ldots$$

$$k_3^2 = k_1^2 - k^2$$

and the $p_{0n}(x_{0n})$ are the various roots of J_0 (Appendix C.1.13). The expressions for the field components can be obtained from Appendix A and are

$$E_\rho = \frac{1}{j\omega\epsilon} \frac{\partial^2 A_z}{\partial\rho \, \partial z}$$

$$E_z = \frac{1}{j\omega\epsilon}\left(k^2 + \frac{\partial^2}{\partial z^2}\right) A_z$$

$$H_\phi = -\frac{\partial A_z}{\partial\rho}$$

Since we have split the interior of the waveguide into two regions in the z-direction (we could have chosen the ρ-direction), the boundary conditions at the source (or source conditions) are

$$H_{\phi 1} = H_{\phi 2}$$

$$E_{\rho 1} - E_{\rho 2} = -I_m \, \delta(\rho - \rho_0) \tag{5-39}$$

If we solve for E_ρ in terms of A_{z1} and A_{z2}, the second condition in (5-39) becomes

$$\sum_{n=1}^{\infty} \frac{k_1 k_3}{j\omega\epsilon}(a_n e^{-k_3 z_0} + b_n e^{k_3 z_0})J_1(k_1\rho) = -I_m \, \delta(\rho - \rho_0)$$

which can be simplified by using the orthogonality of the Bessel functions to obtain the expression

$$a_n e^{-k_3 z_0} + b_n e^{k_3 z_0} = -\frac{2j\omega\epsilon\rho_0 I_m J_1(k_1\rho_0)}{a^2 k_1 k_3 J_1^2(p_{0n})}$$

The boundary condition on H_ϕ gives the second expression

$$a_n e^{-k_3 z_0} = b_n e^{k_3 z_0}$$

so that the coefficients become

$$a_n = b_n e^{2k_3 z_0} = \frac{-j\omega\epsilon p_0 I_m J_1(k_1 p_0) e^{k_3 z_0}}{a^2 k_1 k_3 J_1^2(p_{0n})}$$

and the expression for A_z becomes

$$A_z = -\frac{j\omega\epsilon p_0 I_m}{a^2} \sum_{n=1}^{\infty} \frac{J_1(k_1 p_0) J_0(k_1 p)}{k_1 k_3 J_1^2(p_{0n})} e^{-k_3|z-z_0|} \qquad (5\text{-}40)$$

The expression for E_z becomes

$$E_z = -\frac{p_0 I_m}{a^2} \sum_{n=1}^{\infty} \frac{k_1 J_1(k_1 p_0) J_0(k_1 p)}{k_3 J_1^2(p_{0n})} e^{-k_3|z-z_0|} \qquad (5\text{-}41)$$

The other approach we can use is to find an inhomogeneous scalar equation for $A_{m\phi}$. For this particular case, where there is no ϕ-dependence, one sees by inspecting the expression for the vector Laplacian in cylindrical coordinates in Appendix A that it separates into scalar equations. As a result, the equation satisfied by $A_{m\phi}$ is

$$\left[\frac{1}{p}\frac{\partial}{\partial p}\left(p\frac{\partial}{\partial p}\right) + k^2 - \frac{1}{p^2} + \frac{\partial^2}{\partial z^2}\right] A_{m\phi}(p, z) = -I_m \, \delta(p - p_0) \, \delta(z - z_0)$$
$$(5\text{-}42)$$

and the field components are

$$E_p = \frac{\partial A_{m\phi}}{\partial z}$$

$$E_z = -\frac{1}{p}\frac{\partial}{\partial p}(p A_{m\phi}) \qquad (5\text{-}43)$$

$$H_\phi = -j\omega\epsilon A_{m\phi}$$

The form of the source term is not too obvious but is discussed further in connection with the solution to (5-54) and (5-59). The p-dependent solutions to (5-42) are first order Bessel functions of the first kind and E_z becomes proportional to the zero order Bessel functions of the first kind. We can write the expression for $A_{m\phi}$ as

$$A_{m\phi} = \sum_{n=1}^{\infty} G_z J_1(k_1 p)$$

where $k_1 = p_{0n}/a$ as before and find the equation satisfied by G_z by inserting this expression into (5-42). Again, we use the orthogonality of the first order Bessel functions to get the equation

$$\left(\frac{d^2}{dz^2} - k_3^2\right) G_z = -D \, \delta(z - z_0)$$

where

$$D = \frac{2\rho_0 I_m J_1(k_1\rho_0)}{a^2 J_1^2(p_{0n})}$$

Since this equation is of the same form as (5-33), we can immediately write the solution for G_z as

$$G_z = \frac{D}{2k_3} e^{-k_3|z-z_0|}$$

The expression for $A_{m\phi}$ becomes

$$A_{m\phi}(\rho, z) = \frac{\rho_0 I_m}{a^2} \sum_{n=1}^{\infty} \frac{J_1(k_1\rho_0)J_1(k_1\rho)}{k_3 J_1^2(p_{0n})} e^{-k_3|z-z_0|} \tag{5-44}$$

and the expression for E_z becomes

$$E_z = -\frac{\rho_0 I_m}{a^2} \sum_{n=1}^{\infty} \frac{k_1 J_1(k_1\rho_0)J_0(k_1\rho)}{k_3 J_1^2(p_{0n})} e^{-k_3|z-z_0|} \tag{5-45}$$

For the case where the magnetic ring source coincides with the waveguide wall, the above expression reduces to

$$E_z = -\frac{I_m}{a} \sum_{n=1}^{\infty} \frac{k_1 J_0(k_1\rho)}{k_3 J_1(p_{0n})} e^{-k_3|z-z_0|} \tag{5-46}$$

As an interesting sidelight to the preceding problem, let us consider further the first approach where we started by formulating the problem in terms of a defining equation for A_z, i.e., the expression (5-36). As we mentioned before, when we introduced A_z we replaced the original problem of the magnetic ring source with another problem with an electric dipole as the source. The reason we could do this was because the two sources produced the same field and were therefore equivalent. This is essentially a statement of the equivalence principle[28] which recognizes the fact that any number of different (equivalent) sources can produce the same field within a given region of space. This is not the same as duality since the duality concept is based upon the fact that two problems are represented by the same mathematical form. The concept of an analog is the same as that of a dual except the former term is usually used when one refers to different physical systems. For example, one usually speaks of a fluid problem as being the analog of an electrical problem and a voltage problem as being the dual of a current problem. In

---

[28] Harrington, *Time-Harmonic Electromagnetic Fields*, Sec. 3–5. Also, see R. Harrington, *Electromagnetic Engineering*, McGraw-Hill Book Company, Inc., New York, 1958, Sec. 7–5.

each case, the mathematical equations and boundary conditions are the same.

The reason we wish to discuss this problem further is to emphasize the fact that the two physical problems are different even though the two sources are equivalent. Thus, if we consider the expression (5-36) and add the proper source term we obtain the inhomogeneous equation

$$(\nabla^2 + k^2)A_z = -I\ell\,\delta(\mathbf{\rho} - \mathbf{\rho}_0)\,\delta(z - z_0) \tag{5-47}$$

with the same boundary condition as (5-37). The physical problem posed by (5-47) is that of an infinitesimal z-directed electric dipole of moment $I\ell$ located at the point (ρ_0, ϕ_0, z_0) inside the perfectly conducting cylinder defined by the surface $\rho = a$. However, this general problem is no longer independent of ϕ (unless $\phi_0 = 0$) and, furthermore, the TM_z mode will possess finite values for E_ϕ and H_ρ. Since these two field components must be zero in order for this TM_z mode to be equivalent to the one excited by the uniform ring source, it follows that we must locate the electric dipole along the z-axis. In this case the term

$$\delta(\mathbf{\rho} - \mathbf{\rho}_0) \longrightarrow \delta(\mathbf{\rho}), \qquad \mathbf{\rho}_0 = 0$$

and

$$A_z(\rho, \phi, z) \longrightarrow A_z(\rho, z)$$

Another point worth noting is the fact that even if we solve the problem posed by (5-47), the fields vanish when the dipole is placed against the cylinder whereas the magnetic current source can be placed against the cylinder without the fields vanishing. Although this fact is quite obvious on physical grounds, it is not always quite so obvious when one attempts to replace a given physical problem by another problem that is mathematically equivalent but not physically equivalent.

As another example of a problem in cylindrical coordinates, let us consider a point source located at (ρ_0, ϕ_0, z_0) so that the equation we wish to solve is the scalar Helmholtz equation, i.e.,

$$(\nabla^2 + k^2)G = -\delta(\mathbf{r} - \mathbf{r}_0) = -\frac{\delta(\rho - \rho_0)}{\rho}\,\delta(\phi - \phi_0)\,\delta(z - z_0) \tag{5-48}$$

In practice, the point source could be a z-directed infinitesimal electric or magnetic dipole. This is also the problem that results from Prob. 5.17 when the boundaries of the coaxial cylindrical box vanish. It is also the same problem we solved by a Fourier transform method in Chap. 3 (end of Sec. 3.2). Let us divide the region in the ρ-direction and Fourier transform with respect to z so that we assume a solution

$$G = \frac{1}{\sqrt{2\pi}}\sum_{m=0}^{\infty}\int_{-\infty}^{\infty}\tilde{G}_\rho\cos m(\phi - \phi_0)e^{j\alpha z}\,d\alpha \tag{5-49}$$

where $\tilde{G}_\rho(\rho, \alpha)$ is the direct exponential Fourier transform defined in (3-1) and remains to be found. If we substitute (5-49) into (5-48), we obtain the inhomogeneous ordinary differential equation satisfied by $\tilde{G}_\rho$. Of course, we must utilize the orthogonality of the ϕ-dependent sinusoids and use the Fourier transform representation of $\delta(z - z_0)$. This latter expression is obtained in the same manner as that used to derive (3-6). In any case, the pertinent equation becomes Bessel's equation, i.e.,

$$\left[\frac{1}{\rho}\frac{d}{d\rho}\left(\rho\frac{d}{d\rho}\right) + k_1^2 - \frac{m^2}{\rho^2}\right]\tilde{G}_\rho = -\frac{D\,\delta(\rho - \rho_0)}{\rho} \tag{5-50}$$

where

$$k_1^2 = k^2 - \alpha^2$$

and

$$D = \epsilon_m e^{-j\alpha z_0}(2\pi)^{-3/2}$$

The solution to (5-50) can be obtained by the method of Sec. 2.2 or by reference to Prob. 2.15. Thus, we find that $\tilde{G}_\rho$ is

$$\tilde{G}_\rho = \frac{\pi D}{2j}[H_m^{(2)}(k_1\rho_0)J_m(k_1\rho)u(\rho_0 - \rho) + H_m^{(2)}(k_1\rho)J_m(k_1\rho_0)u(\rho - \rho_0)] \tag{5-51}$$

so that G becomes

$$G = \frac{1}{8\pi j}\int_{-\infty}^{\infty} e^{j\alpha(z-z_0)}\,d\alpha \sum_{m=0}^{\infty} \epsilon_m[H_m^{(2)}(k_1\rho_0)J_m(k_1\rho)u(\rho_0 - \rho)$$
$$+ H_m^{(2)}(k_1\rho)J_m(k_1\rho_0)u(\rho - \rho_0)]\cos m(\phi - \phi_0) \tag{5-52}$$

However this expression can be simplified by using the addition theorem for Hankel functions (Prob. 4.19) so that the final solution[29] to (5-48) is

$$G = \frac{1}{8\pi j}\int_{-\infty}^{\infty} H_0^{(2)}(k_1\rho_1)e^{j\alpha(z-z_0)}\,d\alpha \tag{5-52}$$

The addition theorem for Bessel functions of the first kind is given in Probs. 4.20 and 5.13, and also (3-67) and (3-68). That for Hankel functions[30]

[29]Harrington, *Time-Harmonic Electromagnetic Fields*, p. 244. Also, see Weeks, *Electromagnetic Theory for Engineering Applications*, p. 524.

[30]Harrington, *Time-Harmonic Electromagnetic Fields*, p. 232.

and Bessel functions of the second kind is (see Probs. 4.19 and 4.20)

$$H_0^{(1,2)}(\rho_1) = \sum_{m=-\infty}^{\infty} [H_m^{(1,2)}(\rho_0)J_m(\rho)u(\rho_0 - \rho)$$
$$+ H_m^{(1,2)}(\rho)J_m(\rho_0)u(\rho - \rho_0)]e^{jm(\phi-\phi_0)}$$
$$= \sum_{m=0}^{\infty} \epsilon_m[H_m^{(1,2)}(\rho_0)J_m(\rho)u(\rho_0 - \rho)$$
$$+ H_m^{(1,2)}(\rho)J_m(\rho_0)u(\rho - \rho_0)] \cos m(\phi - \phi_0) \qquad (5\text{-}53)$$
$$N_0(\rho_1) = \sum_{m=0}^{\infty} \epsilon_m[N_m(\rho_0)J_m(\rho)u(\rho_0 - \rho)$$
$$+ N_m(\rho)J_m(\rho_0)u(\rho - \rho_0)] \cos m(\phi - \phi_0)$$

where $\boldsymbol{\rho}_1 = \boldsymbol{\rho} - \boldsymbol{\rho}_0$, $\rho_1 = |\boldsymbol{\rho}_1|$, and the Neumann number is defined as

$$\epsilon_m = 1, \qquad m = 0$$
$$\epsilon_m = 2, \qquad m > 0$$

We could also derive the addition theorem for Hankel functions in (5-53) by recognizing that G in (5-52) is the unbounded space solution to (3-21), i.e., (3-28) for $i = 3$. However, the identity in (3-46) for $i = 3$ leads us directly to (5-53) and the final solution for G in (5-52).

As a final example of a problem in cylindrical coordinates, consider a circular loop of electric current as shown in Fig. 5-4. The axis of the loop is

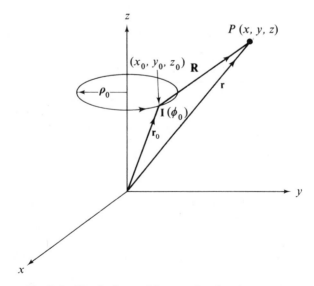

Fig. 5-4 Circular loop of time-varying electric current.

the z-axis and the plane of the loop is defined by $z = z_0$. The current on the loop is assumed to be constant with respect to the coordinates so that only the A_ϕ component of the vector potential is finite. In cylindrical coordinates (Appendix A), A_ϕ satisfies the equation

$$\left(\nabla^2 + k^2 - \frac{1}{\rho^2}\right)A_\phi(\rho, z) = -J_\phi = -I_0\,\delta(\rho - \rho_0)\,\delta(z - z_0) \qquad (5\text{-}54)$$

The general case where the current varies with the source angle ϕ_0 is more difficult since both A_ϕ and A_ρ are finite, depend on all three coordinate variables, and satisfy two coupled partial differential equations. One procedure that can be used in this more general case is to remain in the cartesian coordinate system in order to evaluate A_x and A_y and then later return to the cylindrical system. This case is considered by Fante[31] who studies a current loop with an arbitrary current distribution located inside an infinitely long cylinder of arbitrary complex dielectric constant and radius. The starting point for his work is the expression (4-24) for the magnetic vector potential in integral form. We shall use this viewpoint later on as an alternate method for solving (5-54).

As the first step in solving (5-54), we judiciously guess at a solution for A_ϕ just as we did for G in (5-49). However, there is no summation sign in this case since A_ϕ is independent of ϕ. Instead of guessing at a solution for A_ϕ, i.e., assuming a solution of the proper form according to Method 1 in Sec. 3.1, we could follow the procedure of Method 2 in Sec. 3.1 and multiply both sides of (5-54) by $e^{-j\alpha z}/\sqrt{2\pi}$ and then integrate over z to obtain an ordinary differential equation with ρ as the dependent variable. Naturally, both methods are the same and only differ in viewpoint. The ordinary differential equation that we obtain is Bessel's equation of order one, i.e., (5-50) for $m = 1$ but with $D(\rho) = \rho(I_0 e^{-j\alpha z_0}/\sqrt{2\pi})$. The solution to (5-50) is given by (5-51) so that we can write the solution to (5-54) as

$$A_\phi = \frac{\rho_0 I_0}{4j} \int_{-\infty}^{\infty} [H_1^{(2)}(k_1\rho_0)J_1(k_1\rho)u(\rho_0 - \rho)$$

$$+ H_1^{(2)}(k_1\rho)J_1(k_1\rho_0)u(\rho - \rho_0)]e^{j\alpha(z-z_0)}\,d\alpha \qquad (5\text{-}55)$$

where $k_1^2 = k^2 - \alpha^2$.

If we wish to simplify (5-55) by considering A_ϕ when $k_1\rho \gg 1$, we can replace $H_1^{(2)}(k_1\rho)$ by its asymptotic expansion (large argument value) from Appendix C.3.8 and then use the method of steepest descents to approximate the integral. This procedure is also discussed by Fante. We find that A_ϕ

---

[31]R. Fante, "The Fields of a Current Loop Inside an Infinite Cylindrical Column," *J. Math. Physics*, 3, pp. 325–332, September 1967.

becomes approximately

$$A_\phi = \frac{j\rho_0 I_0}{2r} e^{-jkr} J_1(k\rho_0 \sin \theta), \qquad kr \gg 1 \qquad (5\text{-}56)$$

which for the case of a small loop simplifies further to

$$A_\phi = \frac{jk M_0 e^{-jkr}}{4\pi r} \sin \theta, \qquad k\rho_0 \ll 1 \qquad (5\text{-}57)$$

where $M_0 = I_0 A_0 = I_0 \pi \rho_0^2$.

Both of the expressions (5-56) and (5-57) are derived by Johnson[32] and Wait[33] without resorting to the method of steepest descents. The second expression is also derived by Harrington[34] and Smythe.[35] Unfortunately, none of the expressions (5-55), (5-56), and (5-57) reduce to the usual static expression[36] when k vanishes. The static expression for A_ϕ is often used as a classical example of a problem that leads to elliptic integrals so will not be considered here. However, it is of interest to derive the expression (5-56) from an alternate viewpoint as we mentioned previously. In this case we start with the expression (4-24) which is valid for a current source in unbounded space. Thus,

$$\mathbf{A}(\mathbf{r}) = \frac{1}{4\pi} \int \frac{e^{-jkR}}{R} \mathbf{J}(\mathbf{r}_0) \, dV_0 \qquad (4\text{-}24)$$

By inspecting Fig. 5-4, we see that

$$R = [\rho^2 + \rho_0^2 - 2\rho\rho_0 \cos (\phi - \phi_0) + (z - z_0)^2]^{1/2}$$
$$\mathbf{J}(\mathbf{r}_0) \, dV_0 = I_0 \rho_0 \, d\phi_0 \, \mathbf{a}_{\phi_0}$$
$$\mathbf{A}(\mathbf{r}) = A_\rho \mathbf{a}_\rho + A_\phi \mathbf{a}_\phi$$

Since $\mathbf{a}_{\phi_0}$ is not parallel to $\mathbf{a}_\phi$ and is also a function of the coordinates, we cannot write (4-24) in the form of two scalar expressions until we express $\mathbf{a}_{\phi_0}$ in terms of $\mathbf{a}_\rho$ and $\mathbf{a}_\phi$. One must remember that the unit vectors can be

---

[32]C.C. Johnson, *Field and Wave Electrodynamics*, McGraw-Hill Book Company, Inc., New York, 1965, p. 23.

[33]J. Wait, *Electromagnetic Radiation from Cylindrical Structures*, Pergamon Press, Inc., New York, 1959, p. 194.

[34]Harrington, *Time-Harmonic Electromagnetic Fields*, p. 93.

[35]Smythe, *Static and Dynamic Electricity*, 2nd ed., p. 489.

[36]*Ibid.*, p. 271. Also, see Stratton, *Electromagnetic Theory*, p. 263, and Van Bladel, *Electromagnetic Fields*, p. 155.

taken out of the integral in (4-24) only for the cartesian coordinate system unit vectors. For our case, we find that

$$\mathbf{a}_{\phi_0} = \mathbf{a}_\rho \sin (\phi - \phi_0) + \mathbf{a}_\phi \cos (\phi - \phi_0)$$

so that now (4-24) can be separated into two scalar expressions for A_ρ and A_ϕ, i.e.,

$$A_\rho = \frac{\rho_0 I_0}{4\pi} \int_0^{2\pi} \frac{e^{-jkR}}{R} \sin (\phi - \phi_0) \, d\phi_0 \longrightarrow 0$$

$$A_\phi = \frac{\rho_0 I_0}{4\pi} \int_0^{2\pi} \frac{e^{-jkR}}{R} \cos (\phi - \phi_0) \, d\phi_0$$

(5-58)

The symmetry of the problem shows us that $A_\rho = 0$ so that only a single component of **A** remains. If k vanishes in (5-58), the expression reduces to the usual static expression.[37] If one is interested in an exact solution to (5-58), then Fante's approach can be used. His approach is to replace e^{-jkR}/R in the integrand by the identity (3-43) for $i = 3$ and then use the addition theorem for Hankel functions (5-53) to replace the zero order Hankel function of the second kind by a sum of Hankel functions and Bessel functions. The orthogonality of the sinusoids in ϕ_0 allows one to evaluate the integral over ϕ_0 and finally to obtain the expression (5-55). One advantage to this latter approach, i.e., the use of (4-24), is that it works quite well when I_0 varies with ϕ_0. In this more general case, one should be able to use the expressions in (5-58) provided I_0 outside the integral sign is replaced by $I_0(\phi_0)$ inside the integral sign.

It might also be appropriate to mention that the proper form of the source term in (5-54) is not always obvious. However, we always know that the integral of the current density over the volume where it exists must be equal to the integral of the delta representation over its respective volume, area, or line element. Although this statement is quite obvious, the evaluation of the proper representation is not quite so obvious. For instance, if we consider the source term in (5-54) as an example, the integral of J_ϕ over V_0 is

$$\int_{V_0} J_\phi(\rho_0, z_0) \, dV_0 = \int_{V_0} J_\phi \rho_0 \, d\phi_0 \, d\rho_0 \, dz_0 = 2\pi\rho_0 \int_{A_0} J_\phi \, d\rho_0 \, dz_0 = 2\pi\rho_0 I_0$$

(5-59)

If we use a delta representation for the source, we must obtain the same result. Since the location of the source can be described by the two coordinates ρ_0

---

[37]See Reference 36.

and z_0, we find from Appendix B that

$$\delta(\mathbf{r} - \mathbf{r}_0) = \frac{\delta(\rho - \rho_0)\, \delta(z - z_0)}{2\pi\rho}$$

so that the first integral in (5-59) becomes

$$\int_{V_0} J_\phi \delta(\mathbf{r} - \mathbf{r}_0)\, dV_0 = \int_0^{2\pi} \rho_0\, d\phi_0 \int_{A_0} \frac{J_\phi(\rho_0, z_0)\, \delta(\rho - \rho_0)\, \delta(z - z_0)\, d\rho_0\, dz_0}{2\pi\rho}$$

$$= J_\phi(\rho, z)$$

The result is not what we want even though it agrees with the basic definition of the delta. We must remove the factor $2\pi\rho$ in the denominator of the integral over A_0 and also obtain I_0 instead of J_ϕ. The correct representation thus becomes

$$\int_0^{2\pi} \rho_0\, d\phi_0 \int_{A_0} I_0(\rho_0, z_0)\, \delta(\rho - \rho_0)\, \delta(z - z_0)\, d\rho_0\, dz_0$$

$$= \int_0^{2\pi} \rho_0\, d\phi_0 I_0(\rho, z) = 2\pi\rho_0 I_0(\rho, z) \tag{5-60}$$

All of the expressions in (5-59) except the last one are also valid when J_ϕ is a function of ϕ_0. Since the source representation in (5-54) already includes the integration over ϕ_0, we find that only the integral over A_0 remains as part of the source term.

5.4 Normal Modes in Spherical Coordinates
Source-Free Problems

The normal modes in the spherical coordinate system result from the solutions of Laplace's equation and the scalar wave (Helmholtz) equation. The θ-dependent parts of the two equations are the same and can be represented by associated Legendre functions of the first and second kinds. The ϕ-dependent parts of the two equations are also the same and can be represented by sinusoids (or exponentials). The r-dependent parts of the two equations are different. Laplace's equation yields positive and negative powers of r and the wave equation yields spherical Bessel functions. The various solutions to Laplace's equation are tabulated in Table 5-3A and the solutions to Helmholtz's equation are given in Table 5-3B. In general, the separation constants m and n are not integers, which causes the Legendre functions to become more complicated. This situation is discussed more fully in Chap. 1 in connection with spherical harmonics of nonintegral order. Another point worth noting is the fact that the separation constants m and n are independent

TABLE 5-3A   Solutions to Laplace's Equation
Spherical Coordinates

$$\nabla^2\Phi(r, \theta, \phi) = \left[\frac{1}{r^2}\frac{\partial}{\partial r}\left(r^2\frac{\partial}{\partial r}\right) + \frac{1}{r^2\sin\theta}\frac{\partial}{\partial\theta}\left(\sin\theta\frac{\partial}{\partial\theta}\right) + \frac{1}{r^2\sin^2\theta}\frac{\partial^2}{\partial\phi^2}\right]\Phi(r, \theta, \phi) = 0$$

If

$$\Phi(r, \theta, \phi) = f(r)g(\theta)h(\phi)$$

then

$$\left[\frac{d}{dr}\left(r^2\frac{d}{dr}\right) - n(n+1)\right]f(r) = 0,$$

$$\left[\frac{1}{\sin\theta}\frac{d}{d\theta}\left(\sin\theta\frac{d}{d\theta}\right) + n(n+1) - \frac{m^2}{\sin^2\theta}\right]g(\theta) = 0,$$

$$\left(\frac{d^2}{d\phi^2} + m^2\right)h(\phi) = 0$$

All Combinations

$f(r)$		$g(\theta)$		$h(\phi)$	
$n^2 > 0$	$n = 0$	$m^2 > 0$	$m = 0$	$m^2 > 0$	$m = 0$
r^n $r^{-(n+1)}$		$P_n^m(\cos\theta)$ $Q_n^m(\cos\theta)$		$\cos m\phi$ $\sin m\phi$	
r^n $r^{-(n+1)}$			$P_n(\cos\theta)$ $Q_n(\cos\theta)$		ϕ 1
	1 r^{-1}		$P_0(\cos\theta)$ $Q_0(\cos\theta)$		ϕ 1
	1 r^{-1}	$\tan^m\frac{\theta}{2}$ $\cot^m\frac{\theta}{2}$		$\cos m\phi$ $\sin m\phi$	

of the wave number k. However, these points are best emphasized by considering several problems, which is done in the next two sections.

5.5 Green's Functions for Laplace's Equation in Spherical Coordinates

Let us consider the fairly general problem of the Green's function for a grounded conical box containing a unit dc source at $P(r_0, \theta_0, \phi_0)$. The box is illustrated in Fig. 5-5 and is defined by the spheres $r = a$, b ($a < b$) and the cone $\theta = \theta_1$. A more general problem would result if one excluded the positive z-axis by a cone $\theta = \theta_2$, $\theta_2 < \theta_1$. However, this would affect only the θ-dependent part of G and would necessitate the use of the associated Legendre functions of the second kind, i.e., Q_n^m ($\cos\theta$). This problem would be essentially one of a unit source inside a spherical toroid. The most general problem would be that of a unit source inside a segment of a spherical toroid, i.e., that portion of the toroid between the semi-infinite planes $\phi = \phi_1$, ϕ_2.

TABLE 5-3B   Solutions to Helmholtz's Equation
Spherical Coordinates

$$(\nabla^2 + k^2)\Phi(r, \theta, \phi) = 0$$

If

$$\Phi(r, \theta, \phi) = f(r)g(\theta)h(\phi)$$

then

$$\left[\frac{d}{dr}\left(r^2 \frac{d}{dr}\right) + (kr)^2 - n(n+1)\right]f(r) = 0, \quad k > 0,$$

$$\left[\frac{1}{\sin\theta}\frac{d}{d\theta}\left(\sin\theta \frac{d}{d\theta}\right) + n(n+1) - \frac{m^2}{\sin^2\theta}\right]g(\theta) = 0, \quad \left(\frac{d^2}{d\phi^2} + m^2\right)h(\phi) = 0$$

All Combinations
$b_n(kr)$ is any linear combination of $j_n(kr), n_n(kr), h_n^{(1)}(kr), h_n^{(2)}(kr)$.

$f(r)$		$g(\theta)$		$h(\phi)$	
$n^2 > 0$	$n = 0$	$m^2 > 0$	$m = 0$	$m^2 > 0$	$m = 0$
$b_n(kr)$		$P_n^m(\cos\theta)$ $Q_n^m(\cos\theta)$		$\cos m\phi$ $\sin m\phi$	
$b_n(kr)$			$P_n(\cos\theta)$ $Q_n(\cos\theta)$		ϕ 1
	$b_0(kr)$	$\tan^m \dfrac{\theta}{2}$ $\cot^m \dfrac{\theta}{2}$		$\cos m\phi$ $\sin m\phi$	
	$b_0(kr)$		$P_0(\cos\theta)$ $Q_0(\cos\theta)$		ϕ 1

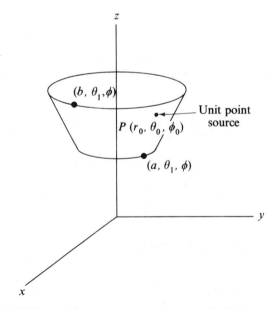

Fig. 5-5 Unit dc point source inside grounded conical conducting box.

Here the ϕ-dependent part of G would be altered to be of the form

$$\sin \frac{m\pi(\phi - \phi_1)}{(\phi_2 - \phi_1)} \sin \frac{m\pi(\phi_0 - \phi_1)}{(\phi_2 - \phi_1)}, \qquad m = 1, 2, 3, \ldots$$

or

$$\sin \frac{m\pi(\phi_2 - \phi)}{\phi_2 - \phi_1} \sin \frac{m\pi(\phi_2 - \phi_0)}{\phi_2 - \phi_1} \qquad (5\text{-}61)$$

The equation for the Green's function is (5-15) except with both sides expressed in spherical coordinates. Thus,

$$\nabla^2 G = \frac{-\delta(r - r_0)\,\delta(\theta - \theta_0)\,\delta(\phi - \phi_0)}{r^2 \sin \theta} \qquad (5\text{-}62)$$

Let us divide the volume inside the spherical box into two regions depending upon whether $r < r_0$ or $r > r_0$. Now we can assume a solution of the form

$$G = \sum_{m=0}^{\infty} \sum_{n} G_r G_{n\theta} G_{m\phi} \qquad (5\text{-}63)$$

where

$$G_{n\theta} = P_n^m(\cos \theta), \qquad P_n^m(\cos \theta_1) = 0$$
$$G_{m\phi} = \cos m(\phi - \phi_0)$$
$$G_r = G_{1r}u(r_0 - r) + G_{2r}u(r - r_0)$$
$$G_{1r} = A_{mn}(r^n - a^{2n+1}r^{-n-1})$$
$$G_{2r} = B_{mn}(r^n - b^{2n+1}r^{-n-1})$$

We should note that in (5-63) m is an integer and is usually an integer in most problems. However, n is not an integer here (it should be replaced by s) but is determined from the relation $P_n^m(\cos \theta_1) = 0$. The index n becomes an integer for $\theta_1 = 0$, $\pi/2$, π. If we substitute (5-63) into (5-62) and perform the indicated operations on $G_{n\theta}$ and $G_{m\phi}$, we obtain

$$\sum_{m=0}^{\infty} \sum_{n} \frac{G_{n\theta} G_{m\phi}}{r^2}\left[\frac{d}{dr}\left(r^2 \frac{d}{dr}\right) - n(n+1)\right]G_r = \frac{-\delta(r - r_0)\,\delta(\theta - \theta_0)\,\delta(\phi - \phi_0)}{r^2 \sin \theta}$$

Now, multiply both sides by $\cos p(\phi - \phi_0)\, d\phi\, P_q^p(\cos \theta) \sin \theta\, d\theta$, integrate over the intervals $(0, 2\pi)$ and $(0, \theta_1)$, respectively, and then replace p by m and q by n to obtain the ordinary inhomogeneous differential equation for G_r,

$$\left[\frac{d}{dr}\left(r^2 \frac{d}{dr}\right) - n(n+1)\right]G_r = \frac{-\epsilon_m}{2\pi N_1(\theta_1)}P_n^m(\cos \theta_0)\,\delta(r - r_0) \qquad (5\text{-}64)$$

where, from Appendix C.4.10,

$$N_1(\theta_1) = (2n + 1)^{-1}\left[\sin\theta\,\frac{\partial P_n^m(\cos\theta)}{\partial\theta}\,\frac{\partial P_n^m(\cos\theta)}{\partial n}\right]_{\theta=\theta_1}\,.$$

and $\epsilon_m = 1$ for $m = 0$, $\epsilon_m = 2$ for $m > 0$. Again, note that n is not an integer for general values of θ. Since G_{1r} and G_{2r} satisfy the homogeneous form of (5-64), the two additional conditions we need in order to find A_{mn} and B_{mn} are derived from the properties of the Green's function at the source. In this case, we find that

$$\mathop{\mathrm{Lim}}_{\epsilon\to0}\frac{dG_r}{dr}\bigg]_{r_0-\epsilon}^{r_0+\epsilon} = \frac{-D}{p(r_0)} \tag{5-65}$$

$$G_{1r}(r_0) = G_{2r}(r_0)$$

where D is the strength of the source, i.e., the coefficient of $-\delta(r - r_0)$ in (5-64), and $p(r)$ is the coefficient of d^2G_r/dr^2. Now, the substitution of G_{1r} and G_{2r} from (5-63) into (5-64) and (5-65) allows us to determine the coefficients A_{mn}, B_{mn} as

$$A_{mn} = -\frac{D(r_0^n - b^{2n+1}r_0^{-n-1})}{(2n + 1)(b^{2n+1} - a^{2n+1})}$$

$$B_{mn} = -\frac{D(r_0^n - a^{2n+1}r_0^{-n-1})}{(2n + 1)(b^{2n+1} - a^{2n+1})} \tag{5-66}$$

Note that the Wronskian of r^n and $r^{-(n+1)}$ is given in Appendix C.3.8, i.e.,

$$W[r^n, r^{-(n+1)}] = -(2n + 1)r^{-2}$$

The solution to (5-63) becomes $G = G_1u(r_0 - r) + G_2u(r - r_0)$ where

$$G_1(r, r_0; \theta, \theta_0; \phi, \phi_0) = -\sum_{m=0}^{\infty}\sum_n\frac{\epsilon_m}{2\pi}\frac{1}{N_1(\theta_1)}$$

$$\times\frac{(r_0^n - b^{2n+1}r_0^{-n-1})(r^n - a^{2n+1}r^{-n-1})}{(2n + 1)(b^{2n+1} - a^{2n+1})}$$

$$\times P_n^m(\cos\theta)P_n^m(\cos\theta_0)\cos m(\phi - \phi_0), \tag{5-67}$$

$$r < r_0$$

$$G_2(r, r_0; \theta, \theta_0; \phi, \phi_0) = G_1(r_0, r; \theta_0, \theta; \phi_0, \phi), \qquad r > r_0$$

Since the expressions in (5-67) are valid for all values of a and b in the interval $(0, \infty)$ and all values θ_1 in the interval $(0, \pi)$, these can be used to generate other Green's functions by letting a, b, and θ_1 take on a variety of specific values. It is also possible to solve for G_r in (5-63) by the general method described in Sec. 2.2. This solution is given in Prob. 2.21. The summation over m in (5-67) can be eliminated by using the addition theorem for Legendre functions from (3-71). This procedure is used in connection with the derivation of (5-71) in the next section.

5.6 Green's Functions for Helmholtz's Equation in Spherical Coordinates

The solutions to the scalar Helmholtz equation in spherical coordinates differ from those for Laplace's equation only in the r-dependent functions as one can see from Tables 5-3A and B. One should also notice that there is no constraint on the separation constants or k. The chief difficulty that one encounters in considering solutions to Maxwell's equations in spherical coordinates is in reducing the vector equations (4-14) and (4-21) to scalar equations. The reason for doing this of course is because we wish to determine the total electromagnetic field in as easy a manner as possible. In the case of cartesian coordinates, we found that the vector equations for the potentials reduced to scalar equations and that the sources entered in a simple manner. As mentioned by several authors[38] and discussed in Sec. 5.3, it can be shown that the total electromagnetic field can be obtained by superposing two partial fields derived from two scalar functions. It is customary to speak of the fields as being transverse magnetic (TM_i) to the direction $\mathbf{a}_i$ when H_i is zero and transverse electric (TE_i) when E_i is zero. Unfortunately, this condition is not true in any coordinate system but only when $i = x, y, z$, and r. Thus, one is restricted to any direction in the cartesian coordinate system, the z-direction in any cylindrical coordinate system (not limited to right circular), and the r-direction in the spherical coordinate system. Also, in the case of the spherical coordinate system, it is not A_r and A_{mr} that satisfy the scalar Helmholtz equation but rather A_r/r and A_{mr}/r. In addition, the Lorentz condition is not valid, which is implicit in the derivations by others[39] of the scalar Helmholtz equation involving A_r/r. This is pointed out in Appendix A. Another point worth noting is the fact that it is very difficult to take into account the sources in a general manner. This point is discussed by Weeks[40] who also mentions that quite often it is expedient to replace the actual source by a dual source in hopes that the dual problem is easier to solve. This is the approach we

---

[38] Stratton, *Electromagnetic Theory*, pp. 351, 415. Also, see Weeks, *Electromagnetic Theory for Engineering Applications*, p. 298.

[39] Jones, *The Theory of Electromagnetism*, p. 486. Also, see Weeks, *Electromagnetic Theory for Engineering Applications*, p. 529, and Harrington, *Time-Harmonic Electromagnetic Fields*, p. 267.

[40] Weeks, *Electromagnetic Theory for Engineering Applications*, p. 301.

employed in obtaining the solution to (5-36) as an alternate to solving (5-42). However, we could have also found the dual to the source term in (5-42) so that (5-36) would be inhomogeneous and thereby display a source term. However, in solving (5-36) we chose instead to introduce the source in connection with the source conditions (5-39), which is even another alternative. We also see by inspecting $\nabla^2\mathbf{A}$ in cylindrical coordinates in Appendix A that the vector equation separates into scalar equations when the field is dependent on only one or two of the coordinates. We made use of this fact in obtaining (5-42). However, the vector equation $\nabla^2\mathbf{A}$ in spherical coordinates in Appendix A does not appear to separate into scalar equations under any conditions, with possibly a few exceptions. One exception occurs when the vector potential has only a ϕ-component and is independent of ϕ. Such a situation results when one considers the fields excited by a uniform current loop and is used as an example in connection with (5-72). With exterior problems in spherical coordinates, it is often useful to introduce Debye potentials, or multipole expansions, or to calculate the fields by a direct integration over the sources.[41]

As an example of the solution of Helmholtz's equation in spherical coordinates, let us consider the interior problem described in Prob. 5.43, i.e., a delta source located between two concentric conducting spheres of radii a and b $(a < b)$ and a Dirichlet boundary condition. The equation we wish to solve is

$$(\nabla^2 + k^2)G(r, \theta, \phi) = -\delta(\mathbf{r} - \mathbf{r}_0), \qquad a < r_0 < b \qquad (5\text{-}68)$$

with the boundary conditions

$$G = 0 \quad \text{at} \quad r = a, b$$

We choose solutions to the homogeneous form of (5-68) from Table 5-3B. If we split the volume between the two concentric spheres in the r-direction, then only the $b_n(kr)$ will satisfy an inhomogeneous equation. We can obtain that equation by choosing the following form for G:

$$G(r, \theta, \phi) = \sum_{n=0}^{\infty} \sum_{m=0}^{n} G_n(r)T_{mn}^i(\theta, \phi) \qquad (5\text{-}69)$$

where the T_{mn}^i are the tesseral harmonics defined in Sec. 1.6. This is the same approach we used in Sec. 5.3 when we solved the scalar Helmholtz equation with a delta source in unbounded space. If we use this procedure here, then according to (3-78) and (3-83), $G_n(r)$ satisfies the inhomogeneous ordinary

---

[41]Van Bladel, *Electromagnetic Fields*, Secs. 7.6, 7.7, 7.9.

differential equation

$$\left[\frac{1}{r^2}\frac{d}{dr}\left(r^2\frac{d}{dr}\right) + k^2 - \frac{n(n+1)}{r^2}\right]G_n(r) = -\frac{\epsilon_m(2n+1)(n-m)!}{4\pi(n+m)!}$$

$$\times\ T_{mn}^i(\theta_0, \phi_0)\frac{\delta(r-r_0)}{r^2} \qquad (5\text{-}70)$$

or

$$\left[\frac{d}{dr}\left(r^2\frac{d}{dr}\right) + (kr)^2 - n(n+1)\right]G_n(r) = -D(r)\,\delta(r-r_0)$$

where

$$D(r) = \frac{\epsilon_m(2n+1)(n-m)!\,T_{mn}^i(\theta_0, \phi_0)}{4\pi(n+m)!}$$

However, the solution to (5-70) for a source of unit magnitude is already given in Prob. 2.17. Thus, the solution to (5-70) becomes

$$G_n(r) = G_{n1}u(r_0 - r) + G_{n2}u(r - r_0)$$

where

$$G_{n1}(r, r_0) = kD(j_{na}n_{nb} - j_{nb}n_{na})^{-1}(j_{nb}n_{n0} - j_{n0}n_{nb})(j_n n_{na} - j_{na}n_n),$$

$$0 < a \leq r \leq r_0$$

$$G_{n2}(r, r_0) = G_{n1}(r_0, r), \qquad r_0 \leq r \leq b < \infty$$

and

$$j_{na} = j_n(ka), \qquad j_n = j_n(kr), \qquad n_{n0} = n_n(kr_0), \qquad \text{etc.}$$

Since (5-69) is valid for $i = o$ and $i = e$ (odd and even solutions in ϕ), the complete solution is the sum of the two and becomes

$$G(r, \theta, \phi) = \frac{k}{4\pi}\sum_{n=0}^{\infty}\sum_{m=0}^{n}\frac{\epsilon_m(2n+1)(n-m)!}{(n+m)!\,(j_{na}n_{nb} - j_{nb}n_{na})}$$

$$\times\ [(j_{nb}n_{n0} - j_{n0}n_{nb})(j_n n_{na} - j_{na}n_n)u(r_0 - r)$$

$$+\ (j_{nb}n_n - j_n n_{nb})(j_{n0}n_{na} - j_{na}n_{n0})u(r - r_0)]$$

$$\times\ [T_{mn}^e(\theta_0, \phi_0)T_{mn}^e(\theta, \phi) + T_{mn}^o(\theta_0, \phi_0)T_{mn}^o(\theta, \phi)]$$

As mentioned in connection with (3-85), the term involving the tesseral harmonics in the second set of square brackets is $P_n^m(\cos\theta)P_n^m(\cos\theta_0)$ $\cos m(\phi - \phi_0)$ so that use of the addition theorem for Legendre functions

(3-71) in the above expression for $G(r, \theta, \phi)$ simplifies it to

$$G(r, \theta, \phi) = \frac{k}{4\pi} \sum_{n=0}^{\infty} \frac{(2n + 1)P_n(\cos \xi)}{(j_{na}n_{nb} - j_{nb}n_{na})}[(j_{nb}n_{n0} - j_{n0}n_{nb})(j_{n}n_{na} - j_{na}n_{n})$$

$$\times u(r_0 - r) + (j_{nb}n_n - j_{n}n_{nb})(j_{n0}n_{na} - j_{na}n_{n0})u(r - r_0)] \qquad (5\text{-}71)$$

where ξ is the angle between $\mathbf{r}$ and $\mathbf{r}_0$.

As an exterior problem of considerable interest, let us consider the uniform electric current ring source illustrated in Fig. 5-4 but in spherical coordinates. Since there is no ϕ-dependence to the problem and the current is ϕ-directed, we would expect A_ϕ to be the only finite component of the vector potential. Of course, this is the same conclusion we arrived at when writing the equation (5-54). If we consider (4-14) and use the expression for $\nabla^2\mathbf{A}$ in spherical coordinates from Appendix A we obtain the equation

$$\left(\nabla^2 + k^2 - \frac{1}{r^2 \sin^2 \theta}\right) A_\phi(r, \theta) = -J_\phi$$

or

$$\left\{\frac{1}{r^2}\frac{\partial}{\partial r}\left(r^2 \frac{\partial}{\partial r}\right) + \frac{1}{r^2}\left[\frac{1}{\sin \theta}\frac{\partial}{\partial \theta}\left(\sin \theta \frac{\partial}{\partial \theta}\right) - \frac{1}{\sin^2 \theta}\right] + k^2\right\} A_\phi(r, \theta)$$

$$= -\frac{I_0 \delta (r - r_0) \delta(\theta - \theta_0)}{r} \qquad (5\text{-}72)$$

Strictly speaking, we should not use (4-14) at all since the Lorentz condition is not valid, but instead use (4-10). However, for this particular case

$$\nabla \times \nabla \times (A_\phi \mathbf{a}_\phi) = -(\nabla^2 A_\phi)\mathbf{a}_\phi$$

and $\nabla\Phi$ does not have a component in the ϕ-direction. Another viewpoint is to realize that $\nabla \cdot \mathbf{A} = 0$ so that the Lorentz condition is satisfied. Since the term in the square brackets is part of the associated Legendre equation (1-32) of degree n and order (superscript) one, we assume that A_ϕ can be written as

$$A_\phi = \sum_{n=1}^{\infty} \Gamma_n(r)P_n^1 (\cos \theta) \qquad (5\text{-}73)$$

The associated Legendre functions of the second kind are omitted since they are not well behaved at $\theta = 0, \pi$. If we substitute (5-73) into (5-72), we obtain the following equation (see Appendix C.4.12):

$$\sum_{n=1}^{\infty}\left[\frac{1}{r^2}\frac{\partial}{\partial r}\left(r^2 \frac{\partial}{\partial r}\right) + k^2 - \frac{n(n + 1)}{r^2}\right]\Gamma_n P_n^1 = -\frac{I_0 \delta(r - r_0) \delta(\theta - \theta_0)}{r}$$

Since the associated Legendre functions are orthogonal over the interval $(0, \pi)$, we multiply both sides of the above equation by $P_m^1 \sin \theta \, d\theta$ and then integrate over the interval $(0, \pi)$ to obtain the inhomogeneous ordinary differential equation for $\Gamma_n(r)$, i.e.,

$$\left[\frac{d}{dr}\left(r^2 \frac{d}{dr}\right) + (kr)^2 - n(n + 1)\right]\Gamma_n(r) = -D(r) \, \delta(r - r_0) \qquad (5\text{-}74)$$

where

$$D(r) = \frac{(2n + 1)rI_0 P_{n0}^1 \sin \theta_0}{2n(n + 1)}$$

$$P_{n0}^1 = P_n^1 (\cos \theta_0)$$

and use was made of an orthogonality relation in Appendix C.4.10. This is a similar equation to that obtained in (5-70) and is the inhomogeneous form of the spherical Bessel function equation (1-35). Moreover, we have already found the Green's function for (1-35) for the same boundary conditions as those relevant to this problem in connection with Prob. 2.20. All we have to do is multiply the solution to Prob. 2.20 by $D(r_0)k$ above to obtain the solution to (5-74). Thus, the solution to (5-72) becomes

$$A_\phi = -\frac{jkr_0 I_0}{2} \sum_{n=1}^{\infty} \frac{(2n + 1)P_n^1 P_{n0}^1 \sin \theta_0}{n(n + 1)} [h_{n0}^{(2)} j_n u(r_0 - r) + h_n^{(2)} j_{n0} u(r - r_0)]$$

$$(5\text{-}75)$$

The electric and magnetic fields are obtained from

$$\mathbf{H} = \nabla \times \mathbf{A} \qquad (4\text{-}7)$$

$$\mathbf{E} = -j\omega\mu\mathbf{A} + \frac{1}{j\omega\epsilon} \nabla\nabla\cdot\mathbf{A} = -j\omega\mu A_\phi \mathbf{a}_\phi \qquad (4\text{-}15)$$

and the only finite component of electric field becomes

$$E_\phi = -\frac{\omega\mu kr_0 I_0}{2} \sum_{n=1}^{\infty} \frac{(2n + 1)P_n^1 P_{n0}^1 \sin \theta_0}{n(n + 1)} [h_{n0}^{(2)} j_n u(r_0 - r) + h_n^{(2)} j_{n0} u(r - r_0)]$$

$$(5\text{-}76)$$

This problem is also considered by others.[42] However, the approach used by Weeks and Harrington is to solve for A_{mr} since the field is TE_r. This approach

---

[42]Weeks, *Electromagnetic Theory for Engineering Applications*, p. 564. Also, see Harrington, *Time-Harmonic Electromagnetic Fields*, p. 315, and Smythe, *Static and Dynamic Electricity*, 2nd ed., p. 489.

is more basic since it is not possible to separate (4-14) in the spherical coordinate system except when

$$A = r\Phi(r, \theta, \phi)a_r \quad \text{or} \quad A_m = r\Phi(r, \theta, \phi)a_r$$

where Φ satisfies the scalar Helmholtz equation in Table 5-3B. With this approach, one solves for E_ϕ and H_θ in terms of A_{mr} and then uses the boundary conditions involving the continuity of E_ϕ and the discontinuity of H_θ at the source. With this viewpoint, one is actually replacing the electric current ring source (magnetic multipoles) by an equivalent source that consists of r-directed electric multipoles.

Still another approach is to use the integral expression for A_ϕ in (5-58), which is valid for a constant current. Next, e^{-jkR}/R is replaced by $-jkh_0^{(2)}(kR)$ according to (3-32) and the addition theorems for spherical Hankel functions from Prob. 5.41 and Legendre functions from Prob. 5.37 are used to replace $h_0^{(2)}(kR)$ by spherical Bessel and Hankel functions and associated Legendre functions. Now, the integral over ϕ can be evaluated and one finds that the expression for A_ϕ reduces to that given by (5-75). The verification of these steps is left for Prob. 5.56.

PROBLEMS

5.1. Show that when G is chosen according to (5-3) that G_1 and G_2 in (5-12) and (5-13), respectively, become

$$G_1(x, x_0; y, y_0)$$

$$= \sum_{n=1}^{\infty} \frac{2}{n\pi} \frac{\sin(n\pi y/b)\sin(n\pi y_0/b)\sinh(n\pi x/b)\sinh[(n\pi/b)(a - x_0)]}{\sinh(n\pi a/b)}$$
$$x < x_0$$

$$G_2(x, x_0; y, y_0) = G_1(x_0, x; y_0, y), \quad x > x_0$$

where

$$G = G_1 u(x_0 - x) + G_2 u(x - x_0)$$

5.2. Show that when $b \rightarrow \infty$ in (5-12) and (5-13), the expression for G_1 becomes

$$G_1(x, x_0; y, y_0) = \sum_{m=1}^{\infty} \frac{2}{m\pi} \sin\frac{m\pi x}{a} \sin\frac{m\pi x_0}{a} \sinh\frac{m\pi y}{a} e^{-m\pi y_0/a}, \quad y < y_0$$

and the expression for G_2 is still given by (5-13).

5.3. Show that the solution of $\nabla^2 G = -\delta(x - x_0)\delta(y - y_0)\delta(z - z_0)$ inside a rectangular box with sides at $x = 0, a; y = 0, b; z = 0, c$ and with the boundary conditions that G vanish on the six walls is

$$G_1(x, x_0; y, y_0; z, z_0) = \sum_{m=1}^{\infty} \sum_{n=1}^{\infty} \frac{4}{ab\Gamma_{mn}}$$

$$\times \frac{\sin(m\pi x/a)\sin(m\pi x_0/a)\sin(n\pi y/b)\sin(n\pi y_0/b)\sinh\Gamma_{mn}z\sinh\Gamma_{mn}(c-z_0)}{\sinh\Gamma_{mn}c},$$
$$z < z_0$$

$$G_2(x, x_0; y, y_0; z, z_0) = G_1(x_0, x; y_0, y; z_0, z), \qquad z > z_0$$

where

$$\Gamma_{mn} = \left[\left(\frac{m\pi}{a}\right)^2 + \left(\frac{n\pi}{b}\right)^2\right]^{1/2}$$

5.4. Verify that (5-18) and (5-19) are solutions to (5-15).

5.5. Find the solution to (5-15) for the problem in Fig. 5-2 when $L \to \infty$, i.e., a semi-infinite cylinder.

5.6. Show that the Green's function in Prob. 5.5 reduces to that given in (5-20) when the cylinder becomes infinite. *Hint:* Use the fact that the solution to Prob. 5.5 can be considered as the superposition of the potentials due to a source at $z = z_0$ and its image at $z = -z_0$.

5.7. Verify that the expression for the Green's function in (5-21) results from (5-20) when $a \to \infty$, i.e., when the Fourier series in p passes to a Fourier integral[43] in p.

5.8. Find the Green's function[44] for a unit dc point source inside an infinite, rectangular tube (Fig. 5-1) with the boundary condition that G vanishes on the wall.

5.9. Show that the Green's function in Prob. 5.8 becomes the following when the unit point source becomes a unit line source:

$$G(x, x_0; y, y_0) = \sum_{m=1}^{\infty} \sum_{n=1}^{\infty} \frac{4}{ab\Gamma_{mn}^2} \sin\frac{m\pi x}{a} \sin\frac{m\pi x_0}{a} \sin\frac{n\pi y}{b} \sin\frac{n\pi y_0}{b}$$

where

$$\Gamma_{mn}^2 = \left(\frac{m\pi}{a}\right)^2 + \left(\frac{n\pi}{b}\right)^2$$

5.10. Show that the Green's function in Prob. 5.9 is equivalent to that in (5-12) and (5-13). *Hint:* Evaluate the sum over n by using the relation

$$\sum_{n=1}^{\infty} \frac{\cos n\beta}{n^2 - \alpha^2} = \frac{1}{2\alpha^2} - \frac{\pi}{2\alpha}\frac{\cos(\beta - \pi)\alpha}{\sin \pi\alpha}, \qquad 0 \le \beta \le 2\pi$$

5.11. Show that the Green's function in Prob. 5.9 can be obtained by assum-

---

[43]Smythe, *Static and Dynamic Electricity*, 2nd ed., p. 179.

[44]*Ibid.*, p. 213.

ing a solution as $G = \sum_{m=1}^{\infty} \sum_{n=1}^{\infty} A_{mn} \sin(m\pi x/a) \sin(n\pi y/b)$ and evaluating the coefficients A_{mn} directly.

5.12. Show that (5-21) can be used to obtain the identity[45]

$$(\rho^2 + z^2)^{-1/2} = \int_0^{\infty} e^{-\alpha|z|} J_0(\alpha\rho)\, d\alpha$$

This expression is valid[46] for $|z| = jx$. Also, see Prob. 3.22 for an alternate derivation.

5.13. Show that (5-21) and the identity in Prob. 5.12 can be used to obtain the addition theorem[47] for Bessel functions of the first kind, i.e.,

$$J_0(\alpha\rho_1) = \sum_{m=0}^{\infty} \epsilon_m \cos m(\phi - \phi_0)[J_m(\alpha\rho_0)J_m(\alpha\rho)u(\rho_0 - \rho)$$

$$+ J_m(\alpha\rho)J_m(\alpha\rho_0)u(\rho - \rho_0) = J_m(\alpha\rho_0)J_m(\alpha\rho)]$$

where $\epsilon_m = 1$, $m = 0$ and $\epsilon_m = 2$, $m > 0$.

5.14. Use the expressions in (3-8) to obtain the Green's function for a unit dc line source in front of a right-angle, grounded, conducting corner. See Prob. 3.2.

5.15. Derive the expression in Prob. 5.12 directly. *Hint:* Use the generating function for $J_n(x)$, i.e., (1-21), and interchange the order of integration.[48]

5.16. Find the Green's function for $\nabla^2 G(\rho, \phi, z) = -\delta(\mathbf{r} - \mathbf{r}_0)$ inside a coaxial cylindrical box[49] defined by $\rho = a$, $\rho = b$ $(b > a)$, $z = 0$, $z = L$ when $G = 0$ on the walls. The result when the region is split in the ρ-direction is

$$G_1 = \frac{1}{\pi L} \sum_{m=0}^{\infty} \sum_{n=1}^{\infty} \frac{\epsilon_m (IK_a - I_aK)(I_0K_b - I_bK_0)}{(I_aK_b - I_bK_a)}$$

$$\times \sin\frac{n\pi z}{L} \sin\frac{n\pi z_0}{L} \cos m(\phi - \phi_0), \qquad 0 < a \le \rho \le \rho_0$$

$$G_2(\rho, \rho_0; \phi, \phi_0; z, z_0) = G_1(\rho_0, \rho; \phi_0, \phi; z_0, z), \qquad \rho_0 \le \rho \le b < \infty$$

where each modified Bessel function is of order m and argument $n\pi\rho/L$, i.e., $I = I_m(n\pi\rho/L)$, $I_a = I_m(n\pi a/L)$, etc. The results from Prob. 2.9 are directly applicable here.

---

[45]*Ibid.*, p. 179.

[46]Van Bladel, *Electromagnetic Fields*, p. 514. Also, see Smythe, *Static and Dynamic Electricity*, p. 182.

[47]Smythe, *Static and Dynamic Electricity*, p. 179.

[48]G. Carrier, M. Krook and C. Pearson, *Functions of a Complex Variable, Theory and Techniques*, McGraw-Hill Book Company, Inc., New York, 1966, p. 88.

[49]See Smythe, *Static and Dynamic Electricity*, p. 191, for the solution when the region is split in the ρ-direction.

5.17. Find the Green's function for $(\nabla^2 + k^2)G(\rho, \phi, z) = -\delta(\mathbf{r} - \mathbf{r}_0)$ for the box and boundary conditions in Prob. 5.16. Split the region in the z-direction. The solution is

$$G_1 = \frac{1}{2\pi} \sum_{m=0}^{\infty} \sum_{n=1}^{\infty} \frac{\epsilon_m}{k_1 N_{mn}} B_m(\beta_{mn}\rho_0)B_m(\beta_{mn}\rho)$$

$$\times \frac{\sinh k_1 z \sinh k_1(L - z_0)}{\sinh k_1 L} \cos m(\phi - \phi_0), \qquad 0 < a \leq \rho \leq \rho_0$$

$$G_2(\rho, \rho_0; \phi, \phi_0; z, z_0) = G_1(\rho_0, \rho; \phi_0, \phi; z_0, z), \qquad \rho_0 \leq \rho \leq b < \infty$$

where $k_1^2 = \beta_{mn}^2 - k^2$ and B_m, β_{mn}, and N_{mn} are defined in (2-43), (2-44), and (2-46), respectively, for $x_1 = a$ and $x_2 = b$. Note that the z-dependent functions satisfy an equation similar to (5-17).

5.18. Repeat Prob. 5.17 when the region is split in the ρ-direction. The solution is

$$G_1 = \frac{1}{2L} \sum_{m=0}^{\infty} \sum_{n=1}^{\infty} \frac{\epsilon_m(J_{mb}N_{m0} - J_{m0}N_{mb})(J_m N_{ma} - J_{ma}N_m)}{J_{ma}N_{mb} - J_{mb}N_{ma}}$$

$$\times \sin \frac{n\pi z}{L} \sin \frac{n\pi z_0}{L} \cos m(\phi - \phi_0), \qquad 0 < a \leq \rho \leq \rho_0$$

$$G_2(\rho, \rho_0; \phi, \phi_0; z, z_0) = G_1(\rho_0, \rho; \phi_0, \phi; z_0, z), \qquad \rho_0 \leq \rho \leq b < \infty$$

where the subscripts on the Bessel functions are defined as in Prob. 2.8 and the argument of each Bessel function is $k_2\rho$, where $k_2^2 = k^2 - (n\pi/L)^2$. Note that the results in Prob. 2.8 are directly applicable except $D = \epsilon_m/\pi L$.

5.19. Consider Prob. 5.17 when the region becomes infinite in the $\pm z$-directions. The result is

$$G = \frac{1}{4\pi} \sum_{m=0}^{\infty} \sum_{n=1}^{\infty} \frac{\epsilon_m}{k_1 N_{mn}} B_m(\beta_{mn}\rho_0)B_m(\beta_{mn}\rho) \cos m(\phi - \phi_0)e^{-k_1|z-z_0|}$$

Note that these results can be obtained from (5-20). Also, B_m, β_{mn}, and N_{mn} are defined as in Prob. 5.17.

5.20. Extend the results of Prob. 5.19 for the case where the origin is included, i.e., $a \to 0$. The result is

$$G = \frac{1}{2\pi b^2} \sum_{m=0}^{\infty} \sum_{n=1}^{\infty} \frac{\epsilon_m}{k_1} \frac{J_m(\beta_{mn}\rho_0)J_m(\beta_{mn}\rho)}{J_m'^2(\beta_{mn}b)} \cos m(\phi - \phi_0)e^{-k_1|z-z_0|}$$

where β_{mn} are the roots of $J_m(\beta_{mn}b) = 0$ (see Appendix C.1.13).

5.21. Consider Prob. 5.19 for $b \to \infty$ and the region split in the ρ-direction. Note that the results in Prob. 5.19 cannot be used nor can the results in Prob. 5.18 be used. However, the results in Prob. 5.18 are applicable since in the limit as $L \to \infty$, the Fourier series in z passes to a Fourier transform. Consequently, one can work this problem by assuming that

$$G = \sum_{m=0}^{\infty} G_{\rho}(\rho) \cos m(\phi - \phi_0) f(z)$$

where

$$f(z) = \frac{1}{\sqrt{2\pi}} \int_{-\infty}^{\infty} \tilde{f}(\alpha) e^{j\alpha z} \, d\alpha$$

The result is

$$G_1 = -\frac{1}{8\pi} \int_{-\infty}^{\infty} \sum_{m=0}^{\infty} \frac{\epsilon_m H_m^{(2)}(k_1 \rho_0)}{H_m^{(2)}(k_1 a)} [J_m(k_1 \rho) N_m(k_1 a) - J_m(k_1 a) N_m(k_1 \rho)]$$

$$\times \cos m(\phi - \phi_0) e^{j\alpha(z - z_0)} \, d\alpha, \qquad 0 < a \le \rho \le \rho_0$$

$$G_2 = G_1(\rho_0, \rho; \phi_0, \phi; z_0, z), \qquad \rho_0 \le \rho \le \infty$$

where $k_1^2 = k^2 - \alpha^2$. Note that the results in Prob. 2.14 can be used directly.

5.22. Extend the results of Prob. 5.21 to the case where $a \to 0$. Now, G_1 and G_2 become

$$G_1 = \frac{1}{8\pi j} \int_{-\infty}^{\infty} \sum_{m=0}^{\infty} \epsilon_m H_m^{(2)}(k_1 \rho_0) J_m(k_1 \rho) \cos m(\phi - \phi_0) e^{j\alpha(z - z_0)} \, d\alpha,$$

$$0 \le \rho \le \rho_0$$

$$G_2 = G_1(\rho_0, \rho; \phi_0, \phi; z_0, z), \qquad \rho_0 \le \rho \le \infty$$

In this case, Prob. 2.15 can be used.

5.23. Show that if one uses the addition theorem for Hankel functions from (5-53) in the expression for G in Prob. 5.22, the integral reduces to the form of (3-46) for $i = 3$ and can be evaluated to give the result

$$G = \frac{e^{-jkR}}{4\pi R}$$

where R is defined in Prob. 5.12. Note that in Probs. 5.17 through 5.23 the source corresponds to that of a z-directed electric dipole and G corresponds to A_z or E_z.

5.24. Find the Green's function for $(\nabla^2 + k^2) G(\rho, \phi, z) = -\delta(\mathbf{r} - \mathbf{r}_0)$ inside a coaxial cylindrical box similar to that described in Prob. 5.16 except that it has sides at $\phi = \phi_1$ and $\phi = \phi_2$. Assume that $G = 0$ on the walls and that $\phi_1 < \phi_0 < \phi_2$. If the result in Prob. 5.18 is used, the correct solution is obtained by the substitution

$$\sum_{m=0}^{\infty} \frac{\epsilon_m}{2L} \cos m(\phi - \phi_0) \longrightarrow \frac{2\pi}{L(\phi_2 - \phi_1)} \sum_{m=1}^{\infty} \sin \frac{m\pi(\phi - \phi_1)}{\phi_2 - \phi_1} \sin \frac{m\pi(\phi_0 - \phi_1)}{\phi_2 - \phi_1}$$

and the order of the Bessel functions is $m\pi/(\phi_2 - \phi_1)$. See (5-61) for the proper form of the ϕ-dependent function.

5.25. Find the Green's function for $(\nabla^2 + k^2)G(\rho, \phi, z) = -(I/\rho)\delta(\rho - \rho_0)$ $\delta(\phi - \phi_0)$ when the electric line source is located outside a conducting wedge[50] defined by the surfaces $\phi = \phi_1$ and $\phi = \phi_2$ ($\phi_0 > \phi_2 > \phi_1$). Assume that $G = 0$ on the wedge and split the region in the ρ-direction. The result is

$$G_1 = \frac{\pi I}{j(2\pi - \phi_2 + \phi_1)} \sum_{m=1}^{\infty} H_s^{(2)}(k\rho_0)J_s(k\rho) \sin s(\phi - \phi_1) \sin s(\phi_0 - \phi_1),$$

$$0 \leq \rho \leq \rho_0$$

$$G_2(\rho, \rho_0; \phi, \phi_0; z, z_0) = G_1(\rho_0, \rho; \phi_0, \phi; z_0, z), \qquad \rho_0 \leq \rho \leq \infty$$

where

$$s = \frac{m\pi}{2\pi - \phi_2 + \phi_1}$$

The results of Prob. 2.15 are also applicable.

5.26. Find the Green's function for $\nabla^2 G(\rho, \phi, z) = -\delta(\mathbf{r} - \mathbf{r}_0)$ inside a grounded cylindrical pipe[51] defined by the surfaces $\phi = \phi_1$, $\phi = \phi_2$ ($\phi_2 > \phi_0 > \phi_1$) and $\rho = b$. The result is

$$G = \frac{4}{b^2(\phi_2 - \phi_1)} \sum_{m=1}^{\infty} \sum_{n=1}^{\infty} \frac{J_s(\beta_{sn}\rho)J_s(\beta_{sn}\rho_0)}{\beta_{sn}J_{s+1}^2(\beta_{sn}b)}$$

$$\times \sin s(\phi - \phi_1) \sin s(\phi_0 - \phi_1)e^{-\beta_{sn}|z-z_0|}, \qquad 0 \leq \rho \leq b$$

where the β_{sn} are defined from $J_s(\beta_{sn}b) = 0$, and the order of the Bessel functions is $s = m\pi/(\phi_2 - \phi_1)$. Note the similarity of this problem to that in the text leading to (5-17).

5.27. Consider Prob. 5.25 when $G = A_z$ and solve for the TM_z modes[52] present.

5.28. Find the Green's function for $(\nabla^2 + k^2)G(\rho, \phi, z) = -(I_m/\rho)\delta(\rho - \rho_0)$ $\delta(\phi - \phi_0)$ when the uniform magnetic line source is located outside an infinitely long perfectly conducting cylinder[53] of radius a located at the origin. Assume that $\partial G/\partial \rho = 0$ on the cylinder. The result is

$$G_1 = \frac{jI_m}{8} \sum_{n=-\infty}^{\infty} \frac{H_n^{(2)}(k\rho_0)}{H_n^{(2)\prime}(ka)}[H_n^{(1)\prime}(ka)H_n^{(2)}(k\rho) - H_n^{(1)}(k\rho)H_n^{(2)\prime}(ka)]$$

$$\times e^{jn(\phi-\phi_0)}, \qquad 0 < a \leq \rho \leq \rho_0$$

$$G_2(\rho, \rho_0; \phi, \phi_0) = G_1(\rho_0, \rho; \phi_0, \phi), \qquad \rho_0 \leq \rho \leq \infty$$

5.29. Consider Prob. 5.28 when $G = A_{mz}$ and solve for the TE_z modes present.

---

[50]See Harrington, *Time-Harmonic Electromagnetic Fields*, p. 239.

[51]See Smythe, *Static and Dynamic Electricity*, p. 213.

[52]See Harrington, *Time-Harmonic Electromagnetic Fields*, p. 237.

[53]*Ibid.*

5.30. Consider Prob. 5.28 when the magnetic line source coincides with the cylindrical surface, i.e., $\rho_0 = a$, to obtain the fields excited by an infinite, narrow, axial slot on a perfectly conducting cylinder.[54]

5.31. Find the Green's function for (5-42) when the uniform magnetic ring source is located concentric with and outside an infinitely long perfectly conducting cylinder[55] of radius a located at the origin. Assume that $(\partial/\partial\rho)(\rho G) = 0$ on the cylinder so that $G \rightarrow A_{mz}$ when $I_m \neq 1$. The result is

$$G_1 = \frac{\rho_0}{4j} \int_{-\infty}^{\infty} \frac{H_1^{(2)}(k_1\rho_0)}{H_0^{(2)}(k_1a)} [J_1(k_1\rho)H_0^{(2)}(k_1a) - J_0(k_1a)H_1^{(2)}(k_1\rho)]$$

$$\times e^{j\alpha(z-z_0)}\,d\alpha, \qquad 0 \leq a \leq \rho \leq \rho_0$$

$$G_2(\rho, \rho_0; z, z_0) = G_1(\rho_0, \rho; z_0, z), \qquad \rho_0 \leq \rho \leq \infty$$

where $k_1^2 = k^2 - \alpha^2$.

5.32. Consider Prob. 5.31 when the magnetic ring source coincides with the cylindrical surface ($\rho_0 = a$) to obtain the fields excited by a narrow, uniform, circumferential slot on a perfectly conducting cylinder. The result[56] for E_z is

$$E_z = \frac{I_m}{2\pi} \int_{-\infty}^{\infty} \frac{k_1 H_0^{(2)}(k_1\rho)}{H_0^{(2)}(k_1a)} e^{j\alpha(z-z_0)}\,d\alpha$$

5.33. Repeat Prob. 5.25 using the cylindrical Fourier (Fourier-Bessel or Hankel) transform defined in (3-59). Assume a continuous solution in ϕ so that the region is split according to $\rho \gtrless \rho_0$ (same procedure as in Prob. 5.25). The inverse[57] cylindrical Fourier transform can be evaluated by equating this solution to that obtained in Prob. 5.25.

5.34. Repeat the example in the text leading to (5-18) and (5-19), i.e., solve (5-15) for $G = 0$ on the boundaries, except divide the cylindrical region according to $\rho \gtrless \rho_0$. Note that this result can be obtained from Prob. 5.16 by letting $a \rightarrow 0$ and using relations from Appendix C.2.8.

5.35. Consider the problem dual to Prob. 5.25, i.e., find the Green's func-

---

[54]*Ibid.*

[55]Note that this problem could be solved by a Weber transform similar to the one described by G. Duff and D. Naylor, *Differential Equations of Applied Mathematics*, John Wiley & Sons, Inc., New York, 1966, p. 337.

[56]Note the similarity with the expression for H_z in the dual problem (except for the boundary conditions) in Van Bladel, *Electromagnetic Fields*, p. 458.

[57]Duff and Naylor, *Differential Equations of Applied Mathematics*, p. 416, have one solution for the inverse transform. However, it is not applicable here since it is the principal value solution and would correspond to the case where $i = 1$ in (3-38). Our problem requires a solution that would correspond to the case $i = 3$ in (3-38).

tion for $(\nabla^2 + k^2)G(\rho, \phi, z) = -(I_m/\rho)\delta(\rho - \rho_0)\delta(\phi - \phi_0)$ with the boundary condition that $\partial G/\partial n = 0$ on the wedge. Note that only the ϕ-dependent part of this problem[58] is different from Prob. 5.25.

5.36. Verify that (5-55) results from the expression for A_ϕ in (5-58) when use is made of the identity (3-43) and the addition theorem for Hankel functions (5-53).

5.37. Show that (5-67) and the first identity in Prob. 1.14 can be used to obtain the addition theorem for Legendre functions (3-71), i.e.,

$$P_n(\cos \xi) = \sum_{m=0}^{\infty} \frac{\epsilon_m(n - m)!}{(n + m)!} P_n^m(\cos \theta) P_n^m(\cos \theta_0) \cos m(\phi - \phi_0)$$

where $\cos \xi = \cos \theta \cos \theta_0 + \sin \theta \sin \theta_0 \cos (\phi - \phi_0)$ and ξ is the angle between $\mathbf{r}$ and $\mathbf{r}_0$. Let $a \to 0$, $b \to \infty$, $\theta_1 \to \pi$, and $N_1(\theta_1) \to 2(n + m)!/(2n + 1)(n - m)!$ in (5-67).

5.38. Use (5-67) to obtain the Green's function for a unit point source outside a grounded conducting sphere of radius a. Use the identity in Prob. 5.37 to show that this result is the same as that obtained by the method of images.

5.39. Use (5-67) to obtain the Green's function for a unit point source above an infinite ground plane with a hemispherical boss[59] of radius a.

5.40. Find the capacitance per unit length between two cones defined by $\theta = \theta_1$ and $\theta = \pi - \theta_2$. The result is

$$C = \frac{2\pi\epsilon}{\ln\left[\cot(\theta_1/2)\cot(\theta_2/2)\right]}$$

5.41. Find the Green's function for $(\nabla^2 + k^2)G(r, \theta, \phi) = -\delta(\mathbf{r} - \mathbf{r}_0)$ in unbounded space in the form of a product solution, i.e., $G(r, \theta, \phi) = G_r(r)G_\theta(\theta)G_\phi(\phi)$. Equate the result to the representation in (3-31) for $i = 3$ to obtain the addition theorem for spherical Hankel functions, i.e.,

$$h_0^{(2)}(kR) = \sum_{n=0}^{\infty} (2n + 1)[h_n^{(2)}(kr_0)j_n(kr)u(r_0 - r)$$

$$+ h_n^{(2)}(kr)j_n(kr_0)u(r - r_0)]P_n(\cos \xi)$$

where $R = |\mathbf{r} - \mathbf{r}_0|$ and ξ is the angle between $\mathbf{r}$ and $\mathbf{r}_0$. Note that the real part of this expression is given in (3-86).

5.42. Consider the real and imaginary part of the addition theorem in Prob. 5.41 to obtain the addition theorem for spherical Bessel functions of the first and second kinds.

---

[58] See Harrington, *Time-Harmonic Electromagnetic Fields*, p. 241.

[59] See Smythe, *Static and Dynamic Electricity*, 2nd ed., p. 200, Problem 13c.

5.43. Find the Green's function for $\nabla^2 G(r, \theta, \phi) = -\delta(\mathbf{r} - \mathbf{r}_0)$ in the region between two concentric perfectly conducting spheres defined by the surfaces $r = a$, $r = b$ ($a < r_0 < b$), when $G = 0$ on the spheres. *Hint:* Split the region in the r-direction so that (5-63) applies. Then, (5-67) is valid except $N_1(\theta_1) \rightarrow 2(n + m)!/(n - m)!\,(2n + 1)$.

5.44. Find the Green's function for $\nabla^2 G = -\delta(\mathbf{r} - \mathbf{r}_0)$ in the region between the cones $\theta = \theta_1$, $\theta = \theta_2$ ($\theta_1 < \theta_0 < \theta_2$), the planes $\phi = \phi_1$, $\phi = \phi_2$ ($\phi_1 < \phi_0 < \phi_2$), and the spheres $r = a$, $r = b$ ($a < r_0 < b$) when $G = 0$ on the surfaces. For the region split in the r-direction, the solution is

$$G_1 = -\sum_{s}^{\infty} \sum_{m=1}^{s} \frac{2}{\phi_2 - \phi_1} \frac{(r_0^s - b^{2s+1}r_0^{-s-1})(r^s - a^{2s+1}r^{-s-1})}{(b^{2s+1} - a^{2s+1})M_1(\theta_1, \theta_2)}$$

$$\times (P_s^m Q_{s1}^m - P_{s1}^m Q_s^m)(P_{s0}^m Q_{s1}^m - P_{s1}^m Q_{s0}^m) \sin\frac{m\pi(\phi - \phi_1)}{\phi_2 - \phi_1} \sin\frac{m\pi(\phi_0 - \phi_1)}{\phi_2 - \phi_1},$$

$$0 < a \leq r \leq r_0$$

$$G_2(r, r_0; \theta, \theta_0; \phi, \phi_0) = G_1(r_0, r; \theta_0, \theta; \phi_0, \phi), \qquad r_0 \leq r \leq b < \infty$$

Also, s is determined from the expression

$$P_{s1}^m Q_{s2}^m - P_{s2}^m Q_{s1}^m = 0$$

and

$$M_1(\theta_1, \theta_2) = \left[\sin\theta \frac{\partial L_s^m(\cos\theta)}{\partial\theta} \frac{\partial L_s^m(\cos\theta)}{\partial s}\right]_{\theta_1}^{\theta_2}, \qquad 0 < \theta_1 < \theta_2 < \pi$$

where $P_s^m = P_s^m(\cos\theta)$, $P_{s1}^m = P_s^m(\cos\theta_1)$, etc., and $L_s = P_s Q_{s1} - P_{s1} Q_s$. *Hint:* Set up the problem as in (5-63) except $G_{m\phi}$ is taken from (5-61) and $G_{n\theta} = L_s^m(\cos\theta)$ is taken from (1-32b). Now, the results from Prob. 2.21 apply except the strength of the source is D, where

$$D = \frac{2(2s + 1)}{\phi_2 - \phi_1} \frac{L_{s0}}{M_1(\theta_1, \theta_2)}$$

Note the similarity to D that is the strength of the source in (5-64).

5.45. Repeat Prob. 5.44 for the equation $(\nabla^2 + k^2)G(r, \theta, \phi) = -\delta(\mathbf{r} - \mathbf{r}_0)$ when the region is split in the r-direction. $G_{s\theta}$ and $G_{m\phi}$ are unchanged from Prob. 5.44 and G_r can be obtained from Prob. 2.17. Note that the expressions for $\Gamma_{1,2}$ in Prob. 2.17 must be multiplied by kD, where D is given in Prob. 5.44. The solution is

$$G_1 = \frac{-2k}{\phi_2 - \phi_1} \sum_s^{\infty} \sum_{m=1}^{s} \frac{(2s + 1)}{M_1(\theta_1, \theta_2)} \frac{(j_{s2}n_{s0} - j_{s0}n_{s2})(j_{s}n_{s1} - j_{s1}n_s)}{(j_{s1}n_{s2} - j_{s2}n_{s1})}$$

$$\times (P_s^m Q_{s1}^m - P_{s1}^m Q_s^m)(P_{s0}^m Q_{s1}^m - P_{s1}^m Q_{s0}^m) \sin\frac{m\pi(\phi - \phi_1)}{\phi_2 - \phi_1} \sin\frac{m\pi(\phi_0 - \phi_1)}{\phi_2 - \phi_1},$$

$$0 < a \leq r \leq r_0$$

$$G_2(r, r_0; \theta, \theta_0; \phi, \phi_0) = G_1(r_0, r; \theta_0, \theta; \phi_0, \phi),$$

$$r_0 \leq r \leq b < \infty, \, 0 < \theta_1 < \theta_2 < \pi$$

5.46. Repeat Prob. 5.43 for the equation $(\nabla^2 + k^2)G(r, \theta, \phi) = -\delta(\mathbf{r} - \mathbf{r}_0)$ and the condition $\partial G/\partial r = 0$ on the spheres. The solution is

$$G_1 = \frac{k}{4\pi} \sum_{m=0}^{\infty} \sum_{n=0}^{m} \frac{\epsilon_m(2n+1)(n-m)!}{(n+m)!} \frac{(j'_{nb}n_{n0} - j_{n0}n'_{nb})(j_n n'_{na} - n_n j'_{na})}{(j'_{na}n'_{nb} - j'_{nb}n'_{na})}$$

$$\times P_n^m(\cos\theta)P_n^m(\cos\theta_0)\cos m(\phi - \phi_0), \qquad 0 < a \leq r \leq r_0$$

$$G_2(r, r_0; \theta, \theta_0; \phi, \phi_0) = G_1(r_0, r; \theta_0, \theta; \phi_0, \phi), \qquad r_0 \leq r \leq b < \infty$$

Note the similarity between G_{r1} above and Γ_1 in Prob. 2.16.

5.47. Find the Green's function for $(\nabla^2 + k^2)G(r, \theta, \phi) = -\delta(\mathbf{r} - \mathbf{r}_0)$ in the region external to the cone $\theta = \theta_2$ $(0 \leq \theta_0 < \theta_2)$ when $G = 0$ on the cone. The solution is

$$G_1 = \frac{-jk}{2\pi} \sum_{s} \sum_{m=0}^{s} \frac{\epsilon_m(2s+1)}{M_2(\theta_2)} h_{s0}^{(2)} j_s P_s^m(\cos\theta)P_s^m(\cos\theta_0)\cos m(\phi - \phi_0),$$

$$0 \leq r \leq r_0$$

$$G_2(r, r_0; \theta, \theta_0; \phi, \phi_0) = G_1(r_0, r; \theta_0, \theta; \phi_0, \phi), \; r_0 \leq r \leq \infty, \; 0 \leq \theta \leq \theta_2 < \pi$$

where s is determined from the condition

$$P_s^m(\cos\theta_2) = 0$$

and

$$M_2(\theta_2) = \left[\sin\theta \frac{\partial P_s^m(\cos\theta)}{\partial\theta} \frac{\partial P_s^m(\cos\theta)}{\partial s}\right]_{\theta=\theta_2}$$

Hint: Choose G in the form of (5-63) except $G_{r1} = kD\Gamma_1$ where Γ_1 is obtained from Prob. 2.20 and

$$D = \frac{\epsilon_m(2s+1)}{2\pi M_2(\theta_2)} P_s^m(\cos\theta_0)$$

5.48. Do Prob. 5.47 for $k = 0$. The solution is

$$G_1 = \sum_{s} \sum_{m=0}^{s} \frac{\epsilon_m r^s r^{-s-1}}{2\pi M_2(\theta_2)} P_s^m(\cos\theta)P_s^m(\cos\theta_0)\cos m(\phi - \phi_0),$$

$$0 \leq r \leq r_0$$

$$G_2(r, r_0; \theta, \theta_0; \phi, \phi_0) = G_1(r_0, r; \theta_0, \theta; \phi_0, \phi),$$

$$r_0 \leq r \leq \infty, \; 0 \leq \theta \leq \theta_2 < \pi$$

where the result for Γ_1 in Prob. 2.21 can be used provided $x_1 = 0$ and $x_2 \to \infty$.

5.49. Repeat Prob. 5.47 using the spherical Fourier transform defined in (3-76). The solution is

$$G = -\sum_{s} \sum_{m=0}^{s} \frac{\epsilon_m(2s+1)}{\pi^2 M_2(\theta_2)} \int_0^\infty \frac{j_s(\alpha r)j_s(\alpha r_0)\alpha^2 \, d\alpha}{k^2 - \alpha^2} P_s^m(\cos\theta)P_s^m(\cos\theta_0)$$

$$\times \cos m(\phi - \phi_0), \qquad 0 \leq r \leq \infty, \; 0 \leq \theta \leq \theta_2 < \pi$$

where s and $M_2(\theta_2)$ are defined in Prob. 5.47. The inverse spherical

Fourier transform can be evaluated by equating this result to that obtained in Prob. 5.47.

5.50. Find the Green's function for

$$\left(\nabla^2 + k^2 - \frac{1}{r^2 \sin^2 \theta}\right) G = \frac{-I_m}{r}\, \delta(r - r_0)\, \delta(\theta - \theta_0)$$

outside a perfectly conducting sphere of radius a located at the origin when $\partial(rG)/\partial r = 0$ on the sphere. The excitation is that of a uniform magnetic ring source concentric with the sphere. The result is

$$G_2 = \frac{-jkr_0 I_m}{2} \sum_{n=1}^{\infty} \frac{2n + 1}{n(n + 1)}\left[j_{n0} - \frac{h_{n0}^{(2)}(ka j_{na})'}{(kah_{na}^{(2)})'}\right] h_n^{(2)} P_n^1 P_{n0}^1 \sin \theta_0,$$

$$r_0 \leq r \leq \infty$$

$$G_1(r, r_0; \theta, \theta_0) = G_2(r_0, r; \theta_0, \theta), \qquad 0 < a \leq r \leq r_0$$

where $j_{n0} = j_n(kr_0), j_{na} = j_n(ka)$, etc., and the prime indicates the derivative with respect to the quantity ka. Also, P_{n0}^1 is $P_n^1(\cos \theta_0)$ and P_n^1 is $P_n^1(\cos \theta)$.

5.51. Consider Prob. 5.50 when the magnetic ring source coincides with the surface of the sphere, i.e., $r_0 = a$. The problem is now that of a conducting sphere with a narrow, uniformly excited, circumferential slot. Let $G = A_{m\phi}$ and find the TM_r fields.[60] For this case,

$$H_\phi = -j\omega\epsilon G \quad \text{and} \quad E_\theta = -\frac{1}{r}\frac{\partial}{\partial r}(rG)$$

where

$$G = -\frac{I_m}{2} \sum_{n=1}^{\infty} \frac{2n + 1}{n(n + 1)} \frac{h_n^{(2)}}{(kah_{na}^{(2)})'} P_n^1 P_{n0}^1 \sin \theta_0, \qquad 0 < a \leq r \leq \infty$$

The result for E_θ is that

$$E_\theta = \frac{I_m}{2r} \sum_{n=1}^{\infty} \frac{2n + 1}{n(n + 1)} \frac{(krh_n^{(2)})'}{(kah_{na}^{(2)})'} P_n^1 P_{n0}^1 \sin \theta_0$$

Note that the expression above does not vanish for $r = a$ but becomes the expression for $(I_m/r)\delta(\theta - \theta_0)$ in terms of a series of associated Legendre functions of the first kind of order 1 and degree n.

5.52. Find the Green's function for

$$\left(\nabla^2 + k^2 - \frac{1}{r^2 \sin^2 \theta}\right) G(r, \theta, \phi) = -\frac{I}{r}\delta(r - r_0)\, \delta(\theta - \theta_0)$$

---

[60]See Harrington, *Time-Harmonic Electromagnetic Fields*, p. 303; also Van Bladel, *Electromagnetic Fields*, p. 340. This result differs from Van Bladel who has a sign error on p. 339 in the expression for $E_\theta(a, \theta)$. The result for E_θ in Problem 5.51 also differs from Harrington, p. 303. It is believed that there is an error of π which is introduced in his expression for a_n in (6-124).

in the region between the cones $\theta = \theta_1$, $\theta = \theta_2$ ($\theta_1 < \theta_0 < \theta_2$) when $G = 0$ on the cones. The result is

$$G_1 = -jkr_0 I \sum_s^\infty \frac{(2s+1)h_{s0}^{(2)}j_s}{M_3(\theta_1, \theta_2)} L_s^1 L_{s0}^1 \sin \theta_0,$$

$$0 \le r \le r_0, \, 0 < \theta_1 \le \theta \le \theta_2 < \pi$$

$$G_2(r, r_0; \theta, \theta_0) = G_1(r_0, r; \theta_0, \theta), \qquad r_0 \le r \le \infty$$

where G_r can be found from Prob. 5.47. Also,

$$L_s^1 = P_s^1 Q_{s1}^1 - P_{s1}^1 Q_s^1, \qquad L_{s0}^1 = P_{s0}^1 Q_{s1}^1 - P_{s1}^1 Q_{s0}^1,$$

s is determined from the expression $P_{s1}^1 Q_{s2}^1 - P_{s2}^1 Q_{s1}^1 = 0$, and

$$M_3(\theta_1, \theta_2) = \left[\sin \theta \, \frac{\partial L_s^1}{\partial \theta} \frac{\partial L_s^1}{\partial s} \right]_{\theta_1}^{\theta_2}$$

Since $E_\phi = -j\omega\mu G$, TE_r modes[61] will be excited.

5.53. Repeat Prob. 5.52 using the spherical Fourier transform defined in (3-76). The use of the spherical Fourier transform is the same as that in Prob. 5.49 and leads to the result

$$G = -\frac{2r_0 I}{\pi} \sum_s^\infty \frac{(2s+1)}{M_3(\theta_1, \theta_2)} L_s^1 L_{s0}^1 \sin \theta_0 \int_0^\infty \frac{j_s(\alpha r_0) j_s(\alpha r) \alpha^2 \, d\alpha}{k^2 - \alpha^2}$$

where L_s^1, L_{s0}^1 and $M_3(\theta_1, \theta_2)$ are defined in Prob. 5.52.

5.54. Find the Green's function for

$$\left(\nabla^2 + k^2 - \frac{1}{r^2 \sin^2 \theta} \right) G(r, \theta, \phi) = -\frac{I_m}{r} \delta(r - r_0) \, \delta(\theta - \theta_0)$$

in the region external to the cone $\theta = \theta_2$ ($0 \le \theta_0 \le \theta_2$) when $(\partial/\partial\theta)$ $(G \sin \theta) = 0$ on the cone. The result is

$$G_1 = -jkr_0 I_m \sum_s^\infty \frac{(2s+1)h_{s0}^{(2)}j_s}{s(s+1)M_4(\theta_2)} P_s^1 P_{s0}^1 \sin \theta_0,$$

$$0 \le r \le r_0, \, 0 \le \theta \le \theta_2 < \pi$$

$$G_2(r, r_0; \theta, \theta_0) = G_1(r_0, r; \theta_0, \theta), \qquad r_0 \le r \le \infty$$

where

$$M_4(\theta_2) = \left[\sin \theta P_s^1 \frac{\partial P_s}{\partial s} \right]_{\theta=\theta_2}$$

---

[61]See Harrington, *Time-Harmonic Electromagnetic Fields*, p. 204, for the case of one cone. Our results agree with his except $L_s^1 \rightarrow P_s^1$. Our expression for $N_1(\theta_1, \theta_2)$ also agrees with his if one uses the identity from Appendix C.4.9 that $(\partial L_s^1/\partial\theta) = -s(s+1) L_s$ for this particular case. On the other hand, one can use the orthogonality integral for $dL_s/d\theta$ instead of L_s^1 and obtain the result in terms of $N_2(\theta_1, \theta_2)$ in Appendix C.4.10. Note that Harrington's expression (6-131) contains a minus sign error from the Wronskian. The same sign error also occurs in his expression (6-140).

and s is determined from the expression

$$\frac{\partial}{\partial\theta}(\sin\theta P_s^1) = 0 \quad \text{at} \quad \theta = \theta_2$$

which is equivalent to the simpler condition, $P_{s2} = 0$, i.e., $P_s = 0$ at $\theta = \theta_2$.

5.55. Consider Prob. 5.54 when the magnetic ring source coincides with the surface of the cone, i.e., $\theta_0 = \theta_2$. Let $G = A_{m\phi}$ and find the TM_r fields. This problem is that of a cone with a narrow circumferential slot[62] which is uniformly excited. The result for the only magnetic field component is

$$H_{\phi 1} = -\omega\epsilon k r_0 I_m \sum_s^\infty \frac{(2s+1)h_{s0}^{(2)}j_s}{s(s+1)M_4(\theta_2)} P_s^1 P_{s2}^1 \sin\theta_2$$

$$0 \leq r \leq r_0, \ 0 \leq \theta \leq \theta_2 < \pi$$

$$H_{\phi 2}(r, r_0; \theta, \theta_0) = H_{\phi 1}(r_0, r; \theta_0, \theta), \qquad r_0 \leq r \leq \infty$$

5.56. Obtain the expression for A_ϕ in (5-75) by using the integral expression for A_ϕ in (5-58). First, replace e^{-jkR}/R according to (3-32) and then use the addition theorems for spherical Hankel and Legendre functions so that the integration with respect to ϕ can be performed.

5.57. Repeat Prob. 5.56 for cylindrical coordinates, i.e., replace e^{-jkR}/R according to (3-43) so that cylindrical coordinates result. This procedure should result in (5-55).

---

[62]Harrington, *Time-Harmonic Electromagnetic Fields*, p. 305; also, Van Bladel, *Electromagnetic Fields*, p. 470.

6

Wave Propagation in Unbounded Space

6.1   Plane Wave Propagation in Simple Media

In many problems involving wave propagation, it is expedient to consider unbounded simple media, i.e., media that are both isotropic and homogeneous. Since the simplest solution to the wave equation is a uniform plane wave, let us consider the formalism involved in solving Maxwell's equations (4-4) in an unbounded simple medium without sources. First, we write the electric and magnetic fields as

$$\mathbf{E} = \mathbf{e}_0 e^{-j\mathbf{k}\cdot\mathbf{r}}$$
$$\mathbf{H} = \mathbf{h}_0 e^{-j\mathbf{k}\cdot\mathbf{r}}$$
$$\mathbf{D} = \mathbf{d}_0 e^{-j\mathbf{k}\cdot\mathbf{r}}$$
$$\mathbf{B} = \mathbf{b}_0 e^{-j\mathbf{k}\cdot\mathbf{r}}$$

(6-1)

where

$$\mathbf{k} = k_1\mathbf{a}_x + k_2\mathbf{a}_y + k_3\mathbf{a}_z$$
$$\mathbf{r} = x\mathbf{a}_x + y\mathbf{a}_y + z\mathbf{a}_z$$

and $\mathbf{e}_0$ and $\mathbf{h}_0$ are constant vectors. Both the electric and magnetic fields in

177

(6-1) satisfy the Helmholtz equation so are obtained from Table 5-1B (cartesian coordinates). The only thing new in (6-1) is the manner of expressing a wave propagating in the general direction given by the propagation vector **k**. However, this procedure is the same as that employed in connection with our representation for the three-dimensional Fourier transform pair in (3-25). We might mention that this wave viewpoint allows one to visualize physically the Fourier transform as a process of summing nonuniform plane waves. Further details on this are given in Stratton.[1]

If we substitute the first two expressions in (6-1) into the first two equations in (4-4), we obtain (no sources)

$$\mathbf{e}_0 \cdot \mathbf{k} = 0$$
$$\mathbf{h}_0 \cdot \mathbf{k} = 0 \qquad (6\text{-}2)$$

where we have used the relation $\nabla e^{-j\mathbf{k}\cdot\mathbf{r}} = -j\mathbf{k}e^{-j\mathbf{k}\cdot\mathbf{r}}$ and the fact that $\mathbf{e}_0$ and $\mathbf{h}_0$ are constant vectors. Next, we substitute the same expressions in (6-1) into the last two equations in (4-4) to obtain

$$\mathbf{h}_0 = (\omega\mu)^{-1}\mathbf{k} \times \mathbf{e}_0$$
$$\mathbf{e}_0 = -(\omega\epsilon)^{-1}\mathbf{k} \times \mathbf{h}_0 \qquad (6\text{-}3)$$

If we substitute the expression for $\mathbf{e}_0$ into that for $\mathbf{h}_0$ in (6-3), we find that $k^2 = \omega^2\mu\epsilon$. On the other hand, if we form the dot product $\mathbf{e}_0 \cdot \mathbf{h}_0$ and substitute $\mathbf{h}_0$ from the first expression in (6-3), we find that

$$\mathbf{e}_0 \cdot \mathbf{h}_0 = 0 \qquad (6\text{-}4)$$

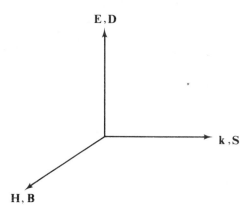

Fig. 6-1   Orientation of field vectors, Poynting vector and propagation vector for simple medium.

[1] J. Stratton, *Electromagnetic Theory*, McGraw-Hill Book Company, Inc., New York, 1941, p. 370.

The two expressions in (6-2) indicate that e_0 and h_0 are both in the plane perpendicular to the direction of propagation while the expression in (6-4) indicates that they are also perpedicular to each other. Since the three vectors **E**, **H**, **k** correspond to a right-hand orthogonal coordinate system, it follows that the Poynting vector $S = E \times H$ is in the direction of **k**. The orientation of these vectors is illustrated in Fig. 6-1.

6.2 Plane Wave Propagation in Anisotropic Electric Media

Let us extend the plane wave analysis carried out in the preceding section to lossless anisotropic media. Also, since anisotropic materials of current interest are anisotropic in either their electric or magnetic properties, it will not be necessary to consider equations so general as those in (4-58) and (4-59). However, we can start with them in their homogeneous form so that we have for electric anisotropy

$$\nabla \times \nabla \times E - k_0^2 \mathcal{K} \cdot E = 0$$
$$\nabla \times (\mathcal{K}^{-1} \cdot \nabla \times H) - k_0^2 H = 0 \tag{6-5}$$

$$D = \epsilon_0 \mathcal{K} \cdot E \tag{4-48}$$

$$B = \mu_0 H \tag{4-2}$$

We have repeated the constitutive relations above to emphasize the fact that for electric anisotropy the relation between **B** and **H** is the same as that for free space. We should also remember that we can solve for either **E** or **H** in (6-5) and find the other fields from Maxwell's equations. Since the representation for $\mathcal{K}$ is given in (4-70) when the magnetostatic field H_0 is in the z-direction, let us leave the direction of the magnetostatic field fixed and assume that a plane wave such as that in (6-1) is propagating at an angle to the z-axis. We define the angle as θ so that it corresponds to the polar angle θ in the spherical coordinate system. If we substitute the expression for **E** from (6-1) into the first equation in (6-5), we obtain

$$k^2 e_0 - k(k \cdot e_0) - k_0^2 \mathcal{K} \cdot e_0 = 0 \tag{6-6}$$

If we substitute the expression for **H** from (6-1) into the second equation in (6-5), we find that we get the more involved expression

$$k \times \mathcal{L} \cdot (k \times h_0) + k_0^2 h_0 = 0$$

In this case, it is easier to solve for e_0 from (6-6) and use Maxwell's equations to solve for **H**. In order to solve (6-6), it is necessary to express the equation in component form. However, let us not follow this approach since it is quite involved. In fact, an easier approach is to substitute $\mathbf{d}_0/\epsilon_0$ for $\mathcal{K} \cdot \mathbf{e}_0$ in the third term of (6-6) and $\mathcal{K}^{-1} \cdot \mathbf{d}_0/\epsilon_0$ for $\mathbf{e}_0$ in the first term of (6-6) which allows one to solve for $\mathbf{d}_0$ in terms of $\mathbf{k} \cdot \mathbf{e}_0$ times an involved expression. This is the procedure followed by Collin[2] and Stratton[3] except they first diagonalize $\mathcal{K}$ and then let the principal axes of the material coincide with the coordinate axes. Nevertheless, let us continue with our procedure, even though the resultant expressions will be more involved, to obtain the desired expression for $\mathbf{d}_0$, i.e.,

$$
\mathbf{d}_0 = \epsilon_0(\mathbf{k} \cdot \mathbf{e}_0)\left[\frac{(k^2\ell_1 - k_0^2)(k_1\mathbf{a}_x + k_2\mathbf{a}_y) + k^2\ell_2(k_2\mathbf{a}_x - k_1\mathbf{a}_y)}{(k^2\ell_1 - k_0^2)^2 + (k^2\ell_2)^2} \right.
$$
$$
\left. + \frac{k_3\mathbf{a}_z}{k^2\ell_3 - k_0^2} \right] \tag{6-7}
$$

We will come back to this expression later after we diagonalize $\mathcal{K}$ and then work the problem in the principal axes coordinate system.

The information that we desire now is the orientation of the field vectors and the propagation vector with respect to each other so that we can construct a diagram similar to Fig. 6-1. If we dot multiply (6-6) from the left by $\mathbf{k}$, the first two terms vanish and we find that $\mathbf{k} \cdot \mathbf{d}_0 = 0$, so that $\mathbf{d}_0$ is in the plane perpendicular to $\mathbf{k}$, the direction of wave propagation. On the other hand, $\mathbf{b}$ and $\mathbf{h}$ are parallel and are both in the plane perpendicular to $\mathbf{k}$ since the second relation in (6-2) is still true. Also, the first relation in (6-3) is still valid but the second relation becomes

$$
-\omega\mathbf{d}_0 = \mathbf{k} \times \mathbf{h}_0 \tag{6-8}
$$

The information that we now have indicates that the three vectors $\mathbf{D}$, $\mathbf{H}$, $\mathbf{k}$ form a right-hand orthogonal coordinate system. We also know from (6-1) that $\mathbf{d}_0$ and $\mathbf{e}_0$ are not parallel unless $\mathbf{k}$ and $\mathbf{e}_0$ are perpendicular, i.e., $\mathbf{k} \cdot \mathbf{e}_0 = 0$. However, the first relation in (6-3) indicates that $\mathbf{E}$ is in the plane containing $\mathbf{k}$ and $\mathbf{D}$, i.e., the plane perpendicular to $\mathbf{H}$. The last item of information we need is the direction of power flow, which we obtain from the Poynting vector. Thus,

$$
\mathbf{S} = \mathbf{E} \times \mathbf{H}^* = \mathbf{e}_0 \times \mathbf{h}_0^* = (\omega\mu_0)^{-1}[\mathbf{k}e_0^2 - \mathbf{e}_0(\mathbf{k} \cdot \mathbf{e}_0)]
$$

where for simplicity we have assumed that all of the quantities in (6-1) are real.[4] If we solve for the angle between $\mathbf{S}$ and $\mathbf{k}$, denoted by α, we find that

---

[2]R. Collin, *Field Theory of Guided Waves*, McGraw-Hill Book Company, Inc., New York, 1960, p. 97.

[3]Stratton, *Electromagnetic Theory*, p. 341.

[4]For the quantities complex, we have
$$
\mathbf{S} = \mathbf{e}_0 \times \mathbf{h}_0^* = (\omega\mu_0)^{-1}[\mathbf{k}^*e_0^2 - \mathbf{e}_0^*(\mathbf{k}^* \cdot \mathbf{e}_0)]
$$

$$\cos \alpha = \frac{\mathbf{S} \cdot \mathbf{k}}{Sk} = \left[1 - \left| \frac{\mathbf{k} \cdot \mathbf{e}_0}{k e_0} \right|^2 \right]^{1/2} \tag{6-9}$$

or

$$\sin \alpha = \frac{\mathbf{k} \cdot \mathbf{e}_0}{|k e_0|}$$

On the other hand, if we solve for the angle between $\mathbf{D}$ and $\mathbf{E}$, we find that it is the same angle α, i.e.,

$$\cos \alpha = \frac{\mathbf{D} \cdot \mathbf{E}}{DE}$$

The orientation of each vector is shown in Fig. 6-2. One sees that the vectors $\mathbf{D}$, $\mathbf{E}$, $\mathbf{k}$, and $\mathbf{S}$ are all in the same plane perpendicular to $\mathbf{H}$ and $\mathbf{B}$. In general, the wave will be a TM_k wave or possibly a TEM_k wave but never a TE_k wave.[5]

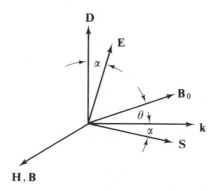

Fig. 6-2 Orientation of field vectors, Poynting vector and propagation vector for anistotropic electric medium. $\mathbf{D}$, $\mathbf{E}$, $\mathbf{k}$, $\mathbf{S}$ are coplanar. The vectors $\mathbf{D}$, $\mathbf{H}$, $\mathbf{k}$ form a right-hand orthogonal coordinate system and the wave can never be a TE_k wave.

In order to get further details concerning plane wave propagation in this anisotropic electric material, let us return to the problem of solving (6-6). As mentioned previously, it will be necessary to express the equation in component form so we let

$$\mathbf{e}_0 = e_x \mathbf{a}_x + e_y \mathbf{a}_y + e_z \mathbf{a}_z$$

---

[5]C. Papas, *Theory of Electromagnetic Wave Propagation*, McGraw-Hill Book Company, Inc., New York, 1965, p. 199.

Now, the vector equation (6-6) reduces to three scalar equations for e_x, e_y, and e_z. Thus, in matrix form

$$
\begin{bmatrix}
k^2 - k_1^2 - k_0^2 g_1 & -k_1 k_2 - k_0^2 g_2 & -k_1 k_3 \\
-k_1 k_2 + k_0^2 g_2 & k^2 - k_2^2 - k_0^2 g_1 & -k_2 k_3 \\
-k_1 k_3 & -k_2 k_3 & k^2 - k_3^2 - k_0^2 g_3
\end{bmatrix}
\begin{bmatrix}
e_x \\
e_y \\
e_z
\end{bmatrix} = 0 \qquad (6\text{-}10)
$$

This equation is too difficult to solve in its present form so we let one of the components of $\mathbf{k}$ in the plane transverse to z vanish, e.g., $k_2 = 0$. We can now make the substitution

$$
\begin{aligned}
k_3 &= k \cos \theta \\
k_1 &= k \sin \theta
\end{aligned}
\qquad (6\text{-}11)
$$

in the above expression and solve for k to obtain the two positive values

$$
k = k_0 \left[\frac{[(g_1^2 + g_2^2 - g_1 g_3) \sin^2 \theta + 2 g_1 g_3] + [(g_1^2 + g_2^2 - g_1 g_3)^2 \sin^4 \theta - 4 g_2^2 g_3^2 \cos^2 \theta]^{1/2}}{2(g_1 - g_3) \sin^2 \theta + g_3} \right]^{1/2} \qquad (6\text{-}12)
$$

The two cases of particular interest occur when $\theta = 0$ or $\pi/2$, so that the wave propagation direction is parallel or perpendicular to $\mathbf{H}_0$, respectively. If we let $\theta = 0$ in (6-12), we obtain

$$
k = k_0 (g_1 \pm j g_2)^{1/2}, \qquad \theta = 0 \qquad (6\text{-}13)
$$

whereas, for $\theta = \pi/2$, we obtain for the plus sign

$$
k = k_0 \left(\frac{g_1^2 + g_2^2}{g_1} \right)^{1/2} \qquad (6\text{-}14)
$$

and for the minus sign, with $\theta = \pi/2$,

$$
k = k_0 (g_3)^{1/2} \qquad (6\text{-}15)
$$

Let us examine the fields for the case where $\theta = 0$. Thus, if we return to (6-10) and (6-11) and use the values for k and θ from (6-13), we find that

$$
e_z = 0
$$

$$
\frac{e_y}{e_x} = \frac{k^2 - k_0^2 g_1}{k_0^2 g_2} = \pm j \qquad (6\text{-}16)
$$

where the $\pm$ signs above coincide with those in (6-13). If we consider the complex electric field

$$
\mathbf{E} = \mathbf{e}_0 e^{-jkz} e^{i\omega t} \qquad (6\text{-}17)
$$

with $\mathbf{e}_0 = = e_x(\mathbf{a}_x \pm j\mathbf{a}_y)$, the real part (for $z = 0$ and e_x real) is

$$\mathbf{E}' = e_x(\mathbf{a}_x \cos \omega t \mp \mathbf{a}_y \sin \omega t)$$

The vector $\mathbf{E}'$ is constant in magnitude as a function of time and rotates in the negative or positive ϕ-direction, according to the upper or lower sign in (6-17), respectively. The angle $\phi = \omega t$ is measured from the positive x-axis toward the positive y-axis. If a direct current were to flow in the positive ϕ-direction in a loop in the x-y plane, it would create a magnetostatic field in the positive z-direction. Thus, we will define our positive circularly polarized wave as one where the electric vector rotates in the positive ϕ-direction. This direction would be counterclockwise looking in the negative z-direction, i.e., the direction from which the wave is coming, and would be clockwise looking in the positive z-direction, i.e., the direction in which the wave is going. Unfortunately, this definition is arbitrary so one should be careful in examining various texts. Our definition of a positive circularly polarized wave corresponds to the definition Collin[6] uses for a right-circular-polarized wave. Moreover, it is the reverse of that used by Stratton[7] but the same as that used by Katz[8] and Lax and Button.[9]

Since $\mathbf{D} = \epsilon_0 \mathfrak{K} \cdot \mathbf{E}$ from (4-48), we can solve for $\mathbf{d}_0$ as

$$\mathbf{d}_0 = \epsilon_0 \mathfrak{K} \cdot \mathbf{e}_0 = d_x \mathbf{a}_x + d_y \mathbf{a}_y$$

where $d_x = \epsilon_0(g_1 \pm jg_2)e_x$

$$d_y = \epsilon_0(g_1 \pm jg_2)e_y = \pm j\epsilon_0(g_1 \pm jg_2)e_x$$

The expression for $\mathbf{d}_0$ now becomes

$$\mathbf{d}_0 = \epsilon_0(g_1 \pm jg_2)(\mathbf{a}_x \pm j\mathbf{a}_y)e_x$$
$$= \epsilon_0(g_1 \pm jg_2)\mathbf{e}_0$$

where the form of $\mathbf{e}_0$ is given in (6-17). From (6-8) we obtain $\mathbf{h}_0$ as

$$\mathbf{h}_0 = h_x \mathbf{a}_x + h_y \mathbf{a}_y$$

[6]Collin, *Field Theory of Guided Waves*, p. 71.

[7]Stratton, *Electromagnetic Theory*, p. 280.

[8]H. Katz, ed., *Solid State Magnetic and Dielectric Devices*, John Wiley & Sons, Inc., New York, 1959, p. 280.

[9]B. Lax and K. Button, *Microwave Ferrites and Ferrimagnetics*, McGraw-Hill Book Company, Inc., New York, 1962, p. 299.

where

$$h_x = -\omega k^{-1} d_y = -\omega k^{-1} \epsilon_0 (g_1 \pm jg_2) e_y$$

$$h_y = \omega k^{-1} d_x = \omega k^{-1} \epsilon_0 (g_1 \pm jg_2) e_x$$

Since k is given by (6-13), we find that

$$\frac{h_x}{h_y} = \frac{-e_y}{e_x} = \frac{b_x}{b_y} = \mp j$$

where we have used (6-16) to obtain the latter expression. Since e_0 above is given by the specific expression in (6-17) and represents a circularly polarized wave, let us indicate this with the $\mp$ superscripts. We can now summarize the preceding results to obtain the following:

$$\mathbf{E}^{\mp} = \mathbf{e}_0^{\mp} e^{j(\omega t - k \mp z)}$$

$$\mathbf{D}^{\mp} = \epsilon_0 (g_1 \pm jg_2) \mathbf{E}^{\mp}$$

$$\mathbf{B}^{\mp} = \mu_0 \mathbf{H}^{\mp}$$

$$\mathbf{H}^{\mp} = \mp j \sqrt{\frac{\epsilon_0}{\mu_0}} (g_1 \pm jg_2)^{1/2} \mathbf{E}^{\mp} = \mp j Y^{\mp} \mathbf{E}^{\mp} \qquad (6\text{-}18)$$

$$Y^{\mp} = Y_0 \sqrt{\epsilon_{\text{eff}}^{\mp}}$$

$$\mathbf{e}_0^{\mp} = e_x (\mathbf{a}_x \pm j\mathbf{a}_y)$$

$$\epsilon_{\text{eff}}^{\mp} = g_1 \pm jg_2 = 1 - \frac{(f_p/f)^2}{1 \pm (f_g/f)}$$

$$k^{\mp} = k_0 \sqrt{\epsilon_{\text{eff}}^{\mp}}$$

The most noticeable feature of the equations in (6-18), other than the fact that the wave is a *TEM* wave, is the relation between $\mathbf{D}^{\mp}$ and $\mathbf{E}^{\mp}$. Rather than the general dyadic relation of (4-48), we now have a scalar relation. However, this scalar relation holds only as long as the electric field transverse to the magnetostatic field is circularly polarized. Since the relation between $\mathbf{D}^{\mp}$ and $\mathbf{E}^{\mp}$ is a scalar one, it means that the dyadic in (4-48) or in (4-70) has been diagonalized. As a result, $\mathbf{D}^{\mp}$ and $\mathbf{E}^{\mp}$ are the normal modes or eigenvectors and the diagonal quantities are the eigenvalues. Also, since the dyadic in (4-70) is hermitian when losses are neglected, the normal modes are orthogonal in the hermitian sense, i.e., $\mathbf{D}^{-} \cdot \mathbf{D}^{+*} = \delta_{-+}$, and the eigenvalues $g_1 \pm jg_2$ are real. This latter point is obvious since g_2 in (4-71) or (4-72) is an imaginary quantity. Another point of interest is the fact that the quantity $g_1 + jg_2$ varies smoothly as a function of frequency whereas $g_1 - jg_2$ becomes infinite when f approaches f_g. This is illustrated in Fig. 6-3 for values of f_p and f_g that correspond to those encountered in the ionosphere. The pole in the quantity $g_1 - jg_2$ is caused by the electrons resonating at the gyrofrequency (cyclotron frequency). This can be seen physically since the orbital

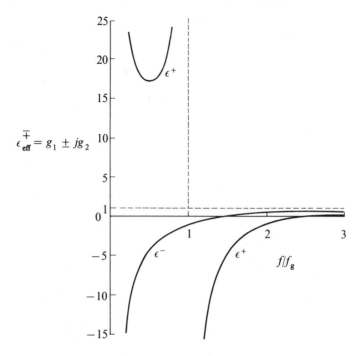

Fig. 6-3   Plot of $\epsilon^{\mp}$ versus f/f_g for plane wave propagation parallel to the dc magnetic field in the ionosphere.

motion of an electron traveling in the $x - y$ plane with a magnetostatic field in the positive z-direction is in the positive ϕ-direction, where ϕ is measured from the positive x-axis toward the positive y-axis (usual cylindrical coordinate system). This is the same direction as the one we used to define a positive circularly polarized wave. This means that the positive circularly polarized component of an incident wave interacts strongly with the electrons in the ionosphere unless the frequency is somewhat larger than the gyrofrequency. If one considers the propagation constants given in (6-18) for the positive and negative circularly polarized waves and the curves of $g_1 \pm jg_2$ in Fig. 6-3, it is clear that both waves propagate when $g_1 \pm jg_2$ is positive and are evanescent (nonabsorptively damped)[10] when $g_1 \pm jg_2$ is negative. The quantity $g_1 \pm jg_2$ becomes zero when

$$f^{\mp} = \frac{f_g}{2}\left\{\left[1 + \frac{2f_p^2}{f_g}\right]^{1/2} \mp 1\right\}$$ (6-19)

---

[10]See Papas, *Theory of Electromagnetic Wave Propagation*, p. 184. The term evanescent is used to describe the spatial damping that occurs when a wave is attenuated in a lossless medium. Common examples are filter action in the stopband and waveguide propagation at frequencies below the modal cutoff frequency.

One sees from Fig. 6-3 that the positively polarized wave possesses a low-frequency passband below f_g whereas both the positively and negatively polarized waves behave as an ordinary waveguide mode at frequencies above f^+ and f^-, respectively. Further information on the propagation characteristics of these waves as well as the ionospheric phenomena of "whistlers" is covered briefly in Johnson.[11]

A procedure that is often used whenever a dyadic quantity is encountered is to diagonalize the dyadic by effecting a rotation of the coordinate axes to the principal axes. In most physical problems the angular rotation is real, e.g., a problem involving a dyadic moment of inertia. However, we will find that with a ferrite or plasma the angular rotation, the principal axes, and the normal modes are all complex. Let us consider in more detail the diagonalization[12] of (4-70). Since the dyadic in (4-70) contains so many zero elements, it is only necessary to diagonalize the part that contains nondiagonal elements. To obtain the two eigenvalues, λ_i, $i = 1, 2$, we solve the equation

$$\begin{vmatrix} g_1 - \lambda & g_2 \\ -g_2 & g_1 - \lambda \end{vmatrix} = 0$$

which gives us

$$\lambda_1 = g_1 - jg_2$$
$$\lambda_2 = g_1 + jg_2 \tag{6-20}$$

To find the principal axes defined by the unit vectors $\mathbf{a}_i$, $i = 1, 2$, we form the dot product $(\mathbf{G} - \lambda_i \mathbf{\mathcal{I}}) \cdot \mathbf{a}_i = 0$. This yields the two equations

$$\begin{bmatrix} g_1 - \lambda_i & g_2 \\ -g_2 & g_1 - \lambda_i \end{bmatrix} \begin{bmatrix} a_{ix} \\ a_{iy} \end{bmatrix} = 0, \qquad i = 1, 2$$

and the solutions

$$\mathbf{a}_1 = a_{1x}(\mathbf{a}_x - j\mathbf{a}_y)$$
$$\mathbf{a}_2 = a_{2x}(\mathbf{a}_x + j\mathbf{a}_y)$$

Since $\mathbf{a}_1$ and $\mathbf{a}_2$ are unit vectors, it follows that a_{1x} and a_{2x} must have the value of $1/\sqrt{2}$. Thus, the final expressions are

$$\mathbf{a}_1 = \frac{1}{\sqrt{2}}(\mathbf{a}_x - j\mathbf{a}_y)$$

$$\mathbf{a}_2 = \frac{1}{\sqrt{2}}(\mathbf{a}_x + j\mathbf{a}_y) \tag{6-21}$$

---

[11]C. C. Johnson, *Field and Wave Electrodynamics*, McGraw-Hill Book Company, Inc., New York, 1965, pp. 389–90.

[12]Collin, *Field Theory of Guided Waves*, p. 569.

$$\mathbf{a}_x = \frac{1}{\sqrt{2}}(\mathbf{a}_1 + \mathbf{a}_2)$$

$$\mathbf{a}_y = \frac{1}{\sqrt{2}}(\mathbf{a}_1 - \mathbf{a}_2)$$

and

$$x_1 = \frac{1}{\sqrt{2}}(x - jy)$$

$$x_2 = \frac{1}{\sqrt{2}}(x + jy)$$

where (x_1, x_2, z) are the principal coordinates.

Before we examine the fields for the case where $\theta = \pi/2$, let us briefly consider the Faraday rotation that takes place when a plane wave propagates in a direction parallel to the magnetostatic field. We can form a suitable plane wave by superposing $\mathbf{E}^-$ and $\mathbf{E}^+$ in (6-18) to obtain

$$\mathbf{E} = e_x[(\mathbf{a}_x + j\mathbf{a}_y)e^{-jk^-z} + (\mathbf{a}_x - j\mathbf{a}_y)e^{-jk^+z}] = E_x\mathbf{a}_x + E_y\mathbf{a}_y$$

When z is zero, E_y is zero and the wave is polarized in the x-direction. For nonzero values of z the angular rotation of the polarization with respect to the z-axis is

$$\phi = (k^- - k^+)\frac{z}{2} \qquad (6\text{-}22)$$

which can be obtained by recognizing that $\tan \phi = E_y/E_x$. Since k^- is always larger than k^+ for frequencies larger than f^+ as one sees by inspection of Fig. 6-3, it is obvious that the angular rotation of the polarization is always positive for positively increasing z. It is also positive for a wave traveling in the negative z-direction since z in (6-22) changes sign for this latter case. This means that a linearly polarized wave passing through the plasma and then being reflected experiences an angular rotation that is double that of a single passage. With Faraday rotation, then, the angular rotation of the polarization of the wave continually increases when the wave is reflected back and forth through the anisotropic medium. Thus, the medium is not reciprocal. We can also expand the terms k^- and k^+ in (6-22) by assuming that f_p/f and f_g/f in (6-18) are much smaller than unity. This gives us

$$\phi \simeq \frac{ze\mu_0}{2mc}\left(\frac{f_p}{f}\right)^2 H_0 \qquad (6\text{-}23)$$

which indicates that the angular rotation of the polarization of the wave increases linearly with the magnetostatic field and reverses when the direction of the magnetostatic field reverses. If we put numbers into (6-23) it becomes

$$\phi = 1.68\left(\frac{f_p}{f}\right)^2 z H_0 \text{ degree/meter-oersted} \qquad (6\text{-}24)$$

If we consider a value for H_0 of 0.5 oe(oersted) and a wave frequency ten times larger than the plasma frequency, the polarization rotates one degree every 120 meters.

Let us now return to the expressions in (6-14) and (6-15) to examine the fields for the case where $\theta = \pi/2$. In particular, it is evident that the propagation constant in (6-15) is independent of the anisotropy of the plasma. As a result, we would expect the fields to behave as if the material were isotropic so that we would have ordinary wave propagation. Since this is true, as we shall now show, this wave is called the ordinary wave. If we substitute the value for k from (6-15) into (6-10) and use the values for k_1 and k_3 from (6-11), we find that e_x and e_y are zero unless $g_1^2 - g_2 g_3 + g_2^2 = 0$. However, this is generally not valid (if at all) so one must conclude that e_x and e_y vanish. Since the coefficient of e_z is zero, it follows that e_z is the only nonzero component of the electric field and $d_z = \epsilon_0 g_3 e_z$ is the only nonzero component of the electric flux density vector. From (6-8) we find that h_y is the only nonzero component of the magnetic field and can be written as

$$h_y = -\sqrt{\frac{\epsilon_0}{\mu_0}} \sqrt{g_3} \, e_z \qquad (6\text{-}25)$$

Thus, the ordinary wave is a linearly polarized TEM_x wave with its electric vector parallel to $\mathbf{H}_0$ and it behaves as if it were propagating in an isotropic medium with constants μ_0 and $\epsilon_0 g_3$. The orientation of the field vectors, Poynting vector, and propagation vector is illustrated in Fig. 6-4.

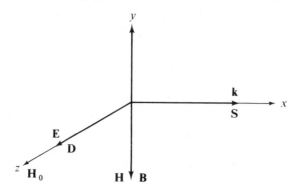

Fig. 6-4   Orientation of field vectors, Poynting vector and propagation vector for the ordinary wave propagating in anisotropic electric medium *Note:* The wave is TEM_x.

The wave that has its propagation constant given by (6-14) is called the extraordinary wave. Again, we find the electric field components from (6-10) and (6-11) and the magnetic field components from (6-8). This gives us the following:

$$e_z = 0$$

$$\frac{e_y}{e_x} = -\frac{g_1}{g_2}$$

$$d_x = d_z = 0$$

$$d_y = \left(\frac{\epsilon_0 k^2}{k_0^2}\right) e_y$$

$$h_x = h_y = b_x = b_y = 0 \qquad\qquad (6\text{-}26)$$

$$h_z = \sqrt{\frac{\epsilon_0}{\mu_0}}\sqrt{\epsilon_{\text{eff}}}\, e_y$$

$$\mathbf{k} = k\mathbf{a}_x$$

$$k = k_0\sqrt{\epsilon_{\text{eff}}}$$

$$\epsilon_{\text{eff}} = \frac{g_1^2 + g_2^2}{g_1} = \frac{f^2 - f_g^2 - 2f_p^2 + (f_p^2/f)^2}{f^2 - f_g^2 - f_p^2}$$

Since propagation is in the x-direction, the extraordinary wave is a TM_x wave. By inspecting Fig. 6-2 one sees that $\mathbf{H}$ is always in the plane perpendicular to the direction of propagation so that the wave for a general angle θ is never a TE wave. Rather, it is a TM wave that reduces to a TEM wave for certain values of θ. The orientation of the field vectors, Poynting vector, and propagation vector of this wave is illustrated in Fig. 6-5. The coefficient of k_0 in (6-26) behaves as an effective dielectric constant and is plotted as a function of f/f_g in Fig. 6-6 for the case where $f_p = 2.8$ MHz and $f_g = 1.4$ MHz. These values of f_p and f_g correspond to typical values for ionospheric propagation and are the same as those used in plotting Fig. 6-3. The pole in the figure is located at $f = (f_g^2 + f_p^2)^{1/2} = \sqrt{5}\, f_g$ and the two zeros at

$$f_{1,2} = \frac{f_g}{2}\left\{\pm 1 + \left[1 + \left(\frac{2f_p}{f_g}\right)^2\right]^{1/2}\right\}$$

or for our particular choice of constants

$$\frac{f_{1,2}}{f_g} = 1.56, 2.56$$

The effective dielectric constant is positive for values of f/f_g between f_1 and the pole at $(f_g^2 + f_p^2)^{1/2}$ and again for values of f/f_g larger than f_2. The former

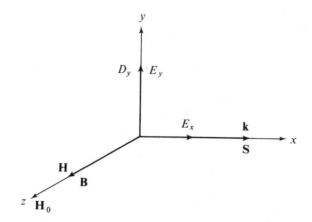

Fig. 6-5 Orientation of field vectors, Poynting vector and propagation vector for the extraordinary wave propagating in anisotropic electric medium *Note:* The wave is TM_x and $D_x = 0$.

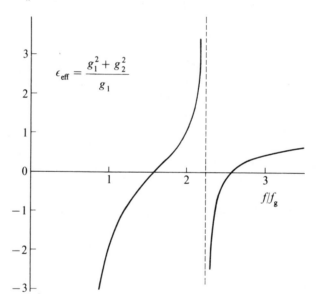

Fig. 6-6 Plot of ϵ_{eff} versus f/f_g for extraordinary plane wave propagating perpendicular to the dc magnetic field in the ionosphere.

constitutes a low-frequency passband and the latter a high-frequency all-pass region.

It should be noted that both Figs. 6-3 and 6-6 are plotted as a function of

f/f_g which is a useful form for the curves when f_g is a constant. However, many laboratory experiments involving plasmas require that the frequency be held constant and f_g be the variable. In this case, it is more useful to plot the preceding figures as a function of f_g/f.

Let us now return to the expression for $\mathbf{d}_0$ in (6-7) and show how it can be considerably simplified when one uses the principal axes coordinate system. In fact, we mentioned previously in connection with the derivation of (6-7) that it is customary in deriving (6-7) to assume that the cartesian coordinate system coincides with the principal axes coordinate system so that the dyadic $\mathcal{K}$ is diagonal, i.e., its three diagonal elements are nonzero and its nondiagonal elements are zero. In order to simplify (6-7), we need to replace the unit vectors $\mathbf{a}_x$ and $\mathbf{a}_y$ by the unit vectors $\mathbf{a}_1$ and $\mathbf{a}_2$ as given by (6-21). However, the expression that we then obtain is still not in its final form since the propagation vector $\mathbf{k}$ also has to be transformed to the new coordinate system. This is easily done by using the definition for $\mathbf{k}$ from (6-1), i.e.,

$$\mathbf{k} = k_1\mathbf{a}_x + k_2\mathbf{a}_y + k_3\mathbf{a}_z = k'_1\mathbf{a}_1 + k'_2\mathbf{a}_2 + k'_3\mathbf{a}_z$$

and solving for k'_1 and k'_2. (We note that the k_i and k'_i are examples of variant scalars mentioned in Sec. 4.3.) When we do this we find that

$$k'_1 = \tfrac{1}{2}(k_1 + jk_2)$$
$$k'_2 = \tfrac{1}{2}(k_1 - jk_2)$$

and the expression for $\mathbf{d}_0$ in (6-7) becomes

$$\mathbf{d}_0 = \epsilon_0(\mathbf{k} \cdot \mathbf{e}_0)\left[\frac{k'_1\mathbf{a}_1}{k^2(\ell_1 + j\ell_2) - k_0^2} + \frac{k'_2\mathbf{a}_2}{k^2(\ell_1 - j\ell_2) - k_0^2} + \frac{k_3\mathbf{a}_z}{k^2\ell_3 - k_0^2}\right]$$

$$\text{(6-27)}$$

In order to simplify this expression further, we need to recognize that $\ell_1 \pm j\ell_2$ are the eigenvalues of the matrix $[L]$, given by (4-73), and are just the inverse of the eigenvalues of the matrix $[G]$, given by (4-70). However, we have already solved for the eigenvalues of (4-70) in (6-20) so that we can write their inverse as

$$\ell_1 + j\ell_2 = \frac{1}{\lambda_1}$$
$$\ell_1 - j\ell_2 = \frac{1}{\lambda_2} \qquad \text{(6-28)}$$
$$\ell_3 = \frac{1}{g_3}$$

If we use the above relations in (6-27) we get the desired expression[13] for d_0, i.e.,

$$d_0 = \epsilon_0(k \cdot e_0)\left[\frac{k_1'\lambda_1 a_1}{k^2 - \lambda_1 k_0^2} + \frac{k_2'\lambda_2 a_2}{k^2 - \lambda_2 k_0^2} + \frac{k_3 g_3 a_z}{k^2 - g_3 k_0^2}\right] \quad (6-29)$$

6.3 Plane Wave Propagation in Anisotropic Magnetic Media

Since we carried through many of the detailed steps in Sec. 6.2, it will not be necessary to repeat them here. Thus, emphasis will be placed on outlining the general procedure and stating the final results.

The homogeneous equations that we wish to solve are obtained from (4-58) and (4-59) so that for the case of magnetic anisotropy we have

$$\nabla \times (\mathfrak{K}_m^{-1} \cdot \nabla \times E) - k_0^2 K E = 0$$
$$\nabla \times \nabla \times H - k_0^2 K \mathfrak{K}_m \cdot H = 0 \quad (6-30)$$

$$B = \mu_0 \mathfrak{K}_m \cdot H \quad (4-49)$$

$$D = \epsilon_0 K E$$

The constitutive relations have been repeated to emphasize that in the case of magnetic anisotropy the relation between D and E is the same as that for a dielectric material. We will use the representation for $\mathfrak{K}_m$ given in (4-70) so that the magnetostatic field H_0 is in the z-direction. Again, we choose a plane wave as defined in (6-1) propagating at the angle θ with respect to the z-axis. If we substitute H into the second expression in (6-30), we obtain

$$k^2 h_0 - k(k \cdot h_0) - k_0^2 K \mathfrak{K}_m \cdot h_0 = 0 \quad (6-31)$$

Inspection of (6-31) indicates that $k \cdot b_0 = 0$ so that b_0 is in the plane perpendicular to k. Since the first relation in (6-2) is still valid, E and D are parallel and in the plane perpendicular to k. The first relation in (6-3) is no longer valid but becomes

$$\omega b_0 = k \times e_0 \quad (6-32)$$

However, the second relation in (6-3) is still valid so that E and D are perpendicular to the plane containing k and h_0. Since E and k are perpendicular, it follows from (6-32) that the vectors B, k, E form a right-hand orthogonal coordinate system and that H is in the plane containing B and k. The orientation of these vectors is shown in Fig. 6-7 along with the Poynting vector S. The Poynting vector S is given as

---

[13]*Ibid.*, Chap. 3, Eq. 105.

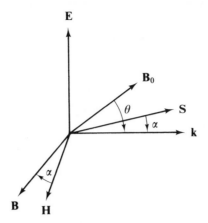

Fig. 6-7 Orientation of field vectors, Poynting vector and propagation vector for anisotropic magnetic medium. **B, H, k, S** are coplanar. The vectors **B, k, E** form a right-hand orthogonal coordinate system and the wave can never be a TM_k wave.

$$\mathbf{S} = \mathbf{E} \times \mathbf{H^*} = \mathbf{e}_0 \times \mathbf{h}_0^* = (\omega \epsilon_0 K)^{-1} [k h_0^2 - \mathbf{h}_0^*(\mathbf{k \cdot h_0})] \qquad (6\text{-}33)$$

and the angle between **S** and **k**, again called α, can be found as

$$\cos \alpha = \frac{\mathbf{S \cdot k}}{Sk} = \left[1 - \left| \frac{\mathbf{k \cdot h_0}}{k h_0} \right|^2 \right]^{1/2} = \frac{\mathbf{B \cdot H}}{BH} \qquad (6\text{-}34)$$

Since **E** is always perpendicular to **k**, the wave will be a TE_k wave or possibly a TEM_k wave but never a TM_k wave. This result is just the dual of that shown in Fig. 6-2 for the wave in an anisotropic electric medium. In fact, as one might expect, the wave in Fig. 6-7 is just the dual of the wave in Fig. 6-2. This fact can be verified by referring to Table 4-1.

In order to solve (6-31) we express $\mathbf{h}_0$ in component form in a cartesian coordinate system. However, rather than proceed in this fashion, it is simpler to use the fact that this wave is the dual of the wave in Sec. 6.2 so that the results can be obtained by inspection. Thus, the expressions dual to (6-18) for the case where $\theta = 0$ are the following:

$$\mathbf{H}^{\mp} = \mathbf{h}_0^{\mp} e^{j(\omega t - k^{\mp} z)}$$

$$\mathbf{B}^{\mp} = \mu_0 \mu^{\mp} \mathbf{H}^{\mp}$$

$$\mathbf{D}^{\mp} = \epsilon_0 K \mathbf{E}^{\mp}$$

$$\mathbf{E}^{\mp} = j \sqrt{\frac{\mu_0}{\epsilon_0}} \sqrt{\mu^{\mp}} \, \mathbf{H}^{\mp} = \pm j Z^{\mp} \mathbf{H}^{\mp}, \qquad Z^{\mp} = Z_0 \sqrt{\mu^{\mp}} \qquad (6\text{-}35)$$

$$\mathbf{h}_0^{\mp} = h_x(\mathbf{a}_x \pm j\mathbf{a}_y)$$

$$k^{\mp} = k_0\sqrt{K}\,\sqrt{\mu^{\mp}}$$

$$\mu^{\mp} = g_1 \pm jg_2 = 1 + f_m(f_g \pm f)^{-1}$$

The effective permeability, $\mu^{\mp} = g_1 \pm jg_2$, is plotted in Fig. 6-8 as a function of f_g/f for a value of f_m that corresponds to that encountered in a polycrystalline YIG ferrite, i.e., $M_0 = 4\pi M_s = 1750$ oe, so that $f_m = 4.9$ GHz (gigahertz). The value of f was chosen as 9.8 GHz. The abscissa in this case

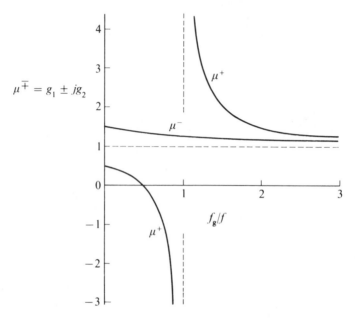

Fig. 6-8   Plot of $\mu^{\mp}$ versus f_g/f for plane wave propagation parallel to the dc magnetic field in YIG ferrite at 9.8 GHz.

was chosen as f_g/f since in many experiments the microwave frequency f is held fixed and the gyrofrequency f_g is varied by changing the applied magnetostatic field H_0. This is in contrast to the effective dielectric constant of an ionospheric plasma plotted in Fig. 6-3 where the abscissa is f/f_g. One sees from inspecting Fig. 6-8 that the effective permeability μ^- for the negative circularly polarized wave is a slowly varying function of f_g/f and decreases from 1.5 to 1 as the abscissa increases from zero to large values. However, the effective permeability μ^+ for the positive circularly polarized wave is initially small and positive and vanishes at $f_g/f = 1 - (f_m/f)$. A stopband then exists until the pole is reached at $f_g/f = 1$. A passband exists for all larger values of f_g/f. This behavior is similar to that encountered with the plasma in Fig. 6-3.

However, the phenomenon is entirely different since the bound electrons in the ferrite precess around the magnetostatic field H_0 whereas in the plasma the free electrons orbit around the field H_0. The former motion is referred to as Larmor precession, whereas the latter motion is referred to as cyclotron resonance.[14]

To consider Faraday rotation in the anisotropic ferrite medium, we use the expression in (6-22) as before and the values of k^+ and k^- from (6-35) for the case where $f_m/f < 1$ and $f_g/f < 1$. Although this situation is not ordinarily fulfilled in practice, it does allow one to determine whether the Faraday rotation encountered by a plane wave in passing through a microwave ferrite is of practical significance. Thus, the desired approximate expression for the angular rotation of the polarization of the plane wave with respect to the x-axis is

$$\phi = \frac{\sqrt{K}\,\omega_m}{2c}z \qquad (6\text{-}36)$$

If we assume that ω_m has the same value as that used in Fig. 6-8 and that $K = 16$, we find that the polarization vector rotates 126° every centimeter.[15] Since this rotation is so large, it indicates that practical devices utilizing Faraday rotation are quite feasible provided that losses within the material are small. These requirements are easily satisfied by modern synthetic microwave ferrites. The first practical device that utilized Faraday rotation was manufactured by Cascade Research Corporation and was termed an "uniline" since it acted as a one-way transmission line. The concept of a one-way transmission line utilizing Faraday rotation was first introduced by Lord Rayleigh with crossed Nicol prisms at optical frequencies and later shown to be feasible with microwave ferrites at microwave frequencies by Hogan.[16] Since then the term isolator has been used to depict any ferrite device that acts as a one-way transmission line. The gyrator is a device whose concept was introduced by Tellegen[17] and refers to a two-port circuit device whose nondiagonal impedance elements are equal but opposite in sign. He suggested several likely materials that could be used to construct such a device and microwave ferrites were one of them. At the present time, ferrites are the only material that have resulted in practical devices, because their losses can be made quite small.

---

[14]C. Kittel, *Introduction to Solid State Physics*, 3rd ed., John Wiley & Sons, Inc., 1966, pp. 318–19.

[15]Lax and Button, *Microwave Ferrites and Ferrimagnetics*, p. 302.

[16]C. Hogan, "The Microwave Gyrator," *Bell System Tech. J.*, Monograph 1959, 1952.

[17]B. Tellegen, "The Gyrator, a New Electric Network Element," *Phillips Res. Rept.*, 3, pp. 81–101, April 1948.

For the case where the direction of propagation is transverse to the direction of the magnetostatic field, i.e., $\theta = \pi/2$ in (6-11), we would expect to obtain results that are dual to those in (6-25) and (6-26). Thus, the final results are the following:

Ordinary wave—TEM_x

$$e_y = \sqrt{\frac{\mu_0 g_3}{\epsilon_0 K}}\, h_z$$

$$\mathbf{k} = k\mathbf{a}_x \tag{6-37}$$

$$k = k_0\sqrt{Kg_3}$$

Extraordinary wave—TE_x

$$h_z = 0$$

$$\frac{h_y}{h_x} = \frac{-g_1}{g_2}$$

$$b_x = b_z = 0$$

$$b_y = \left(\frac{\mu_0 k^2}{k_0^2 K}\right)h_y = \mu_0 \mu_{\text{eff}} h_y \tag{6-38}$$

$$e_x = e_y = d_x = d_y = 0$$

$$e_z = -\sqrt{\frac{\mu_0 \mu_{\text{eff}}}{K\epsilon_0}}\, h_y$$

$$\mathbf{k} = k\mathbf{a}_x$$

$$k = k_0\sqrt{K}\sqrt{\frac{g_1^2 + g_2^2}{g_1}} = k_0\sqrt{K\mu_{\text{eff}}}$$

$$\mu_{\text{eff}} = \left[\frac{f^2 - (f_g + f_m)^2}{f^2 - f_g^2 - f_g f_m}\right]^{1/2}$$

One sees that the ordinary wave is unaffected by the anisotropy of the magnetic material, i.e., it is a linearly polarized TEM_x wave with its magnetic vector parallel to $\mathbf{H}_0$ and it behaves as if it were propagating in an isotropic medium with constants $\epsilon_0 K$ and $\mu_0 g_3$. (We note that g_3 is unity for microwave ferrites.) The orientation of the field vectors, Poynting vector, and propagation vector is illustrated in Fig. 6-9 for the ordinary wave.

The extraordinary wave behaves as though the material were isotropic with a dielectric constant of K and a permeability μ_{eff}. This effective permeability is plotted in Fig. 6-10 as a function of f_g/f for the same material and operating conditions as those used for Fig. 6-8. Inspection of (6-38) and Fig. 6-10 indicates that the effective permeability vanishes at

$$\frac{f_g}{f} = 1 - \frac{f_m}{f}$$

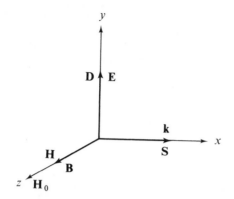

Fig. 6-9 Orientation of field vectors, poynting vector and propagation vector for the ordinary wave propagating in anisotropic magnetic medium. *Note:* The wave is TEM_x.

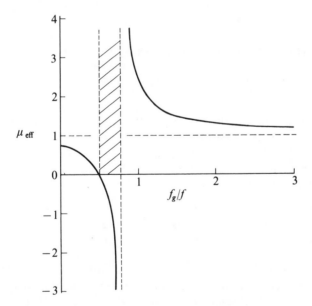

Fig. 6-10 Plot of μ_{eff} versus f_g/f for extraordinary plane wave propagation perpendicular to the dc magnetic field in YIG ferrite at 9.8 GHz. Dashed region indicates the stop-band.

and has a pole at

$$\frac{f_g}{f} = \frac{-f_m}{2f} + \sqrt{\left(\frac{f_m}{2f}\right)^2 + 1}$$

In this case, the stopband exists between the zero and the pole and is indicated by the dashed region in Fig. 6-10. For our choice of constants, the zero occurs at $f_g/f = \frac{1}{2}$ and the pole at $f_g/f = 0.78$. Results similar to those plotted here are also given by Suhl and Walker[18] and Lax and Button.[19] The orientation of the field vectors, Poynting vector, and propagation vector for the extraordinary wave is illustrated in Fig. 6-11.

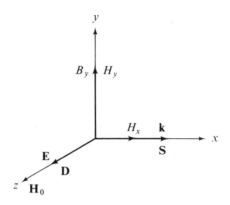

Fig. 6-11   Orientation of field vectors, Poynting vector and propagation vector for the extraordinary wave propagating in anisotropic magnetic medium. *Note:* The wave is TE_x and $B_x = 0$.

The process of diagonalizing the permeability tensor is exactly the same as that for the dielectric tensor in Sec. 6.2. As a result, the eigenvalues are given by (6-20) with g_1 and g_2 given by (4-72); the principal axes and coordinates are given by (6-21); and the expression for $\mathbf{b}_0$ is the dual of that for $\mathbf{d}_0$ in (6-27) and (6-28).

6.4   Wave Propagation in Inhomogeneous Media

Let us now consider wave propagation in a medium that is isotropic but inhomogeneous in either its electric or magnetic properties. Although we could consider the more general problem of a medium that is inhomogeneous in both its electric and magnetic properties and then let the medium be either electrically or magnetically inhomogeneous for two specific problems, it is simpler to consider the two situations separately. This is realistic since most problems of physical interest fall into one or the other of the two cases. Also,

---

[18]H. Suhl and L. Walker, "Topics in Guided Wave Propagation Through Gyromagnetic Media," *Bell System Tech. J.*, Monograph 2322, 1954, p. 91.

[19]Lax and Button, *Microwave Ferrites and Ferrimagnetics*, p. 305.

superposition is valid when the medium is both electrically and magnetically inhomogeneous (or anisotropic, for that matter) so that there is nothing to gain by treating the general problem. Superposition is valid as long as the medium is linear, and this assumption was made when introducing the constitutive relations (4-2). It is interesting to note that superposition does not appear to be valid when the material is both electrically and magnetically inhomogeneous. To see this, let us consider equation (4-46) for two different cases. First, let the field E_a be caused by the source J_a when the material is only electrically inhomogeneous, i.e., $K = K(r)$, $K_m \neq K_m(r)$. Second, let the field E_b be caused by the source J_b when the material is only magnetically inhomogeneous. If these two equations for E_a and E_b are added together, one obtains a different result than that obtained directly from (4-46) for the combined sources $J_a + J_b$ and the total field $E_a + E_b$. Thus, it appears that superposition is not valid. However, this anomaly can be reconciled by noting that the missing term is $-K_m^{-1}\nabla K_m \times \nabla \times E_a$, which vanishes because ∇K_m is zero. This same anomaly occurs if one considers equation (4-47) for the fields H_a and H_b. In this case, the missing term involves ∇K, which is zero. The same procedure can be followed with equations (4-58) and (4-59) when the material is anisotropic but homogeneous to show that superposition is valid. We should note again that the expressions in (4-58) through (4-65) are equally valid when the material is both inhomogeneous and anisotropic.

If we try to attack the problem of wave propagation in an inhomogeneous medium by extending the plane wave analysis that was introduced in Sec. 6.1, and successfully used in Secs. 6.2 and 6.3, we immediately experience difficulties because the propagation vector $\mathbf{k}$ is no longer independent of the coordinates. As a result, the operation $\nabla(e^{-j\mathbf{k}\cdot\mathbf{r}}) = -j\mathbf{k}e^{-\mathbf{k}\cdot\mathbf{r}}$ is no longer valid but has to be left as

$$\nabla(e^{-j\mathbf{k}\cdot\mathbf{r}}) = -j\nabla(\mathbf{k}\cdot\mathbf{r})e^{-j\mathbf{k}\cdot\mathbf{r}}$$

Further, the operation $\nabla(\mathbf{k}\cdot\mathbf{r})$ yields three nonzero terms in general so it does not seem likely that this approach will be too fruitful.

Instead of a general approach, let us consider a specific problem that is two-dimensional in nature.[20] The problem is one where the medium is electrically inhomogeneous and the quantity K in (4-37) is a sinusoidal function of only the z-coordinate. If we were to allow the quantity K to be a function of several spatial variables, then the method of separation of variables would fail.[21] Thus, we restrict ourselves to a simpler problem that we can solve. If

[20]T. Tamir, H. Wang, and A. Oliner, "Wave Propagation in Sinusoidally Stratified Dielectric Media," *IEEE Trans. on Microwave Theory and Techniques*, Vol. MTT-12, pp. 323–336, May 1964.

[21]S. Schelkunoff, *Electromagnetic Fields*, Blaisdell Publishing Company, New York, 1963, p. 292.

the electromagnetic fields are independent of y and we consider a TE_z mode, we can use the results of Prob. 4.9 to write the equation for E_y as

$$\left[\frac{\partial^2}{\partial x^2} + \frac{\partial^2}{\partial z^2} + k_0^2 K(z)\right] E_y(x, z) = 0 \tag{6-39}$$

The equations for H_x and H_z are more involved since they are mixed, i.e., the equation for H_x contains H_z and the equation for H_z contains H_x. As a result, it is easier to solve for E_y and to obtain the magnetic field components from Maxwell's equations in (4-35) or (4-36). Furthermore, if we expand the two curl equations we see that it is still possible for a TE_z mode to propagate independently of a TM_z mode regardless of how K varies with x and z. Let us assume that we can express the sinusoidal variation in K as

$$K(z) = K_0\left(1 - \delta \cos\frac{2\pi z}{L}\right), \qquad 0 \le \delta \le 1 \tag{6-40}$$

and that E_y can be written as $E_y = X(x)Z(z)$, i.e., the separation of variables method works. The partial differential equation in (6-39) now separates into the two ordinary differential equations

$$\frac{X''(x)}{X(x)} = -k_1^2 \tag{6-41}$$

$$\frac{Z''(z)}{Z(z)} + k_0^2 K_0\left(1 - \delta \cos\frac{2\pi z}{L}\right) = k_1^2 \tag{6-42}$$

The solution to (6-41) consists of sinusoids or complex exponentials for $k_1^2 > 0$ whereas the equation (6-42) is of the form of Mathieu's equation (1-42) in Chap. 1. Since we are considering a medium that is unbounded, let us choose a complex exponential for the x-dependent part of the solution from (6-41) and the representation from (1-45) for the z-dependent part from (6-42). The expression for E_y then becomes

$$E_y(x, \zeta) = e^{\pm jv\zeta} \sum_{n=-\infty}^{\infty} a_n e^{-jk_1 x} e^{j2n\zeta} \tag{6-43}$$

where (6-42) was put into the form of (1-42) by using the substitutions

$$\zeta = \frac{\pi z}{L}$$

$$a = \left(\frac{L}{\pi}\right)^2 (k_0^2 K_0 - k_1^2) \tag{6-44}$$

$$h^2 = \left(\frac{K_0 \delta}{2}\right)\left(\frac{k_0 L}{\pi}\right)^2$$

The final form for $E_y(x, z)$ is obtained from (6-43) and (6-44) as

$$E_y(x, z) = e^{-jvz} \sum_{n=-\infty}^{\infty} a_n e^{-j[k_1 x - (2\pi n z / L)]} \tag{6-45}$$

When the material is homogeneous, the so-called modulation index vanishes and the solution to (6-39) is

$$E_y(x, z) = e^{\pm jk_1 x} e^{\pm jz\sqrt{k_0^2 K_0 - k_1^2}} \tag{6-46}$$

whence it is clear that the $\pm$ signs in the exponents determine the direction of wave propagation. Thus, we shall use the minus signs since we wish to consider a wave propagating in the positive x- and z-directions. The general solution in (6-45) reduces to the representation in (6-46) when n vanishes provided

$$v = \sqrt{k_0^2 K_0 - k_1^2}$$

However, for finite value of δ the problem of evaluating v and the coefficients a_n is more involved and is most easily performed graphically by using the Mathieu stability chart in Fig. 1-2 in Chap. 1.

In order to obtain more insight into the filter action that the modulated medium causes, let us consider a specific case where the dielectric material completely fills a rectangular waveguide propagating the TE_{10} mode. The wide dimension of the waveguide is denoted w and is in the x-direction; thus, the solution to (6-41) for the TE_{10} mode is

$$\sin k_1 x$$

with $k_1 = \pi/w$. Next, we solve for a and h^2 from (6-44) in terms of the free space wavelength to obtain

$$a = K_0 \left(\frac{2L}{\lambda_0}\right)^2 - \left(\frac{L}{w}\right)^2$$

$$h^2 = \frac{K_0 \delta}{2} \left(\frac{2L}{\lambda_0}\right)^2$$

If one considers that λ_0 is the only variable in the expressions for a and h^2, then the quantity a vanishes when $\lambda_0 = 2w\sqrt{K_0}$. Also, the ratio $h^2[a + (L/w)^2]^{-1}$ is a constant, i.e.,

$$h^2 \left[a + \left(\frac{L}{w}\right)^2\right]^{-1} = \frac{\delta}{2}$$

Thus, the relation between h^2 and a is the equation of a straight line

$$h^2 = \frac{\delta}{2}a + \frac{\delta}{2}\left(\frac{L}{w}\right)^2 \qquad (6\text{-}47)$$

with a slope of $\delta/2$, an ordinate intercept of $(\delta/2)(L/w)^2$, and an abscissa intercept of $-(L/w)^2$. Now, it becomes clear that the values of a and h^2 that satisfy (6-47) also determine the value of v in (6-45). This situation is illustrated in Fig. 6-12 for a rectangular X-band waveguide with the following values for the parameters:

$$w = 0.9''$$
$$K_0 = 6$$
$$\delta = 0.4$$
$$L = 1.5''$$

The value of frequency that causes the quantity a to vanish is 2.68 GHz, which is the usual cutoff frequency for the TE_{10} mode. However, the point A on the figure corresponds to a frequency of 2.61 GHz, which is slightly lower. Thus, one consequence of the spatial modulation of the dielectric con-

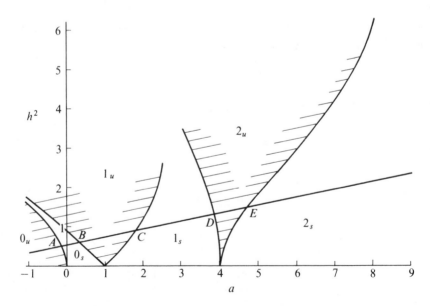

Fig. 6-12 Graphical construction used to determine the behavior of the TE_{10} mode propagating in a sinusoidally modulated dielectric completely filling a rectangular X-band waveguide. Dash regions indicate stopbands.

stant is a reduction of the cutoff frequency by 2.6 %. The point B corresponds to a frequency of 2.835 GHz, so that the first passband is 0.225 GHz wide. A stopband exists between points B and C, which correspond to frequencies of 2.835 and 3.43 GHz, respectively. The next passband exists between points C and D, which correspond to frequencies of 3.43 and 4.14 GHz. The point E corresponds to a frequency of 4.4 GHz and is the beginning of the third passband which extends to the point F, which corresponds to a frequency of 5.56 GHz. The frequency at this point is slightly larger than the cutoff frequency of the TE_{20} mode, i.e., $f_{20} = 5.36$ GHz.

6.5 Wave Propagation in Anisotropic Media

The general problem of wave propagation in a homogeneous anisotropic medium depends upon solving (4-58) and (4-59) for the case of an electric current source or (4-64) and (4-65) for the case of a magnetic current source. As yet, general solutions to these equations have not been found. On the other hand, there is no real need to solve the general problem since practical materials are anisotropic in either their magnetic or electric properties. In most cases, boundary value problems are suggested by practical devices and situations. For instance, propagation at various frequencies through the earth's crust and its atmosphere has been of continuing interest for a considerable number of years. This practical problem was replaced by a large number of boundary value problems, each one slightly more difficult than its predecessor. Zenneck and Sommerfeld were two of the early workers in this area. Apparently they originally felt that long distance radio communication was accomplished through a bound surface wave. This surface wave exists in a region containing two different dielectric constants and is tied closely to the boundary between them. The boundary value problem that Sommerfeld[22] solved was that of a vertical electric dipole situated above a flat, homogeneous earth possessing a complex dielectric constant. He found that the field could be divided into three contributions. The first wave (direct ray) had the dipole as its source, the second wave (reflected ray) had the image dipole as its source, and the third wave was the surface wave. Sommerfeld later decided that the surface wave was not the significant contribution to long distance radio wave propagation. On the other hand, neither was the direct ray. Rather it was the waves that were reflected from the ionosphere. As a result, a new boundary value problem was posed that consisted of a vertical electric dipole in a free space region surrounded by a flat earth below and a flat isotropic ionosphere above. The free space region acts as a parallel-plate waveguide, thus giving rise to the so-called mode theory of

---

[22]A. Sommerfeld, *Partial Differential Equations*, Academic Press, Inc., New York, 1949.

wave propagation. The problem is often simplified from a three-dimensional to a two-dimensional problem (without losing any of the salient features) by replacing the vertical dipole by a horizontal electric or magnetic line source. Unfortunately, the theoretical results did not correlate with all of the experimental results, so that the model was further complicated by making the ionosphere anisotropic or inhomogeneous and the earth inhomogeneous. Many of these topics are considered by Wait,[23] Baños,[24] Budden,[25] and Brekhovskikh.[26]

Now let us return to (4-58) and (4-59), (4-64) and (4-65) and consider several two-dimensional problems. Since we have already assumed that the magnetostatic field is in the z-direction in connection with the tensors $\mathfrak{K}$ and $\mathfrak{K}_m$ defined in (4-70), let us further assume that the electromagnetic fields are independent of z. The problems that we wish to solve involve a plane wave in free space incident obliquely upon either an anisotropic electric (plasma) half

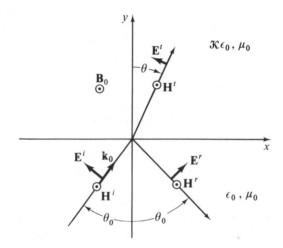

Fig. 6-13   Plane wave incident on anisotropic electric half-space. Incident electric field polarized parallel to plane of incidence. Magnetostatic field parallel to interface and perpendicular to the plane of incidence. Plane of incidence perpendicular to the interface.

[23] J. Wait, *Electromagnetic Waves in Stratified Media*, Pergamon Press, Inc., New York, 1962.

[24] A. Baños, Jr., *Dipole Radiation in the Presence of a Conducting Half-Space*, Pergamon Press, Inc., New York, 1966.

[25] K. Budden, *The Wave-Guide Mode Theory of Wave Propagation*, Prentice-Hall, Inc., Englewood Cliffs, N.J., 1961.

[26] L. Brekhovskikh, *Waves in Layered Media*, Academic Press, Inc., New York, 1960.

space or an anisotropic magnetic (ferrite) halfspace. We shall locate the interface between the two homogeneous regions in the x-z plane at $y = 0$. This
is illustrated in Fig. 6-13, which is identical with Fig. 1 in Wait.[27] The orientation of the field vectors in Fig. 6-13 is also similar to Fig. 6-5 for the anisotropic electric case and Fig. 6-11 for the anisotropic magnetic case. Since the
only finite component of the magnetic field is H_z, for the anisotropic electric
case, we find that (4-65) reduces to

$$\left(\nabla_t^2 + \frac{k_0^2}{\ell_1}\right)H_z = \frac{j\omega\epsilon_0}{\ell_1}J_{mz} \qquad (6\text{-}48)$$

The other equation we need is that involving **E** which we can obtain from
(4-36) and (4-48), i.e.,

$$\mathbf{E} = \frac{1}{j\omega\epsilon_0}\boldsymbol{\mathcal{L}}\cdot\nabla \times \mathbf{H}$$

$$= \frac{1}{j\omega\epsilon_0}\left[\mathbf{a}_x\left(\ell_1\frac{\partial H_z}{\partial y} + \ell_2\frac{\partial H_z}{\partial x}\right) + \mathbf{a}_y\left(\ell_2\frac{\partial H_z}{\partial y} - \ell_1\frac{\partial H_z}{\partial x}\right)\right] \qquad (6\text{-}49)$$

The values of ℓ_1 and ℓ_2 are given in (4-73) and (4-71) and $\nabla_t = \nabla - \mathbf{a}_z(\partial/\partial z)$.
   For the anisotropic magnetic case, the only finite component of electric
field is E_z, as shown in Fig. 6-11. However, the desired equations are just the
duals of (6-48) and (6-49) so we can write them down at once by referring to
Table 4-1. Thus, we obtain

$$\left(\nabla_t^2 + \frac{k_0^2}{\ell_1}\right)E_z = \frac{j\omega\mu_0}{\ell_1}J_z \qquad (6\text{-}50)$$

$$\mathbf{H} = \frac{1}{j\omega\mu_0}\left[\mathbf{a}_x\left(\ell_1\frac{\partial E_z}{\partial y} + \ell_2\frac{\partial E_z}{\partial x}\right) + \mathbf{a}_y\left(\ell_2\frac{\partial E_z}{\partial y} - \ell_1\frac{\partial E_z}{\partial x}\right)\right] \qquad (6\text{-}51)$$

where the values of ℓ_1 and ℓ_2 are given in (4-72) and (4-73).
   We now return to the problem illustrated in Fig. 6-13. If we let the magnitude of the incident magnetic field be unity, we can write it as

$$H_z^i = e^{-j(k_1 x + k_2 y)} \qquad (6\text{-}52)$$

where

$$k_1 = k_0 \sin\theta_0$$

$$k_2 = k_0 \cos\theta_0$$

and k_0 is the free space wave number ($k_0 = 2\pi/\lambda_0$). The reflected field can be
written as

$$H_z^r = R_i e^{-j(k_1 x - k_2 y)} \qquad (6\text{-}53)$$

---

[27]Wait, *Electromagnetic Waves in Stratified Media*, p. 232.

where R_i is the magnetic field reflection coefficient. The transmitted magnetic field differs from the incident field in magnitude and also has a different y-dependence. Since it satisfies the homogeneous form of (6-48), we can express it as

$$H_z^t = T_i e^{-j(k_1 x + k_4 y)} \tag{6-54}$$

and solve for k_4 by inserting the above expression into the homogeneous form of (6-48). The expression for k_4 becomes

$$k_4 = \left(\frac{k_0^2}{\ell_1} - k_1^2\right)^{1/2} = k_0(\ell_1^{-1} - \sin^2 \theta_0)^{1/2} \tag{6-55}$$

We can now use the magnetic field expressions in (6-52), (6-53), and (6-54) to solve for the tangential electric field from (6-49). Thus, we obtain

$$E_x^i = \frac{-k_2}{\omega \epsilon_0} H_z^i$$

$$E_x^r = \frac{k_2}{\omega \epsilon_0} H_z^r \tag{6-56}$$

$$E_x^t = -\frac{1}{\omega \epsilon_0}(\ell_1 k_4 + \ell_2 k_1) H_z^t$$

Since H_z and E_x are continuous at $y = 0$, we obtain the following two equations involving the reflection and transmission coefficients:

$$1 + R_i = T_i$$
$$k_2(1 - R_i) = T_i(\ell_1 k_4 + \ell_2 k_1)$$

Their solution is

$$R_i = \frac{1 - \Delta}{1 + \Delta}$$

$$T_i = \frac{2}{1 + \Delta} \tag{6-57}$$

where

$$\Delta = k_2^{-1}(\ell_1 k_4 + \ell_2 k_1) = \sec \theta_0 [\ell_2 \sin \theta_0 + (\ell_1 - \ell_1^2 \sin^2 \theta_0)^{1/2}]$$

and ℓ_1 and ℓ_2 are given by (4-73) and (4-71).

The problem which is dual to the one just solved is illustrated in Fig. 6-14. The anisotropic electric region is replaced by an anisotropic magnetic region, H_z is replaced by E_z, $\mathbf{E}$ is replaced by $-\mathbf{H}$, and R_i and T_i are replaced by R and T. Since the boundary conditions and the radiation condition are the same for each problem, it follows that the two are exact duals. Consequently, it is possible to write down the expressions for the field components and the electric field reflection and transmission coefficients by referring to Table 4-1.

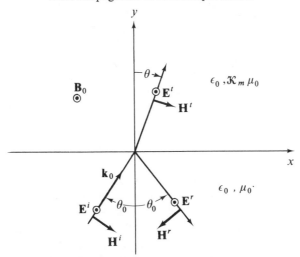

Fig. 6-14 Plane wave incident on anisotropic magnetic half-space. Incident electric field polarized perpendicular to plane of incidence. Magnetostatic field parallel to interface and perpendicular to the plane of incidence. Plane of incidence perpendicular to the interface.

The desired results are the following:

$$E_z^i = e^{-j(k_1 x + k_2 y)}, \qquad H_x^i = \frac{k_2}{\omega\mu_0} E_z^i$$

$$E_z^r = R e^{-j(k_1 x - k_2 y)}, \qquad H_x^r = \frac{-k_2}{\omega\mu_0} E_z^r \qquad (6\text{-}58)$$

$$E_z^t = T e^{-j(k_1 x + k_4 y)}, \qquad H_x^t = \frac{1}{\omega\mu_0}(\ell_1 k_4 + \ell_2 k_1) E_z^t$$

$$R = R_i = \frac{1 - \Delta}{1 + \Delta}$$

$$T = T_i = \frac{2}{1 + \Delta} \qquad (6\text{-}57)$$

where k_1 and k_2 are defined in (6-52), and k_4 and Δ remain the same, i.e.,

$$k_4 = k_0(\ell_1^{-1} - \sin^2 \theta_0)^{1/2}$$

$$\Delta = \sec \theta_0 [\ell_2 \sin \theta_0 + (\ell_1 - \ell_1^2 \sin^2 \theta_0)^{1/2}]$$

and ℓ_1 and ℓ_2 are given by (4-72) and (4-73). We note that the expressions for R and T in this case are the same as those for R_i and T_i in the previous case. However, the values of ℓ_1 and ℓ_2 are not the same since the former is the anisotropic electric material and the latter is the anisotropic magnetic material. In a practical case, ϵ_0 in the anisotropic region in Fig. 6-14 should be replaced by $K\epsilon_0$ and k_0 should be replaced by $\sqrt{K}k_0$.

Another problem that we can consider is the anisotropic electric case illustrated in Fig. 6-13 but with the electric field perpendicular to the plane of incidence, i.e., the fields are oriented as in Fig. 6-14. In this case we wish to express the field components in a form similar to (6-58). However, E_z now satisfies the homogeneous form of (4-58) which is

$$(\nabla_t^2 + k_0^2 g_3)E_z = 0$$

where g_3 is given in (4-71). We see at once that this problem corresponds to that of the ordinary wave propagating in an anisotropic medium as illustrated in Fig. 6-4. Consequently, the problem is the same as that of a plane wave with polarization perpendicular to the plane of incidence propagating from free space into an isotropic dielectric region with constants μ_0 and $g_3\epsilon_0$. If we solve for the electric field reflection and transmission coefficients, we obtain (6-57) again except now the expression for Δ becomes

$$\Delta = \frac{k_5}{k_2} = \sec\theta_0 \, (g_3 - \sin^2\theta_0)^{1/2}$$

where

$$k_5 = (k_0^2 g_3 - k_1^2)^{1/2} = k_0(g_3 - \sin^2\theta_0)^{1/2}$$

and k_5 is the y-directed wave number in the plasma. As one would expect, these last expressions for R and T are found in many references[28] since the corresponding isotropic problem is rather simple.

Let us now consider the more complicated reflection problem illustrated in Fig. 6-15. Although the electric field in the incident wave is perpendicular to the plane of incidence as in Fig. 6-14, the magnetostatic field is perpendicular to the boundary rather than parallel to it. This problem is more complicated than any of the previous ones for two reasons. First, the anisotropy of the material creates a component of magnetic field in the y-direction in both the reflected and transmitted waves. Second, the anisotropy of the material causes two different waves to propagate in the material. Each wave possesses a different wave number and direction of propagation with respect to the z-direction. This is not too obvious at first since the development leading to the dual problem in (6-12) indicated that in an unbounded anisotropic region two waves with different wave numbers would propagate but their direction of propagation would be the same. Further, the two waves would be completely independent of one another and would be TE_k waves. In the

[28]Johnson, *Field and Wave Electrodynamics*, Sec. 3.4. Also, see R. Plonsey and R. Collin, *Principles and Applications of Electromagnetic Fields*, McGraw-Hill Book Company, Inc., New York, 1961, Sec. 10.2.

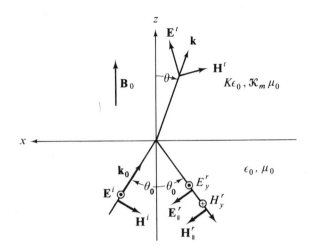

Fig. 6-15 Plane wave incident on anisotropic magnetic half-space. Incident electric field polarized perpendicular to the plane of incidence. Magnetostatic field perpendicular to the interface. Plane of incidence perpendicular to the interface.

present situation the two waves are coupled together by the boundary so that they are not independent of one another. It also follows that the waves will have field components in all three coordinate directions since the direction of propagation does not coincide with one of the coordinate axes. One can visualize the incident wave as exciting both waves in the anisotropic medium which possesses different effective material constants for the two waves. Thus, it is obvious that Snell's law will consist of two relations, one for each wave. The expression in (6-12) will not be too useful now since the angle that it contains is unknown. However, this angle can be eliminated by using Snell's law which we will also derive in what follows.

The fields will be independent of y so we can express them in the following form:

Incident fields

$$E_y^i = e^{jk_0(Sx - Cz)}$$

$$\mathbf{H}^i = -Y_0(C\mathbf{a}_x + S\mathbf{a}_z)E_y^i$$

(6-59)

Reflected fields

$$E_y^r = {}_\perp R_\perp e^{jk_0(Sx + Cz)}$$

$$\mathbf{H}_{||}^r = Y_0(C\mathbf{a}_x - S\mathbf{a}_z)E_y^r$$

$$\mathbf{E}_{||}^r = {}_\perp R_{||}(C\mathbf{a}_x - S\mathbf{a}_z)e^{jk_0(Sx + Cz)}$$

$$H_y^r = -Y_0 {}_\perp R_{||}e^{jk_0(Sx + Cz)}$$

(6-60)

Transmitted fields

$$E_{ya}^t = {}_\perp T_a e^{jk_0 n_a(S_a x - C_a z)}$$

$$E_{xa}^t = \frac{-g_2 P}{Q_a} E_{ya}^t$$

$$E_{za}^t = \frac{-g_2 SP}{q_a Q_a} E_{ya}^t$$

$$H_{xa}^t = \frac{Y_0 P}{q_a} E_{ya}^t \tag{6-61}$$

$$H_{ya}^t = \frac{-Y_0 g_2 KP}{q_a Q_a} E_{ya}^t$$

$$H_{za}^t = \frac{-Y_0 S}{g_3} E_{ya}^t$$

where

$$C = \cos\theta_0, \quad S = \sin\theta_0, \quad C_a = \cos\theta_a, \quad S_a = \sin\theta_a$$

$$Y_0 = \sqrt{\frac{\epsilon_0}{\mu_0}}, \quad \mathbf{k}_a = \mathbf{n}_a k_0, \quad n_a^2 = q_a^2 + S^2$$

$$q_a = n_a C_a, \quad P = \frac{S^2 - Kg_3}{g_3}, \quad Q_a = n_a^2 - Kg_1$$

A similar set of relations is obtained for the other transmitted wave except the subscript a in (6-61) is replaced by b. The total transmitted field in the anisotropic region is just the sum of the two sets. The expressions for the transmitted field components in terms of E_y were obtained by using Prob. 6.7 and the expression $\nabla \times \mathbf{H} = j\omega\epsilon_0 K\mathbf{E}$. Also, if one were to introduce three reflection coefficients R_x, R_y, and R_z instead of the two ${}_\perp R_\perp$ and ${}_\perp R_{\parallel}$, one would find that $R_y = {}_\perp R_\perp$, $R_x = C {}_\perp R_{\parallel}$, and $R_z = -S {}_\perp R_{\parallel}$ so that the two that are used here are sufficient. The notation involving the subscripts on R and T is similar to that used by Budden.[29] The first subscript on R and T indicates whether the electric field in the incident wave is polarized perpendicular ($\perp$) or parallel ($\parallel$) to the plane of incidence. The second subscript on R indicates the same thing for the reflected wave. The second subscript on T refers to the two waves which propagate in the anisotropic region. The reflected fields in (6-60) are two plane waves (with respect to the wave normal), one polarized perpendicular to the plane of incidence and the other polarized parallel to the plane of incidence.

---

[29]K. Budden, *Radio Waves in the Ionosphere*, Cambridge University Press, New York, 1961, Chap. 7.

We can obtain equations involving the reflection and transmission coefficients by invoking the boundary conditions that the tangential components of the fields are continuous at the interface $z = 0$. This yields the following four equations:

$$1 + {}_\perp R_\perp = {}_\perp T_a + {}_\perp T_b$$

$$(-1 + {}_\perp R_\perp)C = P\left(\frac{{}_\perp T_a}{q_a} + \frac{{}_\perp T_b}{q_b}\right)$$

$$-{}_\perp R_\parallel C = g_2 P\left(\frac{{}_\perp T_a}{Q_a} + \frac{{}_\perp T_b}{Q_b}\right) \tag{6-62}$$

$${}_\perp R_\parallel = g_2 K P\left(\frac{{}_\perp T_a}{q_a Q_a} + \frac{{}_\perp T_b}{q_b Q_b}\right)$$

and their solutions:

$${}_\perp R_\perp = -\frac{PA_1 + CA_2}{PA_1 - CA_2}$$

$${}_\perp R_\parallel = \frac{-2g_2 CKP(q_a - q_b)}{PA_1 - CA_2}$$

$${}_\perp T_a = \frac{2q_a Q_a C(CK + q_b)}{PA_1 - CA_2} \tag{6-63}$$

$${}_\perp T_b = \frac{-2q_b Q_b C(CK + q_a)}{PA_1 - CA_2}$$

where

$$A_1 = -CK(Q_a - Q_b) - (q_b Q_a - q_a Q_b)$$

$$A_2 = -CK(q_a Q_a - q_b Q_b) - q_a q_b(Q_a - Q_b)$$

The equations (6-62) also contained exponentials involving x. Since the boundary conditions must be valid for all values of x, it follows that the exponents of the exponentials must be the same. This gives us Snell's law, i.e.,

$$S = n_a S_a = n_b S_b \tag{6-64}$$

which was used to simplify the above expressions. We should note that in this double refracting medium Snell's law applies to the wave normals rather than the rays, i.e., to $\mathbf{k}_a$ and $\mathbf{k}_b$ rather than the Poynting vectors $\mathbf{S}_a$ and $\mathbf{S}_b$ (not to be confused with S_a and S_b above).

All of the quantities involved in the reflection and transmission coefficients are known except the indices of refraction, n_a and n_b. Unfortunately, they cannot be obtained from the expression dual to (6-12) since the angles θ_a and θ_b are unknown. The expressions for k_a and k_b that are dual to (6-12) are obtained from the determinant of the coefficients of **h** in Prob. 6.7. It is customary to solve for q_a and q_b from this determinant rather than k_a and k_b. The equation for q is

$$g_3 q^4 + [S^2(g_1 + g_3) - 2Kg_1g_3]q^2 + (S^2 - Kg_3)[S^2g_1 - K(g_1^2 + g_2^2)] = 0$$
(6-65)

and the solutions for q^2 are

$$q^2 = Kg_1 - \frac{S^2(g_1 + g_3)}{2g_3}$$

$$\pm \frac{1}{2g_3}[S^4(g_1 - g_3)^2 + 4Kg_3(Kg_3 - S^2)(2g_1^2 + g_2^2)]^{1/2} \quad (6\text{-}66)$$

The quadratic equation (6-66) in q^2 gives four values for q. Two of them represent the desired upgoing waves and the other two represent undesired downgoing waves. In the case of electric anisotropy, the equation determining q, (6-65), is called the "Booker quartic" and is discussed rather fully by Budden,[30] who also considers the most general case[31] where both the magnetostatic field and the wave normals are arbitrarily oriented with respect to the coordinate axes. The general Booker quartic contains all powers of q whereas our more specialized case is a quadratic in q^2. The Booker quartic reduces to the Appleton-Hartree formula[32] for the special case of normal incidence and an arbitrarily oriented magnetostatic field. This is the same situation as that leading to (6-12) except we interchanged the directions of the magnetostatic field and the wave normal. However, the results are the same so that (6-12) is the Appleton-Hartree formula provided one divides both sides by k_0 and solves for $n = k/k_0$.

The next problem we wish to consider is shown in Fig. 6-16. This problem differs from that in Fig. 6-15 only in that the electric field is polarized in the plane of incidence rather than perpendicular to it. In this case we use H_y^i as the reference field rather than E_y^i as in the previous problem. Since the transmitted fields are of the same form as those in (6-61), they may be written quite easily by using the relations in (6-61) to express them in terms of H_y. Thus, we write the field components in the following manner:

Incident fields

$$H_y^i = e^{jk_0(Sx - Cz)}$$

$$\mathbf{E}^i = Z_0(C\mathbf{a}_x + S\mathbf{a}_z)H_y^i$$
(6-67)

---

[30]Budden, *Radio Waves in the Ionosphere*, Chap. 8.

[31]*Ibid.*, Chap. 13.

[32]*Ibid.*, Chap. 6.

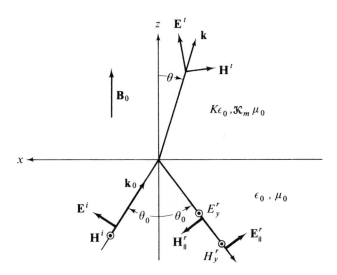

Fig. 6-16 Plane wave incident on anisotropic magnetic halfspace. Incident electric field polarized parallel to the plane of incidence. Magnetostatic field perpendicular to the interface. Plane of incidence perpendicular to the interface.

Reflected fields

$$H_y^r = {}_\|R_\| e^{jk_0(Sx+Cz)}$$

$$E_\|^r = Z_0(-C\mathbf{a}_x + S\mathbf{a}_z)H_y^r$$

$$E_y^r = Z_0 {}_\|R_\perp e^{jk_0(Sx+Cz)}$$ (6-68)

$$H_\|^r = {}_\|R_\perp(C\mathbf{a}_x - S\mathbf{a}_z)e^{jk_0(Sx+Cz)}$$

Transmitted fields

$$H_{ya}^t = {}_\|T_a e^{jk_0 n_a(S_a x - C_a z)}$$

$$H_{xa}^t = -\frac{Q_a}{Kg_2}H_{ya}^t$$

$$H_{za}^t = \frac{q_a Q_a S}{g_2 g_3 KP}H_{ya}^t$$

$$E_{xa}^t = \frac{Z_0 q_a}{K}H_{ya}^t$$ (6-69)

$$E_{ya}^t = -\frac{Z_0 q_a Q_a}{g_2 KP}H_{ya}^t$$

$$E_{za}^t = \frac{Z_0 S}{K}H_{ya}^t$$

As before, a similar set of relations exists for the other transmitted wave with the subscript a replaced by b. The boundary condition on the continuity of the tangential field components at $z = 0$ yields the following four equations involving the reflection and transmission coefficients:

$$1 + {}_{\parallel}R_{\parallel} = {}_{\parallel}T_a + {}_{\parallel}T_b$$

$$-(-1 + {}_{\parallel}R_{\parallel})C = \frac{1}{K}(q_a{}_{\parallel}T_a + q_b{}_{\parallel}T_b)$$

$${}_{\parallel}R_{\perp}C = -\frac{1}{g_2 K}(Q_a{}_{\parallel}T_a + Q_b{}_{\parallel}T_b) \tag{6-70}$$

$${}_{\parallel}R_{\perp} = -\frac{1}{g_2 KP}(q_a Q_a{}_{\parallel}T_a + q_b Q_b{}_{\parallel}T_b)$$

and their solutions:

$${}_{\parallel}R_{\parallel} = \frac{-A_3 + KA_4}{A_3 + CKA_4}$$

$${}_{\parallel}R_{\perp} = \frac{(1 + C)Q_a Q_b(q_a - q_b)}{g_2(A_3 + CKA_4)}$$

$${}_{\parallel}T_a = \frac{-(1 + C)KQ_b(P - Cq_b)}{A_3 + CKA_4} \tag{6-71}$$

$${}_{\parallel}T_b = \frac{(1 + C)KQ_a(P - Cq_a)}{A_3 + CKA_4}$$

where

$$A_3 = P(q_b Q_a - q_a Q_b) - Cq_a q_b(Q_a - Q_b)$$
$$A_4 = P(Q_a - Q_b) - C(q_a Q_a - q_b Q_b)$$

Naturally, Snell's law (6-64) is still valid and has been used to simplify the expressions in (6-71).

It is possible to solve (4-58) and (4-59) or (4-64) and (4-65) in a cartesian coordinate system when the medium is either electrically or magnetically anisotropic and the magnetostatic field is impressed in a direction parallel to one of the coordinate axes. The two cases that are of most interest involve wave propagation in the z-direction with the magnetostatic field either in the z-direction or perpendicular to it. Let us consider first the case where the magnetostatic field is in the z-direction. If we consider the homogeneous forms of (4-58) and (4-59), we find that E_z and H_z satisfy a quartic equation in terms of the operator $\nabla_t = \nabla - \mathbf{a}_z(\partial/\partial z)$. However, it turns out that it is easier to derive two coupled quadratic equations for H_z and E_z by starting with Maxwell's two curl equations in cartesian coordinates with an assumed field variation in the z-direction of the form e^{-jk_3z}. The two curl equations yield six scalar equations, four of them involving the transverse field components directly and two of them involving their transverse derivatives. The transverse field components are obtained in terms of E_z and H_z from the first four equations and then substituted into the last two equations to obtain equations involving E_z and H_z. The results of this lengthy and tedious procedure are the following two coupled quadratic equations when the medium is electrically anisotropic:

$$\nabla_t^2 E_z + A_1 \nabla_t^2 H_z = A_2 H_z$$

$$B_1 \nabla_t^2 E_z + \nabla_t^2 H_z = B_2 H_z$$

which can be simplified further to the more symmetrical form

$$(\nabla_t^2 + T_E^2)E_z = T_{EH}^2 H_z$$

$$(\nabla_t^2 + T_H^2)H_z = T_{HE}^2 E_z$$

$(6\text{-}72)$

where

$$T_E^2 = A_1 B_2 (1 - A_1 B_1)^{-1} = \left(\frac{-g_3}{g_1}\right)(k_3^2 - g_1 k_0^2)$$

$$T_H^2 = A_2 B_1 (1 - A_1 B_1)^{-1} = -k_3^2 + \left(\frac{k_0^2}{g_1}\right)(g_1^2 + g_2^2)$$

$$T_{EH}^2 = A_2 (1 - A_1 B_1)^{-1} = \frac{\omega \mu_0 k_3 g_2}{g_1}$$

$$T_{HE}^2 = B_2 (1 - A_1 B_1)^{-1} = \frac{-\omega \epsilon_0 k_3 g_3 g_2}{g_1}$$

and g_1, g_2, and g_3 are given in (4-71). If we let $k_3 = 0$ in (6-72), the second equation reduces to the homogeneous form of (6-48) as it should. Another point worth noting is the fact that the solutions to (6-72) will be neither TM_z nor TE_z modes but rather linear combinations of both, usually called hybrid modes. Further details concerning these hybrid modes and their application to plasma-filled circular guides can be found elsewhere.[33]

The dual to the problem just considered can be written by referring to Table 4-1. The equations that are dual to (6-72) are the following:

$$(\nabla_t^2 + U_E^2)E_z = U_{EH}^2 H_z$$

$$(\nabla_t^2 + U_H^2)H_z = U_{HE}^2 E_z$$

$(6\text{-}73)$

---

[33]Johnson, *Field and Wave Electrodynamics*, Secs. 11.14–11.18.

where

$$U_E^2 = -k_3^2 + \frac{k^2}{g_1}(g_1^2 + g_2^2)$$

$$U_H^2 = -\left(\frac{g_3}{g_1}\right)(k_3^2 - g_1 k_0^2)$$

$$U_{EH}^2 = \frac{\omega \mu_0 k_3 g_3 g_2}{g_1}$$

$$U_{HE}^2 = \frac{-\omega \epsilon k_3 g_2}{g_1}$$

The quantities g_1, g_2, and g_3 are given in (4-72) and $k^2 = k_0^2 K$, where K is the dielectric constant of the ferrite. The general problem of wave propagation in a gyromagnetic medium was studied very extensively by Epstein.[34] He did not obtain the coupled equations in (6-73) but rather found directly the quartic equation that results from the homogeneous form of (4-58). Although the procedure is straightforward, the calculations are rather tedious so will not be repeated here. Suffice it to state that the resultant quartic can be factored into the product of two quadratics, i.e.,

$$(\nabla_t^2 + \alpha_1^2)(\nabla_t^2 + \alpha_2^2)E_i = 0, \qquad i = x, y, z$$

$$\alpha_1^2 + \alpha_2^2 = U_E^2 + U_H^2$$

$$\alpha_1^2 \alpha_2^2 = -U_{EH}^2 U_{HE}^2$$

The quantities α_1^2 and α_2^2 can be obtained in an explicit manner by solving an algebraic quadratic equation. This same problem was first treated by Kales[35] by the same procedure as ours and later by Gamo.[36] Gamo considered a medium that was both magnetically and electrically anisotropic, i.e., gyrotropic, and transformed the coordinate axes to the principal axes, thereby diagonalizing $\mathcal{K}$ and $\mathcal{K}_m$. He then considered the solutions of the resultant equations in a circular waveguide that contained a gyromagnetic medium. The main conclusion that we can obtain from the above factored equation is that E_i can be found as the superposition of two fields that satisfy the ordinary wave equation, i.e.,

---

[34] P. Epstein, "Wave Propagation in a Gyromagnetic Medium," *Rev. Modern Physics*, Vol. 28, pp. 3–17, January 1956.

[35] M. Kales, *N.R.L. Report No. 4027*, August 1952. Also, see "Modes in Waveguides Containing Ferrites," *J. Appl. Phys.*, Vol. 24, pp. 604–8, May 1953.

[36] H. Gamo, "The Faraday Rotation of Waves in a Circular Waveguide," *J. Phys. Soc. Japan*, Vol. 8, pp. 176–82, March-April 1953.

$$E_i = E_{i1} + E_{i2}$$
$$(\nabla_t^2 + \alpha_1^2)E_{i1} = 0$$
$$(\nabla_t^2 + \alpha_2^2)E_{i2} = 0$$

As mentioned by Epstein, the fields E_1 and E_2 remain independent in unbounded space but are coupled by boundaries and the quartic equation contains spurious solutions that do not satisfy the coupled quadratic equations (6-73).

The case where the magnetostatic field is transverse to the direction of wave propagation in an anisotropic magnetic medium will be considered now. We assume that the field variation in the direction of propagation is of the form e^{-jk_3z} as before but that the magnetostatic field is in the y-direction. The specific magnetic inductive capacity will not be given by (4-70) but by the expression

$$G_{ij} = \begin{bmatrix} g_1 & 0 & -g_2 \\ 0 & g_3 & 0 \\ g_2 & 0 & g_1 \end{bmatrix} \tag{6-74}$$

with the elements still given by (4-72). Rather than transform (4-70) to (6-74), we could equally well let the y-direction be the direction of wave propagation as other authors[37,38] do. However, let us leave the z-direction as the direction of wave propagation. If we expand the two Maxwell curl equations, we obtain the following six equations:

$$\frac{\partial}{\partial y}H_z + jk_3H_y = j\omega\epsilon E_x$$

$$jk_3H_x + \frac{\partial}{\partial x}H_z = -j\omega\epsilon E_y$$

$$\frac{\partial}{\partial y}E_z + jk_3E_y = -j\omega\mu_0(g_1H_x - g_2H_z) \tag{6-75}$$

$$jk_3E_x + \frac{\partial}{\partial x}E_z = j\omega\mu_0 g_3H_y$$

$$\frac{\partial}{\partial x}H_y - \frac{\partial}{\partial y}H_x = j\omega\epsilon E_z$$

$$\frac{\partial}{\partial x}E_y - \frac{\partial}{\partial y}E_x = -j\omega\mu_0(g_2H_x + g_1H_z)$$

---

[37] R. Soohoo, *Theory and Application of Ferrites*, Prentice-Hall, Inc., Englewood Cliffs, N.J., 1960, Chap. 9, Sec. 2.

[38] R. Ghose, *Microwave Circuit Theory and Analysis*, McGraw-Hill Book Company, Inc., New York, 1963, Sec. 13.5.

The first four equations can be used to find expressions for the transverse field components in terms of the longitudinal components. These expressions are then substituted into the last two equations to obtain the equations satisfied by the longitudinal field components. After a considerable amount of manipulation, one obtains the expressions for the transverse components,

$$E_x = \frac{j\omega\mu_0 g_3(\partial/\partial y)H_z + jk_3(\partial/\partial x)E_z}{\alpha_3^2}$$

$$E_y = \frac{\omega\mu_0[g_2 k_3 - jg_1(\partial/\partial x)]H_z + jk_3(\partial/\partial y)E_z}{\alpha_1^2}$$

$$H_x = \frac{-\{j\omega\epsilon(\partial/\partial y)E_z + [k^2 g_2 - jk_3(\partial/\partial x)]H_z\}}{\alpha_1^2}$$

$$H_y = \frac{j\omega\epsilon(\partial/\partial x)E_z + jk_3(\partial/\partial y)H_z}{\alpha_3^2}$$

(6-76)

where $k^2 = \omega^2\mu_0\epsilon$, $\alpha_i^2 = k_3^2 - k^2 g_i$, $i = 1, 3$, and the two coupled equations for the longitudinal field components are

$$\left(\frac{1}{\alpha_3^2}\frac{\partial^2}{\partial x^2} + \frac{1}{\alpha_1^2}\frac{\partial^2}{\partial y^2} - 1\right)E_z + \omega\mu_0\left[\frac{k_3(g_3 - g_1)}{\alpha_1^2\alpha_3^2}\frac{\partial^2}{\partial x\,\partial y} - \frac{jg_2}{\alpha_1^2}\frac{\partial}{\partial y}\right]H_z = 0$$

(6-77)

$$\omega\epsilon\left[\frac{k_3(g_3 - g_1)}{\alpha_1^2\alpha_3^2}\frac{\partial^2}{\partial x\,\partial y} + \frac{jg_2}{\alpha_1^2}\frac{\partial}{\partial y}\right]E_z + \left(\frac{g_1}{\alpha_1^2}\frac{\partial^2}{\partial x^2} + \frac{g_3}{\alpha_3^2}\frac{\partial^2}{\partial y^2} + \frac{k^2 g_2^2}{\alpha_1^2} - g_1\right)H_z = 0$$

Again we see that the solutions to (6-77) will be neither TE_z nor TM_z modes but linear combinations of both. The general solutions to (6-77) have been discussed by Vartanian and Jaynes,[39] but no extensive work has been done since the problem is not yet of practical importance.

The y-dependent part of the field components in (6-76) and (6-77) are unchanged from the isotropic case since the anisotropy is confined to the x-z plane. As a result, it is possible to write the longitudinal field components in the following form:

$$E_z = f_1(x)\sin\frac{n\pi y}{b}$$

$$H_z = f_2(x)\cos\frac{n\pi y}{b}$$

where b is the width of the waveguide in the y-direction (narrow dimension). The functions f_1 and f_2 are determined by substituting the above expressions for E_z and H_z into (6-77). The problem of finding $f_1(x)$ and $f_2(x)$ is quite com-

[39] P. Vartanian and E. Jaynes, "Propagation in Ferrite-Filled Transversely Magnetized Waveguide," *IRE Trans. on Microwave Theory and Techniques*, Vol. MTT-4, pp. 140–43, July 1956.

plicated since they satisfy a quartic differential equation in x of the form

$$\left(\frac{\partial^4}{\partial x^4} + A\frac{\partial^2}{\partial x^2} + B\right)\begin{Bmatrix} f_1(x) \\ f_2(x) \end{Bmatrix} = 0$$

However, the quartic equation can always be factored into a product of two quadratic differential equations. Thus, one can write this product as

$$\left(\frac{\partial^2}{\partial x^2} + C\right)\left(\frac{\partial^2}{\partial x^2} + D\right)\begin{Bmatrix} f_1(x) \\ f_2(x) \end{Bmatrix} = 0$$

and the solution as a linear combination of sinusoids of the form

$$\begin{Bmatrix} \sin px \\ \cos px \end{Bmatrix}\begin{Bmatrix} \sin qx \\ \cos qx \end{Bmatrix}$$

Further details are covered by Vartanian and Jaynes who obtain a formal solution but make no effort to evaluate numerically any of the constants.

PROBLEMS

6.1. Verify the steps leading to (6-7).

6.2. Verify the steps leading to (6-12).

6.3. Verify the steps leading to (6-18).

6.4. Derive (6-22).

6.5. Show that the relation $g_1^2 - g_1g_3 + g_2^2 = 0$, which is necessary for (6-25) to be true, is satisfied only when f_p or f_g vanishes.

6.6. Verify the steps leading to (6-26).

6.7. Show that the expression which is obtained from (6-31) for the case where $k_2 = 0$, i.e., the dual to (6-10), is the following:

$$\begin{bmatrix} k^2 - k_1^2 - k_0^2 Kg_1 & -k_0^2 Kg_2 & -k_1k_3 \\ k_0^2 Kg_2 & k^2 - k_0^2 Kg_1 & 0 \\ -k_1k_3 & 0 & k^2 - k_3^2 - k_0^2 Kg_3 \end{bmatrix}\begin{bmatrix} h_x \\ h_y \\ h_z \end{bmatrix} = 0$$

6.8. Derive the relations in (6-35) by expressing $\mathbf{h}_0$ in a cartesian form and proceeding in the same manner as that leading to (6-18).

6.9. Verify the statement following (6-39), i.e., TE_z and TM_z modes are not coupled[40] when the fields and K are independent of y.

6.10. Repeat Prob. 4.5 but for the vector magnetic potential $\mathbf{A}$, i.e., $\mathbf{H} = \nabla \times \mathbf{A}$, and show that the following results are obtained:

---

[40]Brekhovskikh, *Waves in Layered Media*, p. 170.

$$(\nabla^2 + k^2)\mathbf{A} = -\mathbf{J} + \nabla \cdot \mathbf{A}\left(\frac{\nabla K}{K}\right)$$

$$(\nabla^2 + k^2)\Phi = \frac{-\rho}{K\epsilon_0} - \frac{\nabla K}{K} \cdot (\nabla\Phi + j\omega\mu_0\mathbf{A})$$

$$\mathbf{E} = -j\omega\mu_0\mathbf{A} - \frac{1}{j\omega\epsilon_0}\nabla\left(\frac{\nabla \cdot \mathbf{A}}{k}\right)$$

provided $\nabla \cdot \mathbf{A} + j\omega\epsilon_0 K\Phi = 0$ where $k^2 = \omega^2\mu_0\epsilon_0 K$.

6.11. Consider Prob. 6.10 when $\mathbf{A} = \mathbf{a}_z A_z(x, z)$ and $K = K(z)$. Verify the following results:[41]

$$(\nabla^2 + k^2)A_z = \frac{1}{K}\frac{\partial A_z}{\partial z}\frac{\partial K}{\partial z} - J_z$$

or

$$(\nabla^2 + k_{\text{eff}}^2)\Phi = \frac{-J_z}{\sqrt{K}}$$

where

$$k_{\text{eff}}^2 = k^2 - \frac{3}{4K^2}\left(\frac{\partial K}{\partial z}\right)^2 + \frac{1}{2K}\frac{\partial^2 K}{\partial z^2}$$

and

$$A_z = \Phi\sqrt{k}$$

6.12. Consider a plane wave in free space ($z < 0$) incident obliquely on an electrically inhomogeneous half space ($z > 0$) whose dielectric constant is a function of z. When the electric field is polarized perpendicular to the plane of incidence, i.e., in the y-direction as illustrated in Fig. 6-15, derive the reflection coefficient[42] and show that it is the following:

$$R_\perp = \frac{Y_0 C_0 - Y}{Y_0 C_0 + Y}$$

where

$$Y_0 = \sqrt{\frac{\epsilon_0}{\mu_0}}, \qquad C_0 = \cos\theta_0, \qquad Y = -\frac{H_x^t}{E_y^t}\bigg]_{z=0}$$

Hint: Use the relations in (6-59) and the first two in (6-60).

6.13. Repeat Prob. 6.12 for the case of polarization parallel to the plane of incidence. Use the geometry of Fig. 6-16. The result[43] is

[41]Ibid.

[42]Wait, Electromagnetic Waves in Stratified Media, p. 65.

[43]Ibid., p. 78.

$$R_{\parallel} = \frac{Z_0 C_0 - Z}{Z_0 C_0 + Z}$$

where

$$Z_0 = \frac{1}{Y_0}, \qquad Z = \frac{E_x^t}{H_y^t}\bigg]_{z=0}$$

Hint: Use the relations in (6-67) and the first two in (6-68).

6.14. Consider Prob. 6.12 for the case of a linearly increasing dielectric constant, i.e., $K(z) = 1 + az$, $a > 0$. Define

$$E_y^t = \mathcal{E}(z)e^{jk_1 x}$$

and show that the equation satisfied by $\mathcal{E}(z)$ is the following:

$$\frac{d^2\mathcal{E}}{d\zeta^2} + \left(\frac{k_0}{a}\right)^2 \zeta\mathcal{E} = 0$$

provided one uses the substitution

$$\zeta = C_0^2 + az = \cos^2\theta_0 + az$$

In this case, the expression for $R_\perp$ becomes

$$R_\perp = [H_{1/3}^{(2)}(u_0) - jH_{-2/3}^{(2)}(u_0)][H_{1/3}^{(2)}(u_0) + jH_{-2/3}^{(2)}(u_0)]^{-1}$$

where

$$u = \frac{2}{3}\left(\frac{k_0}{a}\right)\zeta^{3/2}, \qquad u_0 = u]_{z=0}$$

Hint: See Prob. 4.9. The solution[44] to the equation for $\mathcal{E}(\zeta)$ is $u^{1/3}H_{1/3}^{(2)}(u)$ where $H_{1/3}^{(2)}(u)$ is the Hankel function of the second kind of order $\frac{1}{3}$ and argument u.

6.15. Consider Prob. 6.13 and let $H_y^t = \mathcal{3C}(z)e^{jk_1 x}$. Show that the equation[45] for $\mathcal{3C}(z)$ is the following:

$$\frac{d^2\mathcal{3C}(z)}{dz^2} + k_0^2[K(z) - \sin^2\theta_0]\mathcal{3C}(z) - \frac{1}{K(z)}\frac{dK(z)}{dz}\frac{d\mathcal{3C}(z)}{dz} = 0$$

Hint: Use Prob. 4.8. This equation can also be transformed to the wave equation

$$\frac{d^2\Psi}{dz^2} + \left[k_0^2(K - \sin^2\theta_0) + \frac{1}{2K}\frac{d^2K}{dz^2} - \frac{3}{4K^2}\left(\frac{dK}{dz}\right)^2\right]\Psi = 0$$

by the substitution $\mathcal{3C}(z) = \Psi(z)K^{1/2}(z)$. See Prob. 6.11.

6.16. Consider Prob. 4.9 and introduce the vector magnetic potential $\mathbf{C} = \mathbf{a}_y C_y(x, z)$. Use (4-43) and (4-45) and the relation $\mathbf{B} = \nabla \times \mathbf{C}$ to obtain

---

[44]Brekhovskikh, *Waves in Layered Media*, p. 192. Note that he uses $e^{-j\omega t}$ time dependence so that our result is the complex conjugate of his result.

[45]Wait, *Electromagnetic Waves in Stratified Media*, p. 79.

the following:

$$\left(\frac{\partial^2}{\partial x^2} + \frac{\partial^2}{\partial z^2} + k^2\right)C_y = -\mu_0 J_y$$

$$E_y = -j\omega C_y$$

$$B_x = -\frac{\partial C_y}{\partial z}$$

$$B_z = \frac{\partial C_y}{\partial x}$$

where $k^2 = k_0^2 K(z)$. This is an alternate formulation of Prob. 4.9 with E_y and C_y satisfying the same equation.

6.17. Consider the problem formulated in (6-39) and (6-40) when E_y is only a function of z. Show that the TM_z wave reduces to a TEM_z wave and that the results in (6-47) are valid for $w \rightarrow \infty$. The example in Fig. 6-12 is correct except the straight line passes through the origin. This problem is also valid for propagation between parallel, perfectly conducting plates located parallel to the x-z plane.

6.18. Consider the TM_z mode in Prob. 4.8 propagating between parallel, perfectly conducting planes located at $x = 0$ and w when the region in between possesses the dielectric constant specified in (6-40). Show that $H_y(x, z)$ satisfies the following equations:

$$H_y(x, z) = \sum_{m=0}^{\infty} \cos \frac{m\pi x}{w} Z(z)$$

$$\frac{d^2 Z}{dz^2} - \frac{2\pi\delta}{L} \frac{\sin(2\pi z/L)}{1 - \delta \cos(2\pi z/L)} \frac{dZ}{dz} + \left[k_0^2 K_0\left(1 - \delta \cos \frac{2\pi z}{L}\right) - \left(\frac{m\pi}{w}\right)^2\right]Z = 0$$

The substitution

$$\zeta = \frac{\pi z}{L}, \qquad Z(z) = \left(1 - \delta \cos \frac{2\pi z}{L}\right)^{1/2} U(\zeta)$$

reduces the preceding equation to

$$\frac{d^2 U}{d\zeta^2} + \eta U = 0$$

where

$$\eta = \frac{2\delta \cos 2\zeta}{1 - \delta \cos 2\zeta} - \frac{3(\delta \sin 2\zeta)^2}{(1 - \delta \cos 2\zeta)^2} + \left(\frac{L}{w}\right)^2\left[-\left(\frac{m\pi}{w}\right)^2 + k_0^2 K_0(1 - \delta \cos 2\zeta)\right]$$

This last equation can be put into the form of Hill's equation[46] by expanding in a Fourier cosine series.

---

[46]C. Yeh, K. Casey, and Z. Kaprelian, "Transverse Magnetic Wave Propagation in Sinusoidally Stratified Dielectric Media," *IEEE Trans. on Microwave Theory and Techniques*, Vol. MTT-13, pp. 297–303, May 1965. Also, see Schelkunoff, *Electromagnetic Fields*, Sec. 9.4.

6.19. Verify the frequency values that correspond to the points A through F in Fig. 6-12.

6.20. Since H_z^t in (6-54) is the solution to the homogeneous form of (6-48), it follows that an alternate representation is

$$H_z^t = T_i e^{-j\beta \cdot r}$$

where

$$\beta^2 = \frac{k_0^2}{\ell_1}$$

Show that Snell's law is

$$\sin \theta = \sqrt{\ell_1} \sin \theta_0$$

6.21. Carry through the steps leading to (6-57) and verify that the results are correct.

6.22. A uniform plane wave with the electric field polarized parallel to the plane of incidence is incident obliquely on a dielectric half space possessing losses. The geometry is illustrated in Fig. 6-13 with K replaced by $K = K_1 - jK_2$ and $\mathbf{B}_0 = 0$. Show that the fields[47] can be expressed in the following form:

$$H_z^i = e^{-j(k_1 x + k_2 y)}$$

$$\mathbf{E}^i = (-k_2 \mathbf{a}_x + k_1 \mathbf{a}_y) \frac{H_z^i}{\omega \epsilon_0}$$

$$H_z^r = R_i e^{-j(k_1 x - k_2 y)}$$

$$\mathbf{E}^r = (k_2 \mathbf{a}_x + k_1 \mathbf{a}_y) \frac{H_z^r}{\omega \epsilon_0}$$

$$H_z^t = T_i e^{-j(k_1 x + k_6 y)}$$

$$\mathbf{E}^t = (-k_6 \mathbf{a}_x + k_1 \mathbf{a}_y) \frac{H_z^t}{\omega \epsilon_0 K}$$

where $k_6 = k_0 (K - S_0^2)^{1/2} = k_0 (\beta - j\alpha)$, $k_1 = k_0 S_0$, $k_2 = k_0 C_0$, $S_0 = \sin \theta_0$, $C_0 = \cos \theta_0$.

$$\alpha^2 = \frac{1}{2}(K_1 - S_0^2)\left[-1 + \sqrt{1 + \left[\frac{K_2}{K_1 - S_0^2}\right]^2}\right]$$

$$\beta^2 = \frac{1}{2}(K_1 - S_0^2)\left[1 + \sqrt{1 + \left[\frac{K_2}{K_1 - S_0^2}\right]^2}\right]$$

$$R_i = \frac{1 - \Delta}{1 + \Delta}$$

[47]M. Javid and P. Brown, *Field Analysis and Electromagnetics*, McGraw-Hill Book Company, Inc., New York, 1963, Sec. 14–12.

$$T_i = \frac{2}{1 + \Delta}$$

$$\Delta = \frac{k_6}{K k_2}$$

Note that H_z^t and $\mathbf{E}^t$ constitute a nonuniform plane wave, also called an inhomogeneous[48] or nonhomogeneous plane wave.

6.23. In Prob. 6.22 evaluate the time-average Poynting vector in each region and show that the following results are obtained:

$$\mathbf{S}^i = (k_1 \mathbf{a}_x + k_2 \mathbf{a}_y)(2\omega\epsilon_0)^{-1}$$

$$\mathbf{S}^r = (k_1 \mathbf{a}_x - k_2 \mathbf{a}_y) |R_i|^2 (2\omega\epsilon_0)^{-1}$$

$$\mathbf{S}^t = \left[\left(\frac{k_1}{K} + \frac{k_1^*}{K^*} \right) \mathbf{a}_x + \left(\frac{k_6}{K} + \frac{k_6^*}{K^*} \right) \mathbf{a}_y \right] |T_i|^2 e^{-2\alpha k_0 y} (4\omega\epsilon_0)^{-1}$$

where

$$\mathbf{S} = \tfrac{1}{2}\mathrm{Re}\,(\mathbf{E} \times \mathbf{H}^*) = \tfrac{1}{4}(\mathbf{E} \times \mathbf{H}^* + \mathbf{E}^* \times \mathbf{H})$$

Verify these items:

(a) At the interface ($y = 0$):

$$S_y^i + S_y^r = S_y^t$$

$$S_x^i + S_x^r \neq S_x^t$$

(b) For $K_1 = 0$:

$$\alpha^2 = \tfrac{1}{2}(S_0^2 + \sqrt{S_0^4 + K_2^2})$$

$$\beta^2 = \tfrac{1}{2}(-S_0^2 + \sqrt{S_0^4 + K_2^2})$$

$$S_x^t = 0$$

(c) For $K_1 = 0$, $K_2 \gg 1$:

$$\alpha = \beta = \sqrt{\frac{K_2}{2}} = \sqrt{\frac{\sigma}{\omega\epsilon_0}}$$

Item (c) corresponds to the case of a good conductor[49] rather than a lossy dielectric.

6.24. Consider the transmitted wave in Prob. 6.22 as an example of a non-uniform plane wave. Show that the surfaces of constant amplitude are the planes $\alpha y = $ constant and the surfaces of constant phase are the planes $S_0 x + \beta y = $ constant. The surfaces coincide when $S_0 = 0$, i.e., the case of normal incidence, and the transmitted wave becomes uni-

---

[48]Stratton, *Electromagnetic Theory*, p. 501.

[49]*Ibid.*, p. 504.

form. If the angle between the normal to the constant phase plane and the positive y-axis is designated δ, show that $\tan \delta = S_0/\beta$, where δ is a real angle. The wave normal is in the direction of $\mathbf{k}$, which is complex and which is not yet specified in a vector form. Snell's law for real angles becomes

$$\frac{\sin \theta_0}{\sin \delta} = \sqrt{S_0^2 + \beta^2} = \sqrt{K_1 + \alpha^2}$$

6.25. The transmitted wave in Prob. 6.22 can be represented in an alternate form by introducing a complex propagation vector, i.e., $\mathbf{k} = \mathbf{k}_r - j\mathbf{k}_i$ $= k_0(\boldsymbol{\beta}_1 - j\boldsymbol{\alpha}_1)$, $\boldsymbol{\beta}_1 = \beta_1\mathbf{n}_p$, $\boldsymbol{\alpha}_1 = \alpha_1\mathbf{n}_a$, $\mathbf{n}_p = S_p\mathbf{a}_x + C_p\mathbf{a}_y$, $\mathbf{n}_a = S_a\mathbf{a}_x$ $+ C_a\mathbf{a}_y$, where $S_p = \sin \theta_p$, $C_a = \cos \theta_a$, etc., and θ_p and θ_a are the angles between the normals and the positive y-axis. In this formulation, each quantity except $\mathbf{k}$ is real. Derive the following results:[50]

(a) Snell's law, $S_0 = \beta_1 S_p$.
(b) $\mathbf{n}_a = \mathbf{a}_y$ since $S_a = 0$.
(c) $\alpha_1 = \alpha$, $\beta_1^2 = S_0^2 + \beta^2$ where α and β are defined in Prob. 6.22.
(d) $\mathbf{n}_p = (S_0\mathbf{a}_x + \beta\mathbf{a}_y)/\sqrt{S_0^2 + \beta^2}$ so that $\theta_p = \delta$ where δ is defined in Prob. 6.24.
(e) $\beta_1^2 = \left(\dfrac{S_0^2 + K_1}{2}\right)\left[1 + \sqrt{\left[\dfrac{S_0^2 - K_1}{S_0^2 + K_1}\right]^2 + \left[\dfrac{K_2}{S_0^2 + K_1}\right]^2}\right]$

$\alpha_1^2 = \left(\dfrac{S_0^2 + K_1}{2}\right)\left[-1 + \sqrt{\left[\dfrac{S_0^2 - K_1}{S_0^2 + K_1}\right]^2 + \left[\dfrac{K_2}{S_0^2 + K_1}\right]^2}\right]$

(f) $\mathbf{E}^t \cdot \mathbf{n}_p \neq 0$, therefore $\mathbf{E}^t$ is not perpendicular to the planes of constant phase.
(g) $\mathbf{E}^t \cdot \mathbf{k} = 0$, therefore $\mathbf{E}^t$ is perpendicular to the complex propagation vector (wave normal).

6.26. Still another viewpoint[51] on Prob. 6.25 is to introduce the complex propagation vector as $\mathbf{k} = k\mathbf{n}$, $k = k_0(\beta_2 - j\alpha_2)$, $\mathbf{n} = S\mathbf{a}_x + C\mathbf{a}_y$, where both k and $\mathbf{n}$ are complex. Let $\sin \theta = S = a + jb$ and $\cos \theta = C = c + jd$ and show that Snell's law (for complex angles) becomes $S_0 = S(\beta_2 - j\alpha_2)$ and that α_2 and β_2 are related to α and β in Prob. 6.22 as

$$\alpha_2 = \frac{\alpha c + \beta d}{c^2 + d^2}$$

$$\beta_2 = \frac{\beta c - \alpha d}{c^2 + d^2}$$

6.27. An example of a nonuniform plane wave where the constant phase planes are perpendicular to the constant amplitude planes is the case

---

[50]*Ibid.*, p. 340; also see Budden, *Radio Waves in the Ionosphere*, p. 45.

[51]This is the viewpoint used by Stratton, *Electromagnetic Theory*, p. 501.

where the angle of incidence exceeds the critical angle so that the angle of refraction becomes complex. Stratton[52] shows that Snell's law is valid in homogeneous isotropic material whether the material constants and the angles involved are real or complex. Consider a problem with the geometry shown in Fig. 6-13 and material constants $K_0\epsilon_0$, $K_{m0}\mu_0$ in the lower region and $K\epsilon_0$, $K_m\mu_0$ in the upper region. Show that Snell's law is

$$S = \frac{S_0}{S_{0c}}$$

$$S_{0c} = \sqrt{\frac{KK_m}{K_0K_{m0}}} = \sin\theta_{0c}$$

where θ_{0c} is the critical angle. Define the fields as in Prob. 6.22 and show that when $\theta_0 > \theta_{0c}$

$$k_6 = -jk_0(S_0^2 - S_{0c}^2)^{1/2}$$

As a result, one finds that

$$\beta = 0, \quad \alpha = (S_0^2 - S_{0c}^2)^{1/2}$$
$$R_i = e^{j\psi}$$

$$\tan\frac{\psi}{2} = \frac{\alpha K_0}{KC_0}$$

The expressions for R_i and T_i in Prob. 6.22 are still valid except now

$$\Delta = \frac{k_6 K_0}{k_2 K} = -\frac{j\alpha K_0}{KC_0}$$

Since $|R_i| = 1$, it follows that total reflection has occurred so that no average energy flows into the upper region. This can also be seen by noting that E_x^t and H_z^t are in time quadrature. On the other hand, $T_i \neq 0$, since fields (evanescent) still exist in the upper region. This problem is of practical interest since it corresponds to the case of TM_x surface waves.

6.28. Consider Prob. 6.27 from the viewpoint introduced in Prob. 6.26 to show the following:

$$b = c = \alpha_2 = 0$$
$$a = S$$
$$d = \frac{-\alpha}{S_{0c}}$$
$$\beta_2 = S_{0c}$$

The choice of signs is always dictated by the requirement that the fields behave at infinity.

---

[52] *Ibid.*, p. 491.

7

Radiation of Sources in a Half Space

7.1 Electric and Magnetic Line Sources

We considered several problems involving sources in Chap. 5 beginning with solutions to Laplace's equation in two dimensions and finishing with solutions to Helmholtz's equation in two and three dimensions. The emphasis was on presenting the solutions in terms of known one-dimensional problems and their accompanying solutions. We noted that problems that were unbounded in one dimension were most easily solved by introducing a Fourier transform in a variable that corresponded to that dimension. Unfortunately, the introduction of the Fourier transform brings with it the problem of evaluating the inverse Fourier transform. In a few simple cases the inverse can be evaluated by resorting to tables of transform pairs[1] or by using some of the identities derived in Chap. 3. However, in most cases the inverse transform cannot be evaluated in an exact form but must be approximated. In many cases, the quantities of interest are the radiation fields rather than the complete fields everywhere in space so that asymptotic expansions can be

---

[1]G. Campbell and R. Foster, *Fourier Integrals for Practical Applications*, D. Van Nostrand Company, Inc., New York, 1948. Note that many of the available pairs are not useful since the inverse transform is evaluated with a contour that specifies the principal value.

used to replace many of the known functions. When the solution is in the form of an inverse transform, i.e., an integral, then asymptotic methods can be used to approximate the integral.

The first problem that we wish to consider is the simple two-dimensional one shown in Fig. 7-1. It consists of an electric line source in free space situated above a perfectly conducting half space or ground plane. We will assume that the source is time harmonic and uniform in the z-direction so that the

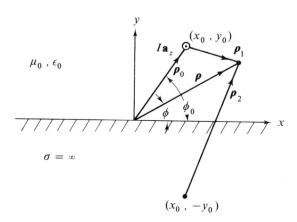

Fig. 7-1   Uniform electric line source in free space above a perfectly conducting ground plane.

two-dimensional Helmholtz equation in x and y applies. The relevant equation is (4-16) which here becomes

$$(\nabla_t^2 + k_0^2)A_z(x, y) = -I\delta(x - x_0)\delta(y - y_0) \qquad (7\text{-}1)$$

with the boundary condition

$$A_z(x, y) = 0, \qquad y = 0$$

where I is the strength of the line source in amperes. We should note that the dimensions of the right side of (7-1) are wrong as the equation now stands although the solution will be correct. This is because a limiting procedure should be used in passing from J_z in (4-16) to I in (7-1). This procedure is noted in the expression preceding (4-28). What we have done already to the right side of (7-1) is passed to the limit to replace J_z by I, i.e., integrated J_z with respect to x_0 and y_0, without disturbing the deltas which indicate the location of the source. We might also recall that the solution to (7-1) in unbounded space is (4-29). As a result, the solution to (7-1) above a perfectly conducting half space is (4-29) plus a similar expression for the image source at $(x_0, -y_0)$ with strength $-I$. Although we already know the solution to the problem illustrated in Fig. 7-1 from the method of images, let us solve the problem by using a one-dimensional exponential Fourier transform in x and α_1, i.e., (3-11). In addition, we could use a one-dimensional sine Fourier transform in y and α_2, i.e., (3-3), but it is easier not to do so. If we apply (3-11) to both sides of (7-1) we obtain the ordinary differential equation

$$\left(\frac{d^2}{dy^2} + k_{40}^2\right)\tilde{A}_z(\alpha_1, y) = -B_1\delta(y - y_0) \tag{7-2}$$

where

$$k_{40}^2 = k_0^2 - \alpha_1^2$$

$$B_1 = \frac{Ie^{-j\alpha_1 x_0}}{\sqrt{2\pi}}$$

Since (7-2) is of the same form as (2-1), we can immediately assume a solution of the form

$$\tilde{A}_z(\alpha_1, y) = A[\sin k_{40}y\, e^{-jk_{40}y_0}u(y_0 - y) + \sin k_{40}y_0 e^{-jk_{40}y}u(y - y_0)] \tag{7-3}$$

At the risk of being repetitious, we should mention that the form of (7-3) was chosen because the first term in y satisfies the boundary condition at $y = 0$, the second term in y satisfies the radiation condition, and both terms satisfy the homogeneous form of (7-2). Only the constant A is present in (7-3) because we have already used the condition that $\tilde{A}_z(\alpha_1, y)$ is continuous at the source, i.e., at $y = y_0$. The constant A can be determined from the strength of the source, i.e.,

$$\frac{d\tilde{A}_z^>}{dy} - \frac{d\tilde{A}_z^<}{dy} = -B_1 \tag{7-4}$$

where $\tilde{A}_z^>$ is the second term in (7-3) and $\tilde{A}_z^<$ is the first term in (7-3). This last condition allows one to determine A as

$$A = \frac{B_1}{k_{40}}$$

The solution to (7-1) becomes

$$A_z(x, y) = \frac{I}{2\pi}\int_{-\infty}^{\infty} \frac{1}{k_{40}}[\sin k_{40}y\, e^{-jk_{40}y_0}u(y_0 - y) + \sin k_{40}y_0 e^{-jk_{40}y}u(y - y_0)]$$
$$\times e^{j\alpha_1(x-x_0)}\, d\alpha_1$$

which can be put into the more meaningful form

$$A_z(x, y) = \frac{I}{2\pi j}\int_{-\infty}^{\infty} \frac{e^{j\alpha_1(x-x_0)}}{k_{40}}[e^{jk_{40}(y-y_0)}u(y_0 - y) + e^{-jk_{40}(y-y_0)}u(y - y_0)]\, d\alpha_1$$
$$- \frac{I}{2\pi j}\int_{-\infty}^{\infty} \frac{e^{j\alpha_1(x-x_0)}}{k_{40}}e^{-jk_{40}(y+y_0)}\, d\alpha_1, \qquad y \geq 0 \tag{7-5}$$

The first term in (7-5) represents a downgoing wave from the source, the second term represents an upgoing wave from the source, and the third term represents an upgoing wave from the image. We should also note that the one-dimensional problem in (7-2) and its solution (7-3) is that of a plane wave source in front of a conducting half space.

The next problem we wish to consider is that shown in Fig. 7-2. This problem differs from that in Fig. 7-1 only in that the conducting half space is replaced by a dielectric half space. As a result, this problem is more complicated than the previous one since fields will exist in both the air and the

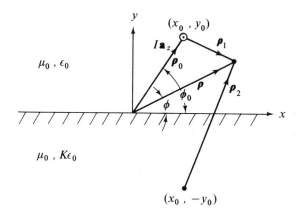

Fig. 7-2 Uniform electric line source in free space above a dielectric half space.

dielectric regions. The equation to be solved in the air region is still (7-1) and the equation to be solved in the dielectric region is the homogeneous two-dimensional Helmholtz equation

$$(\nabla_t^2 + k^2)A_z(x, y) = 0$$

where

$$k^2 = Kk_0^2$$

and K is the dielectric constant (specific electric inductive capacity). Since

$$E_z = -j\omega\mu_0 A_z$$

$$H_x = \frac{\partial A_z}{\partial y}$$

in both regions (we have assumed μ_0 in both regions), the boundary conditions on the continuity of the tangential electric and magnetic fields become

$$\tilde{A}_{z0}^{\lessgtr}(\alpha_1, y) = \tilde{A}_z(\alpha_1, y), \qquad y = 0$$
$$\frac{d\tilde{A}_{z0}^{\lessgtr}(\alpha_1, y)}{dy} = \frac{d\tilde{A}_z(\alpha_1, y)}{dy}, \qquad y = 0 \tag{7-6}$$

where $\tilde{A}_{z0}$ is the solution to (7-2) and A_z is the solution to

$$\left(\frac{d^2}{dy^2} + k_4^2\right)\tilde{A}_z(\alpha_1, y) = 0 \tag{7-7}$$

where $k_4^2 = k^2 - \alpha_1^2 = Kk_0^2 - \alpha_1^2$. We write the solutions to (7-2) and (7-7) in the following form:

$$\tilde{A}_{z0}^{\gtrless}(\alpha_1, y) = Ce^{-jk_{40}(y-y_0)}, \qquad y \geq y_0$$
$$\tilde{A}_{z0}^{\lessgtr}(\alpha_1, y) = Ae^{jk_{40}(y-y_0)} + Be^{-jk_{40}(y+y_0)}, \qquad 0 \leq y \leq y_0 \tag{7-8}$$
$$\tilde{A}_z(\alpha_1, y) = De^{j(k_4 y - k_{40}y_0)}, \qquad y \leq 0$$

The first expression in (7-8) represents an upgoing wave, the second expression consists of a downgoing and an upgoing wave, and the third expression represents a downgoing wave. We now impose the boundary conditions (7-6) and the source condition (7-4) as well as the source condition concerning the continuity of $\tilde{A}_{z0}$ at $y = y_0$. Thus, we obtain the four constants

$$A = \frac{Ie^{-j\alpha_1 x_0}}{j2k_{40}\sqrt{2\pi}}$$
$$B = RA$$
$$C = A(1 + Re^{-2jk_{40}y_0}) \tag{7-9}$$
$$D = \frac{2k_{40}}{k_{40} + k_4}A$$
$$R = \frac{k_{40} - k_4}{k_{40} + k_4} = -1 + \frac{2k_{40}}{k_{40} + k_4}$$

where we have introduced the reflection coefficient R. The second expression for the reflection coefficient is more useful than the first since the second term in that expression vanishes for the case of a perfectly conducting ground, i.e., $k_4 \to \infty$ since $K \to \infty$, and R reduces to -1. The general solution for A_{z0} and A_z can be written as

$$A_{z0}(x, y) = \frac{I}{4\pi j}\int_{-\infty}^{\infty} \frac{e^{j\alpha_1(x-x_0)}}{k_{40}}[e^{-jk_{40}|y-y_0|} + Re^{-jk_{40}(y+y_0)}]\, d\alpha_1$$
$$A_z(x, y) = \frac{I}{2\pi j}\int_{-\infty}^{\infty} \frac{e^{j\alpha_1(x-x_0)}e^{j(k_4 y - k_{40}y_0)}}{k_{40} + k_4}\, d\alpha_1 \tag{7-10}$$

By using the third identity in (3-51) and the second expression for R in (7-9) we can write the first term in (7-10) as

$$A_{z0}(x,y) = \frac{I}{4j}[H_0^{(2)}(k_0\rho_1) - H_0^{(2)}(k_0\rho_2)] + \frac{I}{2\pi j}\int_{-\infty}^{\infty} \frac{e^{j\alpha_1(x-x_0)}e^{-jk_{40}(y+y_0)}}{k_{40} + k_4}\, d\alpha_1$$

(7-11)

which shows more clearly the direct contributions from the source and the image. The problem we have just solved is very similar to the one solved by Collin[2] except he considered a finite thickness dielectric sheet on a perfectly conducting ground plane. Earlier in Sec. 11.1, Collin also considered the dual to our problem but with the line source replaced by a plane wave. Further, he notes that for our problem there is no surface wave if the permeabilities of the two regions are the same. If surface waves exist, they are the contribution of the residues of the poles in the reflection coefficient in (7-10). In the present case, R has a pole when $k_{40} + k_4 = 0$. Since this condition is never realized, R does not have a pole and, therefore, no surface wave exists.

The problem with a source that is dual to the one we have just considered is shown in Fig. 7-3. This case corresponds to a magnetic line source above a dielectric half space. The relevant equation in the free space region for the vector electric potential in terms of the magnetic line source is the dual to (7-1), i.e.,

$$(\nabla_t^2 + k_0^2)A_{mz0}\,(x,y) = -I_m\delta(x - x_0)\delta(y - y_0)$$

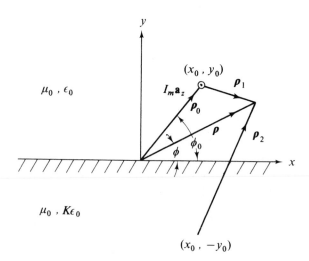

Fig. 7-3   Uniform magnetic line source in free space above a dielectric half space.

[2]R. Collin, *Field Theory of Guided Waves*, McGraw-Hill Book Company, Inc., New York, 1960, Sec. 11.8.

The equation in the dielectric region is

$$(\nabla_t^2 + k^2)A_{mz}(x, y) = 0 \tag{7-12}$$

and the boundary conditions at $y = 0$ are

$$A_{mz0} = KA_{mz}$$

$$\frac{\partial A_{mz0}}{\partial y} = \frac{\partial A_{mz}}{\partial y}$$

If we follow the same procedure as before, i.e., introduce an exponential Fourier transform in x and α_1 and express the potentials in the form of (7-8), we find that the four constants are

$$A = \frac{I_m e^{-j\alpha_1 x_0}}{j2k_{40}\sqrt{2\pi}}$$

$$B = RA$$

$$C = A(1 + Re^{-2jk_{40}y_0}) \tag{7-13}$$

$$D = \frac{2k_{40}}{k_4 + Kk_{40}} A$$

$$R = \frac{-k_4 + Kk_{40}}{k_4 + Kk_{40}} = 1 - \frac{2k_4}{k_4 + Kk_{40}}$$

The second expression for R shows quite clearly that the second term vanishes for $K \rightarrow \infty$, since now $k_4 \rightarrow k_0\sqrt{K}$, and R reduces to 1. The general solution for A_{mz0} and A_{mz} can now be written as

$$A_{mz0}(x, y) = \frac{I_m}{4\pi j} \int_{-\infty}^{\infty} \frac{e^{j\alpha_1(x-x_0)}}{k_{40}} [e^{-jk_{40}|y-y_0|} + Re^{-jk_{40}(y+y_0)}] \, d\alpha_1 \tag{7-14}$$

$$A_{mz}(x, y) = \frac{I_m}{2\pi j} \int_{-\infty}^{\infty} \frac{e^{j\alpha_1(x-x_0)}}{k_4 + Kk_{40}} e^{j(k_4 y - k_{40}y_0)} \, d\alpha_1$$

The reflection coefficient R in (7-13) now has a pole at

$$\alpha_1 = |\alpha_p| = k_0\left(\frac{K}{1 + K}\right)^{1/2}$$

which is just the value of α_1 that corresponds to the Brewster[3] angle θ_B, i.e.,

$$\alpha_1 = k_0 \sin \theta_B$$

$$\sin \theta_B = \left(\frac{K}{1 + K}\right)^{1/2}$$

---

[3]Collin, *Field Theory of Guided Waves*, p. 455.

Now, the fact that no pole existed in the previous problem meant that no Brewster angle existed because the permeabilities were the same in both regions. This is the well-known fact that no Brewster angle exists for perpendicular polarization unless the regions have different permeabilities.[4]

We can complicate the two preceding problems by letting the dielectric possess losses. In this case the quantity K must be considered complex with a negative imaginary part. The quantities k_{40}, k_4, and α_1 also become complex in such a fashion that the integrals containing these quantities are well behaved at infinity.

Another way in which we can complicate the two problems is by replacing the isotropic dielectric by an anisotropic dielectric. Let us do this with the problem shown in Fig. 7-3. We shall also assume that the magnetostatic field is in the x-direction rather than arbitrarily oriented. The first equation in (7-12) is still valid in the air region but the second equation is not valid for the anisotropic dielectric. Instead, in the anisotropic region, we should use the homogeneous form of (4-63) for C_{mz} or the homogeneous form of (4-65) for H_z. Since the electric vector potential is still in the z-direction, the relevant equation in the anisotropic dielectric region is

$$(\ell_1 \nabla_t^2 + k_0^2)C_{mz}(x, y) = 0 \quad \text{or} \quad (\ell_1 \nabla_t^2 + k_0^2)H_z(x, y) = 0 \qquad (7\text{-}15)$$

where ℓ_1 is given in (4-73) and (4-71). The equation involving H_z in (7-15) can also be obtained from (6-5) by letting $k_3 = 0$. In the free space region, $C_{mz0} = \epsilon_0 A_{mz0}$ so that the first equation in (7-12) becomes

$$(\nabla_t^2 + k_0^2)C_{mz0}(x, y) = -\epsilon_0 I_m \delta(x - x_0)\delta(y - y_0) \qquad (7\text{-}16)$$

Since it follows from (4-60) that

$$H_z = -j\omega C_{mz}$$

and from the definition of $\mathbf{C}_m$ in (4-36) that

$$E_x = -\frac{1}{\epsilon_0}\left(\ell_1 \frac{\partial C_{mz}}{\partial y} + \ell_2 \frac{\partial C_{mz}}{\partial x}\right)$$
$$E_y = -\frac{1}{\epsilon_0}\left(\ell_2 \frac{\partial C_{mz}}{\partial y} - \ell_1 \frac{\partial C_{mz}}{\partial x}\right) \qquad (7\text{-}17)$$

the boundary conditions concerning the continuity of the tangential electric

[4] *Ibid.*, p. 90.

and magnetic fields at $y = 0$ become, respectively, in terms of the electric vector potential

$$\frac{\partial C_{mz0}}{\partial y} = \ell_1 \frac{\partial C_{mz}}{\partial y} + \ell_2 \frac{\partial C_{mz}}{\partial x}$$

$$C_{mz0} = C_{mz}$$

(7-18)

After introducing a one-dimensional exponential Fourier transform in x and α_1, the relevant equations and boundary conditions become

$$\left(\frac{d^2}{dy^2} + k_{40}^2\right) \tilde{C}_{mz0}(\alpha_1, y) = \frac{-\epsilon_0 I_m}{\sqrt{2\pi}} e^{-j\alpha_1 x_0} \delta(y - y_0), \qquad y \geq 0$$

$$\left(\frac{d^2}{dy^2} + k_{41}^2\right) \tilde{C}_{mz}(\alpha_1, y) = 0, \qquad y \leq 0$$

(7-19)

$$\frac{d\tilde{C}_{mz0}}{dy} = \left(\ell_1 \frac{d}{dy} + j\alpha_1 \ell_2\right) \tilde{C}_{mz}, \qquad y = 0$$

$$\tilde{C}_{mz0} = \tilde{C}_{mz}, \qquad y = 0$$

where

$$k_{40}^2 = k_0^2 - \alpha_1^2$$

as before and

$$k_{41}^2 = \frac{k_0^2}{\ell_1} - \alpha_1^2$$

The only difference between the one-dimensional problem in (7-19) and the two previous ones is that one of the boundary conditions is more complicated, although still of the homogeneous type. If we introduce the four constants, A, B, C, D, as in (7-8) and then apply the boundary conditions and the source conditions, we obtain the expressions

$$A = \frac{\epsilon_0 I_m e^{-j\alpha_1 x_0}}{j 2 k_{40} \sqrt{2\pi}}$$

$$B = RA$$

$$C = A(1 + R e^{-2jk_{40}y_0})$$

$$D = \frac{2k_{40}}{k_{40} + k_{41}\ell_1 + \alpha_1 \ell_2} A$$

(7-20)

$$R = \frac{k_{40} - k_{41}\ell_1 - \alpha_1 \ell_2}{k_{40} + k_{41}\ell_1 + \alpha_1 \ell_2} = 1 - \frac{2(k_{41}\ell_1 + \alpha_1 \ell_2)}{k_{40} + k_{41}\ell_1 + \alpha_1 \ell_2}$$

Again, the second expression for R is written so that the second term in it vanishes when the lower region becomes a perfect conductor. The general solutions for C_{mz0} and C_{mz} can now be written as

$$C_{mz0}(x, y) = \frac{\epsilon_0 I_m}{4\pi j} \int_{-\infty}^{\infty} \frac{e^{j\alpha_1(x-x_0)}}{k_{40}} [e^{-jk_{40}|y-y_0|} + Re^{-jk_{40}(y+y_0)}] \, d\alpha_1$$

$$C_{mz}(x, y) = \frac{\epsilon_0 I_m}{2\pi j} \int_{-\infty}^{\infty} \frac{e^{j\alpha_1(x-x_0)} e^{j(k_{41}y - k_{40}y_0)}}{k_{40} + k_{41}\ell_1 + \alpha_1\ell_2} \, d\alpha_1$$

(7-21)

and the fields in both regions can be obtained from (7-17).

Although the inverse transform expressions in (7-5), (7-10), (7-14), and (7-21) represent formally the solutions to our problems, there are only a few of the simpler cases where the inverse transform is known and tabulated. In all other cases, the inverse transform must be evaluated by resorting to complex variable theory, i.e., by replacing the integral by a suitable contour integral and then using Cauchy's integral formula and the residue theorem to evaluate the contour integral. Even then, there are many cases where the integrand is so complicated that no attempt is made to obtain an exact solution but rather attention is directed toward obtaining an approximate solution. In the case of electromagnetic problems, one is usually interested in fields at distances that are many wavelengths removed from finite sources, i.e., the fields in the far zone or radiation zone, so that asymptotic methods are often employed. In simple cases, exact functions are replaced by large argument expressions whereas in more complicated cases the exact integrals are evaluated by asymptotic methods. The general nature of these asymptotic methods is covered quite well by Carrier, Krook, and Pearson[5] whereas many specific examples are considered in detail by others.[6]

7.2   Electric and Magnetic Dipoles

We now turn our attention to several three-dimensional problems. These problems will be similar to the two-dimensional problems we have already discussed as far as the nature of the two half spaces is concerned and will differ only in that the source will be an infinitesimal electric or magnetic dipole located at a point. The two-dimensional dipoles, or uniform line sources, that we considered previously can be visualized as an array of equal amplitude three-dimensional dipoles laid end to end. We shall also be con-

---

[5]G. Carrier, M. Krook, and C. Pearson, *Functions of a Complex Variable, Theory and Techniques*, McGraw-Hill Book Company, Inc., New York, 1966, Chap. 6.

[6]A. Baños, Jr., *Dipole Radiation in the Presence of a Conducting Half-Space*, Pergamon Press, Inc., New York, 1966. Also, see D. Jones, *The Theory of Electromagnetism*, Pergamon Press, Inc., New York, 1964, and J. Wait, *Electromagnetic Waves in Stratified Media*, Pergamon Press, Inc., New York, 1962.

cerned with dipoles that are either vertical or horizontal, i.e., perpendicular or parallel to the interface, respectively. The first example we wish to consider is the vertical electric dipole shown in Fig. 7-4. We will assume that k_0 and k are complex with a negative imaginary part and that the permeabilities and dielectric constants are different in both regions. The equation in the upper

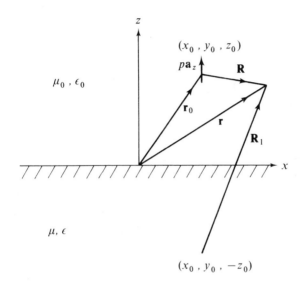

Fig. 7-4   Vertical electric dipole above a half space. The constants of the upper region are μ_0, ϵ_0 (not necessarily free space) whereas those of the lower region are μ, ϵ.

region for the only finite component of the magnetic vector potential is (4-16) with $i = z$. If a limiting procedure is employed on the source term, the relevant equations become

$$(\nabla^2 + k_0^2)A_{z0}(\mathbf{r}) = -p\delta(\mathbf{r} - \mathbf{r}_0), \quad z \geq 0$$
$$(\nabla^2 + k^2)A_z(\mathbf{r}) = 0, \quad z \leq 0$$

$$(7\text{-}22)$$

where $k_0^2 = \omega^2\mu_0\epsilon_0$, $k^2 = \omega^2\mu\epsilon$, and p is the electric dipole moment, i.e., $p = I\,\Delta P$. The constants μ_0 and ϵ_0 are those of the upper region for this problem and are not necessarily those of free space. Also, the dimensions on both sides of the first equation in (7-22) are not the same since we have already passed to the limit with the source term just as we did with (7-1). The field components can be written down by using the relevant equations in (4-7). Thus, one obtains for this TM_z wave the expressions

$$H_x = \frac{\partial A_z}{\partial y}$$

$$H_y = -\frac{\partial A_z}{\partial x}$$

$$E_x = \frac{1}{j\omega\epsilon}\frac{\partial^2 A_z}{\partial x\,\partial z}$$

$$E_y = \frac{1}{j\omega\epsilon}\frac{\partial^2 A_z}{\partial y\,\partial z}$$

$$E_z = \frac{1}{j\omega\epsilon}\left(k^2 + \frac{\partial^2}{\partial z^2}\right)A_z$$

The boundary conditions on the continuity of the tangential electric and magnetic fields at the interface give us the following relations between the magnetic vector potentials in the two regions:

$$A_{z0} = A_z, \qquad z = 0$$

$$\epsilon\frac{\partial A_{z0}}{\partial z} = \epsilon_0\frac{\partial A_z}{\partial z}, \qquad z = 0 \tag{7-23}$$

We now introduce the double exponential Fourier transform pair in x, y and α_1, α_2 that is defined in (3-16) and apply the direct transform to the equations in (7-22) and (7-23). The results of this process are the following equations:

$$\left(\frac{d^2}{dz^2} + k_{30}^2\right)\tilde{A}_{z0}(\alpha_1, \alpha_2, z) = \frac{-p}{2\pi}e^{-j(\alpha_1 x_0 + \alpha_2 y_0)}\delta(z - z_0)$$

$$\left(\frac{d^2}{dz^2} + k_3^2\right)\tilde{A}_z(\alpha_1, \alpha_2, z) = 0$$

$$\tilde{A}_{z0}(\alpha_1, \alpha_2, z) = \tilde{A}_z(\alpha_1, \alpha_2, z), \qquad z = 0 \tag{7-24}$$

$$\epsilon\frac{d\tilde{A}_{z0}(\alpha_1, \alpha_2, z)}{dz} = \epsilon_0\frac{d\tilde{A}_z(\alpha_1, \alpha_2, z)}{dz}, \qquad z = 0$$

where

$$k_{30}^2 = k_0^2 - \alpha_1^2 - \alpha_2^2 = k_0^2 - \lambda^2$$

$$k_3^2 = k^2 - \alpha_1^2 - \alpha_2^2 = k^2 - \lambda^2$$

It is interesting to note that the one-dimensional problem defined in (7-24) is the same as the one-dimensional problem that resulted from the previous examples of line sources in isotropic media. Also, the one-dimensional problems we have considered thus far are just as simple as many of the one-dimensional Green's function problems introduced in Chap. 2. The complexity in the exterior boundary value problems considered here arises when one attempts to evaluate the inverse transforms. It might be well to recall that any problem that results in a partial differential equation has infinitely many solutions. In the case of interior problems, i.e., problems with most dimensions of finite extent, the boundary and source conditions allow one to pick out the relevant solutions which can still be infinite in number. In the case of exterior problems, one has to add the radiation condition to the boundary and source conditions in order to try to pick out the relevant solutions. Unfortunately, when using transform techniques the general solution is not in a form where one can proceed very far in picking out the relevant solutions until actual attempts are made to evaluate the inverse transform. As a result, the reason these transform problems seem so simple is because we have deferred the major part of the effort of obtaining the solution by introducing the transform technique in the first place. Thus, the transform technique allows one to replace a partial differential equation by an ordinary differential equation which is easy to solve but the task of finding the solutions to the rest of the partial differential equation is merely delayed since these solutions are contained in the inverse transform. The problem of evaluating the inverse transform is quite complicated in general since the integrands are often transcendental or multivalued functions. For instance, the book by Baños[7] is concerned for the most part with the asymptotic evaluation of the inverse transforms of the four problems that result from horizontal and vertical electric and magnetic dipoles situated above a half space.

Let us now return to the problem in (7-24) and write down the solution in a form similar to that in (7-8). Thus, we obtain

$$\tilde{A}_{z0}^{>}(\alpha_1, \alpha_2, z) = Ce^{-jk_{30}(z-z_0)}, \qquad z \geq z_0$$

$$\tilde{A}_{z0}^{<}(\alpha_1, \alpha_2, z) = Ae^{jk_{30}(z-z_0)} + Be^{-jk_{30}(z+z_0)}, \qquad 0 \leq z \leq z_0 \qquad (7\text{-}25)$$

$$\tilde{A}_z(\alpha_1, \alpha_2, z) = De^{j(k_3 z - k_{30} z_0)}, \qquad z \leq 0$$

The four constants are evaluated by imposing the two boundary conditions in (7-24) and the two source conditions from the first equation in (7-24). The constants are

$$A = \frac{pe^{-j(\alpha_1 x_0 + \alpha_2 y_0)}}{4\pi jk_{30}}$$

$$B = RA$$

$$C = A(1 + Re^{-2jk_{30}z_0}) \qquad (7\text{-}26)$$

$$D = \frac{2\epsilon k_{30}}{\epsilon k_{30} + \epsilon_0 k_3} A$$

$$R = \frac{\epsilon k_{30} - \epsilon_0 k_3}{\epsilon k_{30} + \epsilon_0 k_3} = 1 - \frac{2\epsilon_0 k_3}{\epsilon k_{30} + \epsilon_0 k_3}$$

---

[7]Baños, *Dipole Radiation in the Presence of a Conducting Half-Space.*

We can now write the general solution to (7-22) in terms of the inverse transforms as

$$A_{z0}(x, y, z) = \frac{p}{8\pi^2 j} \int\!\!\!\int_{-\infty}^{\infty} \frac{e^{j[\alpha_1(x-x_0)+\alpha_2(y-y_0)]}}{k_{30}} [e^{-jk_{30}|z-z_0|} + Re^{-jk_{30}(z+z_0)}]\, d\alpha_1\, d\alpha_2$$

$$= \frac{P}{8\pi^2 j} \int\!\!\!\int_{-\infty}^{\infty} \frac{e^{j[\alpha_1(x-x_0)+\alpha_2(y-y_0)]}}{k_{30}} \Bigg[e^{-jk_{30}|z-z_0|} + e^{-jk_{30}(z+z_0)} \tag{7-27}$$

$$- \frac{2\epsilon_0 k_3}{\epsilon k_{30} + \epsilon_0 k_3} e^{-jk_{30}(z+z_0)} \Bigg]\, d\alpha_1\, d\alpha_2$$

$$A_z(x, y, z) = \frac{\epsilon p}{4\pi^2 j} \int\!\!\!\int_{-\infty}^{\infty} \frac{1}{\epsilon k_{30} + \epsilon_0 k_3} e^{j[\alpha_1(x-x_0)+\alpha_2(y-y_0)+k_3 z-k_{30}z_0]}\, d\alpha_1\, d\alpha_2$$

The results in (7-27) are very similar in form to those given by Jones[8] since he uses the same definition of the Fourier transform pair. We can also use the last identity in (3-54) to simplify A_{z0} in (7-27). Thus, it can be written as

$$A_{z0}(x, y, z) = \frac{pe^{-jkR}}{4\pi R} + \frac{pe^{-jkR_1}}{4\pi R_1} - \frac{p\epsilon_0}{4\pi^2 j} \int\!\!\!\int_{-\infty}^{\infty} \frac{k_3 e^{j[\alpha_1(x-x_0)+\alpha_2(y-y_0)-k_{30}(z+z_0)]}}{k_{30}(\epsilon k_{30} + \epsilon_0 k_3)}\, d\alpha_1\, d\alpha_2$$

$$\tag{7-28}$$

which shows more clearly that the magnetic vector potential in the upper region can be considered as the superposition of three contributions. We also note the existence of a pole in both integrands in (7-27). The contribution of this pole appears in the form of a surface wave[9] just as we saw with the line source in connection with (7-14). The value of λ at the pole is

$$\lambda = \lambda_p = k_0 k \left[\frac{k_0^2 \mu^2 - k^2 \mu_0^2}{k_0^4 \mu^2 - k^4 \mu_0^2} \right]^{1/2} \tag{7-29}$$

When the permeabilities of the two regions are the same, i.e., $\mu = \mu_0$, (7-29) reduces to

$$\lambda = \lambda_p = \frac{k_0 k}{(k^2 + k_0^2)^{1/2}} \tag{7-30}$$

which agrees with Sommerfeld.[10] This last expression finally reduces to (7-14) for the case of a dielectric half space.

---

[8] Jones, *The Theory of Electromagnetism*, Sec. 6.25.

[9] Sommerfeld, *Partial Differential Equations in Physics*, p. 253. Also, see Baños, *Dipole Radiation in the Presence of a Conducting Half-Space*, Sec. 2.64.

[10] Sommerfeld, *Partial Differential Equations in Physics*, p. 253.

When the electric dipole is oriented horizontally, i.e., parallel to the interface, the problem becomes more complicated because the boundary conditions cannot be satisfied with a single component of the magnetic vector potential. The geometry of the problem is shown in Fig. 7-5 which differs from Fig. 7-4 only in that the dipole is in the x-direction. As mentioned by

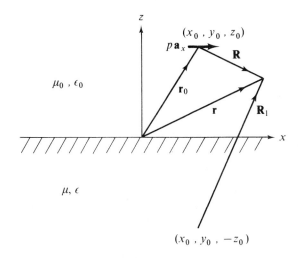

Fig. 7-5   Horizontal electric dipole above a half space. The constants of the upper region are μ_0, ϵ_0 (not necessarily free space) whereas those of the lower region are μ, ϵ.

Sommerfeld,[11] an x-component of the magnetic vector potential is sufficient to satisfy the boundary conditions only when the dipole is located in unbounded space or above a perfectly conducting surface. In the case we wish to consider, it is necessary to include a z-component of the magnetic vector potential in order to satisfy the boundary conditions. As before, A_{x0} in the upper region is the only component of the vector potential that satisfies an inhomogeneous equation. The relevant equations for the components of the magnetic vector potentials are

$$(\nabla^2 + k_0^2)A_{x0}(\mathbf{r}) = -p\delta(\mathbf{r} - \mathbf{r}_0), \quad z \geq 0$$

$$(\nabla^2 + k_0^2)A_{z0}(\mathbf{r}) = 0, \quad z \geq 0$$

$$(\nabla^2 + k^2)A_x(\mathbf{r}) = 0, \quad z \leq 0$$

$$(\nabla^2 + k^2)A_z(\mathbf{r}) = 0, \quad z \leq 0$$

$$(7\text{-}31)$$

---

[11] *Ibid.*, Sec. 33.

and the field components in terms of the potentials are

$$H_x = \frac{\partial A_z}{\partial y}$$

$$H_y = \frac{\partial A_x}{\partial z} - \frac{\partial A_z}{\partial x}$$

$$H_z = -\frac{\partial A_x}{\partial y}$$

$$E_x = \frac{1}{j\omega\epsilon}\left[\left(k^2 + \frac{\partial^2}{\partial x^2}\right)A_x + \frac{\partial^2 A_z}{\partial x\,\partial z}\right] = -j\omega\mu A_x + \frac{1}{j\omega\epsilon}\frac{\partial}{\partial x}\nabla\cdot\mathbf{A}$$

$$E_y = \frac{1}{j\omega\epsilon}\frac{\partial}{\partial y}\left(\frac{\partial A_x}{\partial x} + \frac{\partial A_z}{\partial z}\right) = \frac{1}{j\omega\epsilon}\frac{\partial}{\partial y}\nabla\cdot\mathbf{A}$$

$$E_z = \frac{1}{j\omega\epsilon}\left[\left(k^2 + \frac{\partial^2}{\partial z^2}\right)A_z + \frac{\partial^2 A_x}{\partial x\,\partial z}\right] = -j\omega\mu A_z + \frac{1}{j\omega\epsilon}\frac{\partial}{\partial z}\nabla\cdot\mathbf{A}$$

The boundary conditions in this case are even more involved than those in (7-18). From the continuity of E_y we find that

$$\frac{1}{\epsilon_0}\nabla\cdot\mathbf{A}_0 = \frac{1}{\epsilon}\nabla\cdot\mathbf{A}$$

Now, the continuity of E_x gives the relation

$$\mu_0 A_{x0} = \mu A_x$$

The continuity of H_x is satisfied for $A_{z0} = A_z$ and the continuity of H_y for

$$\frac{\partial A_{x0}}{\partial z} - \frac{\partial A_{z0}}{\partial x} = \frac{\partial A_x}{\partial z} - \frac{\partial A_z}{\partial x}$$

These four boundary conditions at $z = 0$ can be summarized in the following form:

$$\mu_0 A_{x0} = \mu A_x$$

$$\frac{\partial A_{x0}}{\partial z} = \frac{\partial A_x}{\partial z}$$

$$A_{z0} = A_z \tag{7-32}$$

$$\epsilon\left(\frac{\partial A_{x0}}{\partial x} + \frac{\partial A_{z0}}{\partial z}\right) = \epsilon_0\left(\frac{\partial A_x}{\partial x} + \frac{\partial A_z}{\partial z}\right)$$

If we introduce the double exponential Fourier transform (3-16) as before, we obtain the one-dimensional problem below:

$$\left(\frac{d^2}{dz^2} + k_{30}^2\right) \tilde{A}_{x0}(\alpha_1, \alpha_2, z) = \frac{-p}{2\pi} e^{-j(\alpha_1 x_0 + \alpha_2 y_0)} \delta(z - z_0), \qquad z \geq 0$$

$$\left(\frac{d^2}{dz^2} + k_{30}^2\right) \tilde{A}_{z0}(\alpha_1, \alpha_2, z) = 0, \qquad z \geq 0$$

$$\left(\frac{d^2}{dz^2} + k_3^2\right) \tilde{A}_x(\alpha_1, \alpha_2, z) = 0, \qquad z \leq 0$$ (7-33)

$$\left(\frac{d^2}{dz^2} + k_3^2\right) \tilde{A}_z(\alpha_1, \alpha_2, z) = 0, \qquad z \leq 0$$

and the boundary conditions at $z = 0$

$$\mu_0 \tilde{A}_{x0} = \mu \tilde{A}_x$$

$$\frac{d\tilde{A}_{x0}}{dz} = \frac{d\tilde{A}_x}{dz}$$

$$\tilde{A}_{z0} = \tilde{A}_z$$

$$\frac{d}{dz}(\epsilon \tilde{A}_{z0} - \epsilon_0 \tilde{A}_z) = j\alpha_1(\epsilon_0 \tilde{A}_x - \epsilon \tilde{A}_{x0})$$

The problem defined in (7-33) differs from that defined in (7-24) in that there are two extra unknown functions. However, there are two additional boundary conditions so that the solutions are completely specified. We write the solutions in the following form:

$$\tilde{A}_{x0}^{>}(\alpha_1, \alpha_2, z) = Ce^{-jk_{30}(z-z_0)}, \qquad z \geq z_0$$

$$\tilde{A}_{x0}^{<}(\alpha_1, \alpha_2, z) = Ae^{jk_{30}(z-z_0)} + Be^{-jk_{30}(z+z_0)}, \qquad 0 \leq z \leq z_0$$

$$\tilde{A}_x(\alpha_1, \alpha_2, z) = De^{j(k_3 z - k_{30} z_0)}, \qquad z \leq 0 \qquad (7\text{-}34)$$

$$\tilde{A}_{z0}(\alpha_1, \alpha_2, z) = Ee^{-jk_{30}(z+z_0)}, \qquad z \geq 0$$

$$\tilde{A}_z(\alpha_1, \alpha_2, z) = Fe^{j(k_3 z - k_{30} z_0)}, \qquad z \leq 0$$

When we impose the boundary and source conditions to evaluate the six constants, we obtain the following six equations:

$$A + Be^{-2jk_{30}z_0} - C = 0$$

$$A - Be^{-2jk_{30}z_0} + C = \frac{pe^{-j(\alpha_1 x_0 + \alpha_2 y_0)}}{2\pi jk_{30}}$$

$$A + B = \frac{\mu}{\mu_0} D$$

$$A - B = \frac{k_3}{k_{30}} D$$

$$E = F$$

$$\epsilon k_{30} E + \epsilon_0 k_3 F = \alpha_1[\epsilon(A + B) - \epsilon_0 D]$$

and their solutions:

$$A = \frac{pe^{-j(\alpha_1 x_0 + \alpha_2 y)}}{4\pi j k_{30}}$$

$$B = RA$$

$$C = A(1 + Re^{-2jk_{30}z_0})$$

$$D = \frac{2\mu_0 k_{30}}{\mu k_{30} + \mu_0 k_3} A \tag{7-35}$$

$$R = \frac{\mu k_{30} - \mu_0 k_3}{\mu k_{30} + \mu_0 k_3} = -1 + \frac{2\mu k_{30}}{\mu k_{30} + \mu_0 k_3}$$

$$E = F = \frac{2\alpha_1(\mu\epsilon - \mu_0\epsilon_0)k_{30}}{(\mu k_{30} + \mu_0 k_3)(\epsilon k_{30} + \epsilon_0 k_3)} A$$

The general solution to (7-31) becomes

$$A_{x0}(x, y, z) = \frac{p}{8\pi^2 j} \int\int_{-\infty}^{\infty} \frac{e^{j[\alpha_1(x-x_0) + \alpha_2(y-y_0)]}}{k_{30}} [e^{-jk_{30}|z-z_0|} + Re^{-jk_{30}(z+z_0)}]\, d\alpha_1\, d\alpha_2$$

$$= \frac{p}{8\pi^2 j} \int\int_{-\infty}^{\infty} \frac{e^{j[\alpha_1(x-x_0) + \alpha_2(y-y_0)]}}{k_{30}} \Bigg[e^{-jk_{30}|z-z_0|} - e^{-jk_{30}(z+z_0)}$$

$$+ \frac{2\mu k_{30}}{\mu k_{30} + \mu_0 k_3} e^{-jk_{30}(z+z_0)} \Bigg] d\alpha_1\, d\alpha_2 \tag{7-36}$$

$$A_x(x, y, z) = \frac{\mu_0 p}{4\pi^2 j} \int\int_{-\infty}^{\infty} \frac{e^{j[\alpha_1(x-x_0) + \alpha_2(y-y_0) + k_3 z - k_{30}z_0]}}{\mu k_{30} + \mu_0 k_3}\, d\alpha_1\, d\alpha_2$$

$$A_{z0}(x, y, z) = \frac{(\mu\epsilon - \mu_0\epsilon_0)p}{4\pi^2 j} \int\int_{-\infty}^{\infty} \frac{e^{j[\alpha_1(x-x_0) + \alpha_2(y-y_0) - k_{30}(z+z_0)]}}{(\mu k_{30} + \mu_0 k_3)(\epsilon k_{30} + \epsilon_0 k_3)} \alpha_1\, d\alpha_1\, d\alpha_2$$

$$A_z(x, y, z) = \frac{(\mu\epsilon - \mu_0\epsilon_0)p}{4\pi^2 j} \int\int_{-\infty}^{\infty} \frac{e^{j[\alpha_1(x-x_0) + \alpha_2(y-y_0) + k_3 z - k_{30}z_0]}}{(\mu k_{30} + \mu_0 k_3)(\epsilon k_{30} + \epsilon_0 k_3)} \alpha_1\, d\alpha_1\, d\alpha_2$$

Since the form of A_{x0} in (7-36) is very similar to that for A_{z0} in (7-27), we can rewrite it as

$$A_{x0}(x, y, z) = \frac{pe^{-jkR}}{4\pi R} - \frac{pe^{-jkR_1}}{4\pi R_1} + \frac{\mu p}{4\pi^2 j} \int\int_{-\infty}^{\infty} \frac{e^{j[\alpha_1(x-x_0) + \alpha_2(y-y_0) - k_{30}(z+z_0)]}}{\mu k_{30} + \mu_0 k_3}\, d\alpha_1\, d\alpha_2$$

$$\tag{7-37}$$

so that the three contributions are shown in an explicit manner. As before, R is the distance from the source to the field point and R_1 is the distance from the image to the field point, i.e., $R^2 = (x - x_0)^2 + (y - y_0)^2 + (z - z_0)^2$ and $R_1^2 = (x - x_0)^2 + (y - y_0)^2 + (z + z_0)^2$, respectively. Both of the integrands in the expressions for A_{x0} and A_x in (7-36) contain a pole so that there will be a surface wave. The value of λ at the pole is

$$\lambda = \lambda_p = k_0 k \left[\frac{k_0^2 \epsilon^2 - k^2 \epsilon_0^2}{k_0^4 \epsilon^2 - k^4 \epsilon_0^2} \right]^{1/2} \tag{7-38}$$

When the permeabilities of the two regions are the same, no pole occurs and, therefore, no surface wave exists.

The integrands in the expressions for A_{z0} and A_z in (7-36) are much more complicated than before since two poles exist. One pole is the same as that in (7-29) whereas the other one is the same as that in (7-38). When the permeabilities of the two regions are the same, the expressions for A_{z0} and A_z simplify to the following:

$$A_{z0}(x, y, z) = \frac{\epsilon_0 p}{4\pi^2 jk_0^2} \int\!\!\!\int\limits_{-\infty}^{\infty} \frac{(k_3 - k_{30}) e^{j[\alpha_1(x-x_0) + \alpha_2(y-y_0) - k_{30}(z+z_0)]}}{\epsilon k_{30} + \epsilon_0 k_3} \alpha_1 \, d\alpha_1 \, d\alpha_2 \tag{7-39}$$

$$A_z(x, y, z) = \frac{\epsilon_0 p}{4\pi^2 jk_0^2} \int\!\!\!\int\limits_{-\infty}^{\infty} \frac{(k_3 - k_{30}) e^{j[\alpha_1(x-x_0) + \alpha_2(y-y_0) + k_3 z - k_{30} z_0]}}{\epsilon k_{30} + \epsilon_0 k_3} \alpha_1 \, d\alpha_1 \, \alpha \alpha_2$$

with the pole defined in (7-30). When the dielectric constants of the two regions are the same, the expressions for A_{z0} and A_z simplify to the following:

$$A_{z0}(x, y, z) = \frac{\mu_0 p}{4\pi^2 jk_0^2} \int\!\!\!\int\limits_{-\infty}^{\infty} \frac{(k_3 - k_{30}) e^{j[\alpha_1(x-x_0) + \alpha_2(y-y_0) - k_{30}(z+z_0)]}}{\mu k_{30} + \mu_0 k_3} \alpha_1 \, d\alpha_1 \, d\alpha_2 \tag{7-40}$$

$$A_z(x, y, z) = \frac{\mu_0 p}{4\pi^2 jk_0^2} \int\!\!\!\int\limits_{-\infty}^{\infty} \frac{(k_3 - k_{30}) e^{j[\alpha_1(x-x_0) + \alpha_2(y-y_0) + k_3 z - k_{30} z_0]}}{\mu k_{30} + \mu_0 k_3} \alpha_1 \, d\alpha_1 \, d\alpha_2$$

with the pole again defined in (7-30). Although the pole in (7-39) and that in (7-40) is defined by the same relation, one should remember that in (7-39) $k_0^2 = \omega^2 \mu_0 \epsilon_0$ and $k^2 = \omega^2 \mu_0 \epsilon$ whereas in (7-40) $k_0^2 = \omega^2 \mu_0 \epsilon_0$ as before but $k^2 = \omega^2 \mu \epsilon_0$.

One other point that is worth noting is the fact that the integrals that occur for the four different cases of dipole orientation are all related to several basic integrals. This point is emphasized by Baños.[12] For instance, the quantity α_1 in the numerator of (7-40) can be obtained by taking the x-derivative of an integral such as the one that appears in the expression for A_{x0} or A_x in (7-36). Similarly, the quantity $k_3 - k_{30}$ that also appears in the numerator of (7-40) can be obtained by combining a z-derivative and a z_0-derivative of a similar expression.

PROBLEMS

7.1. Verify the steps leading to (7-5).

7.2. Repeat the problem in Fig. 7-1 using a one-dimensional sine Fourier transform in y and α_2 to obtain the result

$$A_z(x, y) = \frac{I}{\pi^2} \int_0^\infty \sin \alpha_2 y_0 \sin \alpha_2 y \, d\alpha_2 \int_{-\infty}^\infty \frac{e^{j\alpha_1(x-x_0)}}{\alpha_1^2 - k_{50}^2} \, d\alpha_1$$

where $k_{50}^2 = k_0^2 - \alpha_2^2$.

7.3. Use the identities in (3-51) to show that (7-5) can be put into the form of (4-29) plus a similar term for the image. *Hint:* The term in the square brackets is just $e^{jk_{40}|y-y_0|}$.

7.4. Evaluate the integrals in Prob. 7.2 and show that the results are the same as those obtained in Prob. 7.3. *Hint:* Use the results of (3-48) and (3-49) to evaluate the second integral in Prob. 7.2 and the last identity in (3-51) to evaluate the first integral.

7.5. Verify the constants in (7-9).

7.6. Consider the problem in Fig. 7-2 for the case where the dielectric half space is replaced by one possessing both electric and magnetic properties, i.e., $k^2 = KK_m k_0^2$. Since (7-8) is still valid, except k_4 is replaced by k_{41}, show that the four constants become the following:

$$A, B, C \text{ unchanged}$$

$$D = \frac{2k_{40}}{K_m k_{40} + k_{41}} A$$

$$R = \frac{K_m k_{40} - k_{41}}{K_m k_{40} + k_{41}}$$

where $k_{41}^2 = KK_m k_0^2 - \alpha_1^2$.
Note: A pole now exists at

$$|\alpha_p| = k_0 k \left[\frac{k^2 - K^2 k_0^2}{k^4 - K^2 k_0^4}\right]^{1/2} = k_0 \left[\frac{K_m(K_m - K)}{K_m^2 - 1}\right]^{1/2}$$

[12]Baños, *Dipole Radiation in the Presence of a Conducting Half-Space*, Sec. 2.22.

7.7. Verify that the equations in (7-15) follow from (4-63) and (4-65).

7.8. Repeat Prob. 7.6 for the magnetic line source in Fig. 7-3. In this case, the results in (7-13) and (7-14) are valid with k_4 replaced by k_{41}. The expression for the pole is the dual of the one in Prob. 7.6.

7.9. Consider the problem defined by (7-15) and (7-16) but with an electric line source. Show that the relevant equations and boundary conditions are the following:

$$(\nabla_t^2 + k_0^2 g_3)C_z = 0, \qquad y < 0$$
$$(\nabla_t^2 + k_0^2)C_{z0} = -\mu_0 I \delta(x - x_0)\delta(y - y_0), \qquad y > 0$$

or

$$(\nabla_t^2 + k_0^2 g_3)E_z = 0, \qquad y < 0$$
$$(\nabla_t^2 + k_0^2)E_z = j\omega\mu_0 I \delta(x - x_0)\delta(y - y_0), \qquad y > 0$$
$$C_{z0} = C_z, \qquad y = 0$$
$$\frac{\partial C_{z0}}{\partial y} = \frac{\partial C_z}{\partial y}, \qquad y = 0$$

where g_3 is defined in (4-71).

7.10. Continue Prob. 7.9 with the electric line source. Define the four constants for **C** in the same manner as in (7-8) and show that they are the same as those defined in (7-9) provided K is replaced by g_3. *Note:* The anisotropic dielectric behaves as an isotropic material with constants μ_0 and $g_3\epsilon_0$.

7.11. Repeat Prob. 7.10 when the anisotropic region is magnetic. *Note:* This problem is the dual of that defined in (7-15) and (7-16) and is that of an electric line source above a magnetically anisotropic half space.

7.12. Repeat the problem defined by (7-15) and (7-16) when the anisotropic region is magnetic. *Note:* This is the dual of Prob. 7.10 and is that of a magnetic line source above a magnetically anisotropic half space.

7.13. Verify that the expressions in (7-19) follow from those in (7-15) through (7-18).

7.14. Verify that the four constants in (7-26) follow from (7-24) and (7-25).

7.15. Show that if one replaces the second expression for R in (7-26) by

$$R = -1 + 2\epsilon k_{30}(\epsilon k_{30} + \epsilon_0 k_3)^{-1}$$

the expression[13] for A_{z0} in (7-28) becomes

---

[13]This is of the form obtained by J. Stratton, *Electromagnetic Theory*, McGraw-Hill Book Company, Inc., New York, 1941, p. 581, and Baños, *Dipole Radiation in the Presence of a Conducting Half-Space*, p. 35. The expression in (7–28) is of the form obtained by Sommerfeld, *Partial Differential Equations in Physics*, p. 249.

$$A_{z0} = \frac{pe^{-jk_0R}}{4\pi R} - \frac{pe^{-jk_0R_1}}{4\pi R_1} + \frac{p\epsilon}{4\pi^2 j} \int\limits_{-\infty}^{\infty}\int \frac{e^{j[\alpha_1(x-x_0)+\alpha_2(y-y_0)-k_{30}(z+z_0)]}}{\epsilon k_{30} + \epsilon_0 k_3} d\alpha_1 \, d\alpha_2$$

One sees that the integrand above is simpler than that in (7-28).

7.16. Derive the expressions in (7-29) and (7-30).

7.17. Verify the boundary conditions in (7-32).

7.18. Carry through the steps leading to the four constants in (7-35).

7.19. Consider the problem in Fig. 7-4 but with a vertical magnetic dipole. Write $\mathbf{A}_m$ in the form shown in (7-25) and evaluate the four constants. *Note:* This problem is the dual of that in Fig. 7-4.

7.20. Consider Prob. 7.19 when the permeabilities of the two regions are the same. *Note:* No pole exists in R or D so that no surface wave exists. As mentioned by Baños[14], this is the simplest type of "Sommerfeld problem." The dual to this problem is the vertical electric dipole and two media with the same dielectric constants. See Prob. 7.23.

7.21. Repeat Prob. 7.20 but for the case where the dielectric constants are the same. In this case, a pole exists and is defined by (7-30) except now $k^2/k_0^2 = \mu/\mu_0$.

7.22. Consider the problem which is dual to that shown in Fig. 7-5, i.e., a horizontal magnetic dipole, and solve for $\mathbf{A}_m$ in the form of (7-34). Since the problem is an exact dual, the results can be written down at once. It is also simple to extend these results to specific cases where several of the constants of the media are the same.

7.23. Consider the problem in Fig. 7-4 when the dielectric constants in the two regions are the same and $z = z_0 = 0$. The expressions for A_z in (7-27) and A_{z0} in (7-28) become the following:

$$A_z(x, y, 0) = \frac{p}{4\pi^2 j} \int\limits_{-\infty}^{\infty}\int \frac{e^{j[\alpha_1(x-x_0)+\alpha_2(y-y_0)]}}{k_{30} + k_3} d\alpha_1 \, d\alpha_2$$

$$A_{z0}(x, y, 0) = \frac{pe^{-jk_0R}}{4\pi R} + \frac{pe^{-jk_0R_1}}{4\pi R_1} - \frac{p}{4\pi^2 j} \int\limits_{-\infty}^{\infty}\int \frac{k_3 e^{j[\alpha_1(x-x_0)+\alpha_2(y-y_0)]}}{k_{30}(k_{30} + k_3)} d\alpha_1 \, d\alpha_2$$

$$= \frac{pe^{-jk_0R}}{4\pi R} - \frac{pe^{-jk_0R_1}}{4\pi R_1} + \frac{p}{4\pi^2 j} \int\limits_{-\infty}^{\infty}\int \frac{e^{j[\alpha_1(x-x_0)+\alpha_2(y-y_0)]}}{k_{30} + k_3} d\alpha_1 \, d\alpha_2$$

$$= A_z(x, y, 0)$$

since $R = R_1 = \rho_1$; $\rho_1^2 = (x - x_0)^2 + (y - y_0)^2$ at $z = z_0 = 0$.

---

[14]Baños, *Dipole Radiation in the Presence of a Conducting Half-Space*, p. 37.

7.24. Use the identity

$$\frac{e^{-jk\rho_1}}{\rho_1} = \int_0^\infty \frac{J_0(\lambda\rho_1)\lambda\,d\lambda}{\sqrt{\lambda^2 - k^2}}$$

which can be obtained from (3-52) or Stratton[15] to evaluate the integral

$$\int_0^\infty \sqrt{\lambda^2 - k^2}\,J_0(\lambda\rho)\lambda\,d\lambda = \frac{1}{\rho^3} - \frac{e^{-jk\rho}}{\rho^2}\left(jk + \frac{1}{\rho}\right)$$

Hint: Consider $(d/dk)\int_0^\infty \sqrt{\lambda^2 - k^2}\,J_0(\lambda\rho)\lambda\,d\lambda$.

7.25. Evaluate the integral in Prob. 7.23 by using (3-54), (3-52), and Prob. 7.24. The result[16] is exact and is

$$A_z(x, y, 0) = \frac{p}{2\pi(k^2 - k_0^2)\rho_1}\frac{\partial}{\partial\rho_1}\left(\frac{e^{-jk_0\rho_1}}{\rho_1} - \frac{e^{-jk\rho_1}}{\rho_1}\right)$$

Hint: $(k_{30} + k_3)^{-1} = (k_{30} - k_3)(k_0^2 - k^2)^{-1}$.

---

[15]Stratton, *Electromagnetic Theory*, p. 576.

[16]Baños, *Dipole Radiation in the Presence of a Conducting Half-Space*, p. 103.

Vector And Electromagnetic Field Relations

Vector Identities

$$\mathbf{a \cdot b \times c} = \mathbf{b \cdot c \times a} = \mathbf{c \cdot a \times b}$$

$$\mathbf{a \times b \times c} = \mathbf{b(a \cdot c)} - \mathbf{c(a \cdot b)}$$

$$\nabla(\phi\psi) = \phi\nabla\psi + \psi\nabla\phi$$

$$\nabla \cdot (\phi\mathbf{a}) = \mathbf{a} \cdot \nabla\phi + \phi\nabla \cdot \mathbf{a}$$

$$\nabla \times (\phi\mathbf{a}) = \nabla\phi \times \mathbf{a} + \phi\nabla \times \mathbf{a}$$

$$\nabla(\mathbf{a \cdot b}) = (\mathbf{a} \cdot \nabla)\mathbf{b} + (\mathbf{b} \cdot \nabla)\mathbf{a} + \mathbf{a} \times \nabla \times \mathbf{b} + \mathbf{b} \times \nabla \times \mathbf{a}$$

$$\nabla \cdot (\mathbf{a \times b}) = \mathbf{b} \cdot \nabla \times \mathbf{a} - \mathbf{a} \cdot \nabla \times \mathbf{b}$$

$$\nabla \times (\mathbf{a \times b}) = \mathbf{a}\nabla \cdot \mathbf{b} - \mathbf{b}\nabla \cdot \mathbf{a} + (\mathbf{b} \cdot \nabla)\mathbf{a} - (\mathbf{a} \cdot \nabla)\mathbf{b}$$

$$\nabla \times \nabla \times \mathbf{a} = \nabla\nabla \cdot \mathbf{a} - \nabla^2\mathbf{a}$$

$$\nabla \times \nabla\phi = 0$$

$$\nabla \cdot \nabla \times \mathbf{a} = 0$$

Gauss' Theorems

For the closed surface S enclosing the volume V with the positive unit normal $\mathbf{n}$ directed out of V, the following Gauss' theorems hold:

$$\int_V \nabla\phi \, dV = \oint \phi\mathbf{n} \, dS$$

$$\int_V \nabla\cdot\mathbf{a} \, dV = \oint_S \mathbf{a}\cdot\mathbf{n} \, dS$$

$$\int_V \nabla\times\mathbf{a} \, dV = \oint_S \mathbf{n}\times\mathbf{a} \, dS$$

Stokes' Theorems

For the open surface S enclosed by the contour C with the positive unit normal $\mathbf{n}$ directed away from that side of S which is enclosed by a positive traversal of the contour, the following Stokes' theorems hold:

$$\int_S \mathbf{n}\times\nabla\phi \, dS = \oint_C \phi \, d\mathbf{C}$$

$$\int_S \nabla\times\mathbf{a}\cdot\mathbf{n} \, dS = \oint_C \mathbf{a}\cdot d\mathbf{C}$$

Green's Second Identity

The following identities can be obtained from Gauss' theorems:

$$\int_V (\psi\nabla^2\phi - \phi\nabla^2\psi) \, dV = \oint_S (\psi\nabla\phi - \phi\nabla\psi)\cdot\mathbf{n} \, dS = \oint_S \left(\psi\frac{\partial\phi}{\partial n} - \phi\frac{\partial\psi}{\partial n}\right) dS$$

$$\int_V (\mathbf{B}\cdot\nabla\times\nabla\times\mathbf{A} - \mathbf{A}\cdot\nabla\times\nabla\times\mathbf{B}) \, dV = \oint_S (\mathbf{A}\times\nabla\times\mathbf{B} - \mathbf{B}\times\nabla\times\mathbf{A})\cdot\mathbf{n} \, dS$$

Vector Relations for Different Coordinate Systems

Cartesian coordinates

$$\mathbf{a}_x = \mathbf{a}_\rho\cos\phi - \mathbf{a}_\phi\sin\phi = \mathbf{a}_r\cos\phi\sin\theta + \mathbf{a}_\theta\cos\phi\cos\theta - \mathbf{a}_\phi\sin\phi$$

$$\mathbf{a}_y = \mathbf{a}_\rho\sin\phi + \mathbf{a}_\phi\cos\phi = \mathbf{a}_r\sin\phi\sin\theta + \mathbf{a}_\theta\sin\phi\cos\theta + \mathbf{a}_\phi\cos\phi$$

$$\mathbf{a}_z = \mathbf{a}_r\cos\theta - \mathbf{a}_\theta\sin\theta$$

$$\nabla f = \frac{\partial f}{\partial x}\mathbf{a}_x + \frac{\partial f}{\partial y}\mathbf{a}_y + \frac{\partial f}{\partial z}\mathbf{a}_z$$

$$\nabla\cdot\mathbf{A} = \frac{\partial A_x}{\partial x} + \frac{\partial A_y}{\partial y} + \frac{\partial A_z}{\partial z}$$

$$\nabla\times\mathbf{A} = \left(\frac{\partial A_z}{\partial y} - \frac{\partial A_y}{\partial z}\right)\mathbf{a}_x + \left(\frac{\partial A_x}{\partial z} - \frac{\partial A_z}{\partial x}\right)\mathbf{a}_y + \left(\frac{\partial A_y}{\partial x} - \frac{\partial A_x}{\partial y}\right)\mathbf{a}_z$$

$$\nabla^2 f = \frac{\partial^2 f}{\partial x^2} + \frac{\partial^2 f}{\partial y^2} + \frac{\partial^2 f}{\partial z^2}$$

$$\nabla^2 \mathbf{A} = \mathbf{a}_x \nabla^2 A_x + \mathbf{a}_y \nabla^2 A_y + \mathbf{a}_z \nabla^2 A_z$$

$$\nabla \nabla \cdot \mathbf{A} = \left(\frac{\partial^2 A_x}{\partial x^2} + \frac{\partial^2 A_y}{\partial x\, \partial y} + \frac{\partial^2 A_z}{\partial x\, \partial z} \right) \mathbf{a}_x + \left(\frac{\partial^2 A_x}{\partial x\, \partial y} + \frac{\partial^2 A_y}{\partial y^2} + \frac{\partial^2 A_z}{\partial y\, \partial z} \right) \mathbf{a}_y$$
$$+ \left(\frac{\partial^2 A_x}{\partial x\, \partial z} + \frac{\partial^2 A_y}{\partial y\, \partial z} + \frac{\partial^2 A_z}{\partial z^2} \right) \mathbf{a}_z$$

$$\nabla \times \nabla \times \mathbf{A} = \left(-\frac{\partial^2 A_x}{\partial y^2} - \frac{\partial^2 A_x}{\partial z^2} + \frac{\partial^2 A_y}{\partial x\, \partial y} + \frac{\partial^2 A_z}{\partial x\, \partial z} \right) \mathbf{a}_x + \left(-\frac{\partial^2 A_y}{\partial x^2} - \frac{\partial^2 A_y}{\partial z^2} \right.$$
$$\left. + \frac{\partial^2 A_x}{\partial x\, \partial y} + \frac{\partial^2 A_z}{\partial y\, \partial z} \right) \mathbf{a}_y + \left(-\frac{\partial^2 A_z}{\partial x^2} - \frac{\partial^2 A_z}{\partial y^2} + \frac{\partial^2 A_x}{\partial x\, \partial z} + \frac{\partial^2 A_y}{\partial y\, \partial z} \right) \mathbf{a}_z$$

Cylindrical coordinates

$$\mathbf{a}_\rho = \mathbf{a}_x \cos \phi + \mathbf{a}_y \sin \phi = \mathbf{a}_r \sin \theta + \mathbf{a}_\theta \cos \theta$$
$$\mathbf{a}_\phi = \mathbf{a}_x \sin \phi + \mathbf{a}_y \cos \phi$$
$$\mathbf{a}_z = \mathbf{a}_r \cos \theta - \mathbf{a}_\theta \sin \theta$$

The finite derivatives and operations involving the unit vectors are the following:

$$\frac{\partial \mathbf{a}_\rho}{\partial \phi} = \mathbf{a}_\phi, \qquad \frac{\partial \mathbf{a}_\phi}{\partial \phi} = -\mathbf{a}_\rho, \qquad \nabla \cdot \mathbf{a}_\rho = \frac{1}{\rho}, \qquad \nabla \times \mathbf{a}_\phi = \frac{\mathbf{a}_z}{\rho}$$

$$\nabla f = \frac{\partial f}{\partial \rho} \mathbf{a}_\rho + \frac{1}{\rho} \frac{\partial f}{\partial \phi} \mathbf{a}_\phi + \frac{\partial f}{\partial z} \mathbf{a}_z$$

$$\nabla \cdot \mathbf{A} = \frac{1}{\rho} \frac{\partial}{\partial \rho} (\rho A_\rho) + \frac{1}{\rho} \frac{\partial A_\phi}{\partial \phi} + \frac{\partial A_z}{\partial z}$$

$$\nabla \times \mathbf{A} = \left(\frac{1}{\rho} \frac{\partial A_z}{\partial \phi} - \frac{\partial A_\phi}{\partial z} \right) \mathbf{a}_\rho + \left(\frac{\partial A_\rho}{\partial z} - \frac{\partial A_z}{\partial \rho} \right) \mathbf{a}_\phi + \left[\frac{1}{\rho} \frac{\partial(\rho A_\phi)}{\partial \rho} - \frac{1}{\rho} \frac{\partial A_\rho}{\partial \phi} \right] \mathbf{a}_z$$

$$\nabla^2 f = \frac{1}{\rho} \frac{\partial}{\partial \rho} \left(\rho \frac{\partial f}{\partial \rho} \right) + \frac{1}{\rho^2} \frac{\partial^2 f}{\partial \phi^2} + \frac{\partial^2 f}{\partial z^2} = \frac{\partial^2 f}{\partial \rho^2} + \frac{1}{\rho} \frac{\partial f}{\partial \rho} + \frac{1}{\rho^2} \frac{\partial^2 f}{\partial \phi^2} + \frac{\partial^2 f}{\partial z^2}$$

$$\nabla^2 \mathbf{A} = \left(\nabla^2 A_\rho - \frac{A_\rho}{\rho^2} - \frac{2}{\rho^2} \frac{\partial A_\phi}{\partial \phi} \right) \mathbf{a}_\rho + \left(\nabla^2 A_\phi - \frac{A_\phi}{\rho^2} + \frac{2}{\rho^2} \frac{\partial A_\rho}{\partial \phi} \right) \mathbf{a}_\phi + \nabla^2 A_z \mathbf{a}_z$$

$$\nabla \nabla \cdot \mathbf{A} = \left(\frac{\partial^2 A_\rho}{\partial \rho^2} + \frac{\partial^2 A_z}{\partial \rho\, \partial z} + \frac{1}{\rho} \frac{\partial^2 A_\phi}{\partial \rho\, \partial \phi} + \frac{1}{\rho} \frac{\partial A_\rho}{\partial \rho} - \frac{1}{\rho^2} \frac{\partial A_\phi}{\partial \phi} - \frac{A_\rho}{\rho^2} \right) \mathbf{a}_\rho$$
$$+ \left(\frac{1}{\rho} \frac{\partial^2 A_z}{\partial \phi\, \partial z} + \frac{1}{\rho^2} \frac{\partial^2 A_\phi}{\partial \phi^2} + \frac{1}{\rho} \frac{\partial^2 A_\rho}{\partial \rho\, \partial \phi} + \frac{1}{\rho^2} \frac{\partial A_\rho}{\partial \phi} \right) \mathbf{a}_\phi$$
$$+ \left(\frac{\partial^2 A_z}{\partial z^2} + \frac{1}{\rho} \frac{\partial^2 A_\phi}{\partial \phi\, \partial z} + \frac{\partial^2 A_\rho}{\partial \rho\, \partial z} + \frac{1}{\rho} \frac{\partial A_\rho}{\partial z} \right) \mathbf{a}_z$$

$$\nabla \times \nabla \times \mathbf{A} = \left(-\frac{1}{\rho^2}\frac{\partial^2 A_\rho}{\partial\phi^2} - \frac{\partial^2 A_\rho}{\partial z^2} + \frac{\partial^2 A_z}{\partial\rho\,\partial z} + \frac{1}{\rho}\frac{\partial^2 A_\phi}{\partial\rho\,\partial\phi} + \frac{1}{\rho^2}\frac{\partial A_\phi}{\partial\phi}\right)\mathbf{a}_\rho$$

$$+ \left(-\frac{\partial^2 A_\phi}{\partial z^2} + \frac{1}{\rho}\frac{\partial A_z}{\partial\phi\,\partial z} - \frac{\partial^2 A_\phi}{\partial\rho^2} - \frac{1}{\rho}\frac{\partial A_\phi}{\partial\rho} + \frac{A_\phi}{\rho^2} - \frac{1}{\rho^2}\frac{\partial A_\rho}{\partial\phi}\right.$$

$$\left. + \frac{1}{\rho}\frac{\partial^2 A_\rho}{\partial\phi\,\partial\rho}\right)\mathbf{a}_\phi + \left(-\frac{\partial^2 A_z}{\partial\rho^2} - \frac{1}{\rho^2}\frac{\partial^2 A_z}{\partial\phi^2} + \frac{\partial^2 A_\rho}{\partial\rho\,\partial z} + \frac{1}{\rho}\frac{\partial^2 A_\phi}{\partial\phi\,\partial z}\right.$$

$$\left. + \frac{1}{\rho}\frac{\partial A_\rho}{\partial z} - \frac{1}{\rho}\frac{\partial A_z}{\partial\rho}\right)\mathbf{a}_z$$

Spherical coordinates

$$\mathbf{a}_r = \mathbf{a}_x \cos\phi \sin\theta + \mathbf{a}_y \sin\phi \sin\theta + \mathbf{a}_z \cos\theta = \mathbf{a}_\rho \sin\theta + \mathbf{a}_z \cos\theta$$

$$\mathbf{a}_\theta = \mathbf{a}_x \cos\phi \cos\theta + \mathbf{a}_y \sin\phi \cos\theta - \mathbf{a}_z \sin\theta = \mathbf{a}_\rho \cos\theta - \mathbf{a}_z \sin\theta$$

$$\mathbf{a}_\phi = -\mathbf{a}_x \sin\phi + \mathbf{a}_y \cos\phi$$

The finite derivatives and operations involving the unit vectors are the following:

$$\frac{\partial\mathbf{a}_r}{\partial\phi} = \sin\theta\,\mathbf{a}_\phi, \qquad \frac{\partial\mathbf{a}_r}{\partial\theta} = \mathbf{a}_\theta, \qquad \frac{\partial\mathbf{a}_\theta}{\partial\theta} = -\mathbf{a}_r, \qquad \frac{\partial\mathbf{a}_\theta}{\partial\phi} = \cos\theta\,\mathbf{a}_\phi$$

$$\frac{\partial\mathbf{a}_\phi}{\partial\phi} = -\mathbf{a}_r \sin\theta - \mathbf{a}_\theta \cos\theta, \qquad \nabla\cdot\mathbf{a}_r = \frac{2}{r}, \qquad \nabla\cdot\mathbf{a}_\theta = \frac{1}{r\tan\theta}$$

$$\nabla\times\mathbf{a}_\phi = \frac{1}{r\tan\theta}\mathbf{a}_r - \frac{1}{r}\mathbf{a}_\theta, \qquad \nabla\times\mathbf{a}_\theta = \frac{1}{r}\mathbf{a}_\phi$$

$$\nabla f = \frac{\partial f}{\partial r}\mathbf{a}_r + \frac{1}{r}\frac{\partial f}{\partial\theta}\mathbf{a}_\theta + \frac{1}{r\sin\theta}\frac{\partial f}{\partial\phi}\mathbf{a}_\phi$$

$$\nabla\cdot\mathbf{A} = \frac{1}{r^2}\frac{\partial}{\partial r}(r^2 A_r) + \frac{1}{r\sin\theta}\frac{\partial}{\partial\theta}(\sin\theta\,A_\theta) + \frac{1}{r\sin\theta}\frac{\partial A_\phi}{\partial\phi}$$

$$= \frac{\partial A_r}{\partial r} + \frac{2A_r}{r} + \frac{1}{r}\frac{\partial A_\theta}{\partial\theta} + \frac{A_\theta}{r\tan\theta} + \frac{1}{r\sin\theta}\frac{\partial A_\phi}{\partial\phi}$$

$$\nabla\times\mathbf{A} = \frac{1}{r\sin\theta}\left[\frac{\partial}{\partial\theta}(\sin\theta\,A_\phi) - \frac{\partial A_\theta}{\partial\phi}\right]\mathbf{a}_r + \frac{1}{r}\left[\frac{1}{\sin\theta}\frac{\partial A_r}{\partial\phi} - \frac{\partial}{\partial r}(rA_\phi)\right]\mathbf{a}_\theta$$

$$+ \frac{1}{r}\left[\frac{\partial}{\partial r}(rA_\theta) - \frac{\partial A_r}{\partial\theta}\right]\mathbf{a}_\phi$$

$$= \left(\frac{1}{r}\frac{\partial A_\phi}{\partial\theta} + \frac{A_\phi}{r\tan\theta} - \frac{1}{r\sin\theta}\frac{\partial A_\theta}{\partial\phi}\right)\mathbf{a}_r + \left(\frac{1}{r\sin\theta}\frac{\partial A_r}{\partial\phi} - \frac{\partial A_\phi}{\partial r} - \frac{A_\phi}{r}\right)\mathbf{a}_\theta$$

$$+ \left(\frac{\partial A_\theta}{\partial r} + \frac{A_\theta}{r} - \frac{1}{r}\frac{\partial A_r}{\partial\theta}\right)\mathbf{a}_\phi$$

$$\nabla^2 f = \frac{1}{r^2}\frac{\partial}{\partial r}\left(r^2\frac{\partial f}{\partial r}\right) + \frac{1}{r^2\sin\theta}\frac{\partial}{\partial\theta}\left(\sin\theta\frac{\partial f}{\partial\theta}\right) + \frac{1}{r^2\sin^2\theta}\frac{\partial^2 f}{\partial\phi^2}$$

$$= \frac{\partial^2 f}{\partial r^2} + \frac{2}{r}\frac{\partial f}{\partial r} + \frac{1}{r^2}\frac{\partial^2 f}{\partial\theta^2} + \frac{1}{r^2\tan\theta}\frac{\partial f}{\partial\theta} + \frac{1}{r^2\sin^2\theta}\frac{\partial^2 f}{\partial\phi^2}$$

$$\nabla^2\mathbf{A} = \left(\nabla^2 A_r - \frac{2A_r}{r^2} - \frac{2\cot\theta}{r^2}A_\theta - \frac{2}{r^2}\frac{\partial A_\theta}{\partial\theta} - \frac{2}{r^2\sin\theta}\frac{\partial A_\phi}{\partial\phi}\right)\mathbf{a}_r$$

$$+ \left(\nabla^2 A_\theta + \frac{2}{r^2}\frac{\partial A_r}{\partial\theta} - \frac{A_\theta}{r^2\sin^2\theta} - \frac{2\cos\theta}{r^2\sin^2\theta}\frac{\partial A_\phi}{\partial\phi}\right)\mathbf{a}_\theta$$

$$+ \left(\nabla^2 A_\phi + \frac{2}{r^2\sin\theta}\frac{\partial A_r}{\partial\phi} - \frac{1}{r^2\sin^2\theta}A_\phi + \frac{2\cos\theta}{r^2\sin^2\theta}\frac{\partial A_\theta}{\partial\phi}\right)\mathbf{a}_\phi$$

$$\nabla\nabla\cdot\mathbf{A} = \left(\frac{\partial^2 A_r}{\partial r^2} + \frac{2}{r}\frac{\partial A_r}{\partial r} - \frac{2A_r}{r^2} - \frac{A_\theta}{r^2\tan\theta} + \frac{1}{r\tan\theta}\frac{\partial A_\theta}{\partial r} + \frac{1}{r}\frac{\partial^2 A_\theta}{\partial\theta\,\partial r} - \frac{1}{r^2}\frac{\partial A_\theta}{\partial\theta}\right.$$

$$+ \left.\frac{1}{r\sin\theta}\frac{\partial^2 A_\phi}{\partial\phi\,\partial r} - \frac{1}{r^2\sin\theta}\frac{\partial A_\phi}{\partial\phi}\right)\mathbf{a}_r$$

$$+ \left(\frac{1}{r}\frac{\partial^2 A_r}{\partial r\,\partial\theta} + \frac{2}{r^2}\frac{\partial A_r}{\partial\theta} - \frac{A_\theta}{r^2\sin^2\theta} + \frac{1}{r^2\tan\theta}\frac{\partial A_\theta}{\partial\theta} + \frac{1}{r^2}\frac{\partial^2 A_\theta}{\partial\theta^2}\right.$$

$$+ \left.\frac{1}{r^2\sin\theta}\frac{\partial^2 A_\phi}{\partial\phi\,\partial\theta} - \frac{\cos\theta}{r^2\sin^2\theta}\frac{\partial A_\phi}{\partial\phi}\right)\mathbf{a}_\theta$$

$$+ \left(\frac{1}{r\sin\theta}\frac{\partial^2 A_r}{\partial r\,\partial\phi} + \frac{2}{r^2\sin\theta}\frac{\partial A_r}{\partial\phi} + \frac{\cos\theta}{r^2\sin^2\theta}\frac{\partial A_\theta}{\partial\phi} + \frac{1}{r^2\sin\theta}\frac{\partial^2 A_\phi}{\partial\phi\,\partial\theta}\right.$$

$$+ \left.\frac{1}{r^2\sin^2\theta}\frac{\partial^2 A_\phi}{\partial\phi^2}\right)\mathbf{a}_\phi$$

$$\nabla\times\nabla\times\mathbf{A} = \left(\frac{1}{r}\frac{\partial^2 A_\theta}{\partial r\,\partial\theta} + \frac{1}{r^2}\frac{\partial A_\theta}{\partial\theta} - \frac{1}{r^2}\frac{\partial^2 A_r}{\partial\theta^2} + \frac{1}{r\tan\theta}\frac{\partial A_\theta}{\partial r} + \frac{1}{r\tan\theta}\frac{A_\theta}{r}\right.$$

$$- \left.\frac{1}{r^2\tan\theta}\frac{\partial A_r}{\partial\theta} - \frac{1}{r^2\sin^2\theta}\frac{\partial^2 A_r}{\partial\phi^2} + \frac{1}{r\sin\theta}\frac{\partial^2 A_\phi}{\partial r\,\partial\phi} + \frac{1}{r^2\sin\theta}\frac{\partial A_\phi}{\partial\phi}\right)\mathbf{a}_r$$

$$+ \left(\frac{1}{r^2\sin^2\theta}\frac{\partial^2 A_\phi}{\partial\phi\,\partial\theta} + \frac{\cos\theta}{r^2\sin^2\theta}\frac{\partial A_\phi}{\partial\phi} - \frac{1}{r^2\sin^2\theta}\frac{\partial^2 A_\phi}{\partial\phi^2} - \frac{2}{r}\frac{\partial A_\theta}{\partial r}\right.$$

$$+ \left.\frac{1}{r}\frac{\partial^2 A_r}{\partial r\,\partial\theta} - \frac{\partial^2 A_\theta}{\partial r^2}\right)\mathbf{a}_\theta + \left(\frac{1}{r\sin\theta}\frac{\partial^2 A_r}{\partial\phi\,\partial r} - \frac{2}{r}\frac{\partial A_\phi}{\partial r} - \frac{1}{r^2}\frac{\partial^2 A_\phi}{\partial\theta^2}\right.$$

$$- \frac{\partial^2 A_\phi}{\partial r^2} - \frac{1}{r^2\tan\theta}\frac{\partial A_\phi}{\partial\theta} + \frac{A_\phi}{r^2\sin^2\theta} + \frac{1}{r^2\sin^2\theta}\frac{\partial^2 A_\theta}{\partial\theta\,\partial\phi}$$

$$- \left.\frac{\cos\theta}{r^2\sin^2\theta}\frac{\partial A_\theta}{\partial\phi}\right)\mathbf{a}_\phi$$

Electromagnetic field relations

$$\mathbf{E} = -j\omega\mu\mathbf{A} + \frac{1}{j\omega\epsilon}\nabla\nabla\cdot\mathbf{A} \qquad\qquad \mathbf{H} = -j\omega\epsilon\mathbf{A}_m + \frac{1}{j\omega\mu}\nabla\nabla\cdot\mathbf{A}_m$$

$$\mathbf{H} = \nabla\times\mathbf{A} \qquad\qquad\qquad\qquad\qquad \mathbf{E} = -\nabla\times\mathbf{A}_m$$

Cartesian coordinates

TM_z mode

$$(\nabla^2 + k^2)A_z = -J_z$$

$$A_z(x, y, z)$$

$$E_x = \frac{1}{j\omega\epsilon}\frac{\partial^2 A_z}{\partial x\,\partial z}$$

$$E_y = \frac{1}{j\omega\epsilon}\frac{\partial^2 A_z}{\partial y\,\partial z}$$

$$E_z = \frac{1}{j\omega\epsilon}\left(k^2 + \frac{\partial^2}{\partial z^2}\right)A_z$$

$$H_x = \frac{\partial A_z}{\partial y}$$

$$H_y = -\frac{\partial A_z}{\partial x}$$

$$H_z = 0$$

TE_z mode

$$(\nabla^2 + k^2)A_{mz} = -J_{mz}$$

$$A_{mz}(x, y, z)$$

$$H_x = \frac{1}{j\omega\mu}\frac{\partial^2 A_{mz}}{\partial x\,\partial z}$$

$$H_y = \frac{1}{j\omega\mu}\frac{\partial^2 A_{mz}}{\partial y\,\partial z}$$

$$H_z = \frac{1}{j\omega\mu}\left(k^2 + \frac{\partial^2}{\partial z^2}\right)A_{mz}$$

$$E_x = -\frac{\partial A_{mz}}{\partial y}$$

$$E_y = \frac{\partial A_{mz}}{\partial x}$$

$$E_z = 0$$

Cylindrical coordinates

$$A_z(\rho, \phi, z)$$

$$E_\rho = \frac{1}{j\omega\epsilon}\frac{\partial^2 A_z}{\partial\rho\,\partial z}$$

$$E_\phi = \frac{1}{j\omega\epsilon}\frac{\partial^2 A_z}{\rho\,\partial\phi\,\partial z}$$

$$E_z = \frac{1}{j\omega\epsilon}\left(k^2 + \frac{\partial^2}{\partial z^2}\right)A_z$$

$$H_\rho = \frac{1}{\rho}\frac{\partial A_z}{\partial\phi}$$

$$H_\phi = -\frac{\partial A_z}{\partial\rho}$$

$$H_z = 0$$

$$\mathbf{A}(\rho, z) \neq f(\phi)$$

$$\left(\nabla^2 - \frac{1}{\rho^2} + k^2\right)A_\rho = -J_\rho$$

$$\left(\nabla^2 - \frac{1}{\rho^2} + k^2\right)A_\phi = -J_\phi$$

$$(\nabla^2 + k^2)A_z = -J_z$$

$$A_\rho \sim B_1(k_1\rho)h(k_3 z) \sim A_{m\rho}$$

$$A_{mz}(\rho, \phi, z)$$

$$H_\rho = \frac{1}{j\omega\mu}\frac{\partial^2 A_{mz}}{\partial\rho\,\partial z}$$

$$H_\phi = \frac{1}{j\omega\mu}\frac{\partial^2 A_{mz}}{\rho\,\partial\phi\,\partial z}$$

$$H_z = \frac{1}{j\omega\mu}\left(k^2 + \frac{\partial^2}{\partial z^2}\right)A_{mz}$$

$$E_\rho = -\frac{1}{\rho}\frac{\partial A_{mz}}{\partial\phi}$$

$$E_\phi = \frac{\partial A_{mz}}{\partial\rho}$$

$$E_z = 0$$

$$\mathbf{A}_m(\rho, z) \neq f(\phi)$$

$$\left(\nabla^2 - \frac{1}{\rho^2} + k^2\right)A_{m\rho} = -J_{m\rho}$$

$$\left(\nabla^2 - \frac{1}{\rho^2} + k^2\right)A_{m\phi} = -J_{m\phi}$$

$$(\nabla^2 + k^2)A_{mz} = -J_{mz}$$

$$A_\phi \sim B_1(k_1\rho)h(k_3z) \sim A_{m\phi}$$

$$A_z \sim B_0(k_1\rho)h(k_3z) \sim A_{mz}$$

$$k_1^2 + k_3^2 = k^2$$

TM$_\phi$ Mode (or *TE$_z$*)

$A_\phi(\rho, z)$

$E_\rho = 0$

$E_\phi = -j\omega\mu A_\phi$

$E_z = 0$

$H_\rho = -\dfrac{\partial A_\phi}{\partial z}$

$H_\phi = 0$

$H_z = \dfrac{1}{\rho}\dfrac{\partial}{\partial\rho}(\rho A_\phi)$

TE$_\phi$ Mode (or *TM$_z$*)

$A_{m\phi}(\rho, z)$

$H_\rho = 0$

$H_\phi = -j\omega\epsilon A_{m\phi}$

$H_z = 0$

$E_\rho = \dfrac{\partial A_{m\phi}}{\partial z}$

$E_\phi = 0$

$E_z = -\dfrac{1}{\rho}\dfrac{\partial}{\partial\rho}(\rho A_{m\phi})$

TM$_\rho$ Mode (or *TM$_z$*)

$A_\rho(\rho, z)$

$E_\rho = \dfrac{1}{j\omega\epsilon}\left[\dfrac{1}{\rho}\dfrac{\partial}{\partial\rho}\left(\rho\dfrac{\partial}{\partial\rho}\right) - \dfrac{1}{\rho^2} + k^2\right]A_\rho$

$E_\phi = 0$

$E_z = \dfrac{1}{j\omega\epsilon}\dfrac{1}{\rho}\dfrac{\partial}{\partial\rho}\left(\rho\dfrac{\partial A_\rho}{\partial z}\right)$

$H_\rho = 0$

$H_\phi = \dfrac{\partial A_\rho}{\partial z}$

$H_z = 0$

TE$_\rho$ Mode (or *TE$_z$*)

$A_{m\rho}(\rho, z)$

$H_\rho = \dfrac{1}{j\omega\mu}\left[\dfrac{1}{\rho}\dfrac{\partial}{\partial\rho}\left(\rho\dfrac{\partial}{\partial\rho}\right) - \dfrac{1}{\rho^2} + k^2\right]A_{m\rho}$

$H_\phi = 0$

$H_z = \dfrac{1}{j\omega\mu}\dfrac{1}{\rho}\dfrac{\partial}{\partial\rho}\left(\rho\dfrac{\partial A_{m\rho}}{\partial z}\right)$

$E_\rho = 0$

$E_\phi = \dfrac{-\partial A_{m\rho}}{\partial z}$

$E_z = 0$

Spherical coordinates

$\mathbf{H} = \nabla \times \mathbf{A}$

$\mathbf{E} = -j\omega\mu\mathbf{A} - \nabla\Phi$

where $\Phi = -\dfrac{1}{j\omega\epsilon}\dfrac{\partial A_r}{\partial r}$

or $\mathbf{E} = \dfrac{1}{j\omega\epsilon}\nabla \times \nabla \times \mathbf{A}$

$\mathbf{A} = A_r\mathbf{a}_r$

$\mathbf{E} = -\nabla \times \mathbf{A}_m$

$\mathbf{H} = -j\omega\epsilon\mathbf{A}_m - \nabla\Phi_m$

where $\Phi_m = -\dfrac{1}{j\omega\mu}\dfrac{\partial A_{mr}}{\partial r}$

or $\mathbf{H} = \dfrac{1}{j\omega\mu}\nabla \times \nabla \times \mathbf{A}_m$

$\mathbf{A}_m = A_{mr}\mathbf{a}_r$

TM$_r$ Mode

$$(\nabla^2 + k^2)\frac{A_r}{r} = 0$$

$$A_r(r, \theta, \phi)$$

$$E_r = \frac{1}{j\omega\epsilon}\left(k^2 + \frac{\partial^2}{\partial r^2}\right)A_r$$

$$E_\theta = \frac{1}{j\omega\epsilon}\frac{1}{r}\frac{\partial^2 A_r}{\partial r\,\partial\theta}$$

$$E_\phi = \frac{1}{j\omega\epsilon}\frac{1}{r\sin\theta}\frac{\partial^2 A_r}{\partial r\,\partial\phi}$$

$$H_r = 0$$

$$H_\theta = \frac{1}{r\sin\theta}\frac{\partial A_r}{\partial\phi}$$

$$H_\phi = -\frac{1}{r}\frac{\partial A_r}{\partial\theta}$$

TE$_r$ Mode

$$(\nabla^2 + k^2)\frac{A_{mr}}{r} = 0$$

$$A_{mr}(r, \theta, \phi)$$

$$H_r = \frac{1}{j\omega\mu}\left(k^2 + \frac{\partial^2}{\partial r^2}\right)A_{mr}$$

$$H_\theta = \frac{1}{j\omega\mu}\frac{1}{r}\frac{\partial^2 A_{mr}}{\partial r\,\partial\theta}$$

$$H_\phi = \frac{1}{j\omega\mu}\frac{1}{r\sin\theta}\frac{\partial^2 A_{mr}}{\partial r\,\partial\phi}$$

$$E_r = 0$$

$$E_\theta = -\frac{1}{r\sin\theta}\frac{\partial A_{mr}}{\partial\phi}$$

$$E_\phi = \frac{1}{r}\frac{\partial A_{mr}}{\partial\theta}$$

Note: The Lorentz condition is not valid for this spherical case.

B

The Dirac Delta

The Dirac delta is often defined by its "sifting" property,[1] i.e.,

$$\int_V f(\mathbf{r})\delta(\mathbf{r} - \mathbf{r}_0) \, dV = \begin{cases} f(\mathbf{r}_0), \, P_0(x_0, y_0, z_0) \text{ in } V \\ 0, \, P_0(x_0, y_0, z_0) \text{ not in } V \end{cases} \tag{B-1}$$

There is no restriction on the number of dimensions involved and $f(\mathbf{r})$ can be a scalar function or a vector function. However, it is rather obvious that $f(\mathbf{r})$ must be defined at the point $P_0(x_0, y_0, z_0)$. If the function $f(\mathbf{r})$ is a constant, e.g., unity, then one sees that the delta is normalized. As a result, it is customary to speak of the delta as a symbolic representation for a unit source. However, the source is of a unit magnitude in the sense that the integral of the delta over the coordinates involved is unity. This is most easily seen by referring to Table B-1 which lists the representations for the Dirac delta in three different coordinate systems. If we consider a three-dimensional orthogonal curvilinear coordinate system with elements of length $h_i u_i$, where the h_i are the scale factors and the u_i are the curvilinear coordinates, then we

---

[1] R. Collin, *Field Theory of Guided Waves*, McGraw-Hill Book Company, Inc., New York, 1960, p. 44. Also, see D. Jones, *The Theory of Electromagnetism*, Pergamon Press, Inc., New York, 1964, p. 35, and J. Van Bladel, *Electromagnetic Fields*, McGraw-Hill Book Company, Inc., New York, 1964, Appendix 6.

can replace the one three-dimensional Dirac delta by three one-dimensional deltas as follows:

$$\delta(\mathbf{r} - \mathbf{r}_0) = \delta(\mathbf{R}) = \frac{\delta(u_1 - u_{10})}{h_1} \frac{\delta(u_2 - u_{20})}{h_2} \frac{\delta(u_3 - u_{30})}{h_3} \qquad \text{(B-2)}$$

where $\mathbf{r}$ is the vector from the origin to some point $P(x, y, z)$ and $\mathbf{r}_0$ is the vector from the origin to another point $P_0(x_0, y_0, z_0)$. Thus, $\mathbf{R}$ is the vector from P_0 toward P. When the delta is written in two dimensions, one of the terms on the right side of (B-2) vanishes. However, this does not allow one to arbitrarily omit one of the terms since the integral of the delta must be unity. As mentioned by Van Bladel,[2] the proper procedure is to replace the denominator of the right side of (B-2) by the integral of the three scale factors over the coordinate that is to be ignored. For instance, if one is considering spherical coordinates with no ϕ-dependence, the denominator becomes

$$\int_0^{2\pi} r^2 \sin \theta \, d\phi = 2\pi r^2 \sin \theta$$

If the problem involves spherical coordinates but with no dependence on either ϕ or θ, the denominator becomes

$$\int_0^{\pi} d\theta \int_0^{2\pi} d\phi \, r^2 \sin \theta = 4\pi r^2$$

TABLE B-1   Representations of the Dirac Delta in Three
Different Coordinate Systems

$\delta(\mathbf{R})$

	$\delta(x, y, z)$	$\delta(x, y)$	$\delta(x)$
Cartesian	$\delta(x - x_0)\delta(y - y_0)\delta(z - z_0)$	$\delta(x - x)_0(y - y_0)$	$\delta(x - x_0)$
	$\delta(\rho, \phi, z)$	$\delta(\rho, z)$	$\delta(\rho)$
Cylindrical	$\frac{1}{\rho}\delta(\rho - \rho_0)\delta(\phi - \phi_0)\delta(z - z_0)$	$\frac{1}{2\pi\rho}\delta(\rho - \rho_0)\delta(z - z_0)$	$\frac{1}{2\pi\rho}\delta(\rho - \rho_0)$
	$\delta(r, \theta, \phi)$	$\delta(r, \theta)$	$\delta(r)$
Spherical	$\frac{1}{r^2 \sin \theta}\delta(r - r_0)\delta(\theta - \theta_0)\delta(\phi - \phi_0)$	$\frac{1}{2\pi r^2 \sin \theta}\delta(r - r_0)\delta(\theta - \theta_0)$	$\frac{1}{4\pi r^2}\delta(r - r_0)$

---

[2]Van Bladel, *Electromagnetic Fields*, Appendix 6.

C

Special Functions

C.1 Cylindrical Bessel Functions

1. Bessel's equation is

$$\frac{d^2f}{dx^2} + \frac{1}{x}\frac{df}{dx} + \left(\alpha^2 - \frac{s^2}{x^2}\right)f = \frac{1}{x}\frac{d}{dx}\left(x\frac{df}{dx}\right) + \left(\alpha^2 - \frac{s^2}{x^2}\right)f = 0, \qquad s = n$$

where $f = B_s(\alpha x)$ and $B_s(\alpha x)$ is any linear combination of Bessel functions of the first and second kinds, including Hankel functions. The order of the Bessel functions is represented by s, a noninteger, or n, an integer. Thus, Bessel's equation is valid for nonintegers s and integers n.

2. Two linearly independent solutions of Bessel's equation can be represented as the series

$$J_{\pm s}(\alpha x) = \left(\frac{\alpha x}{2}\right)^{\pm s}\sum_{m=0}^{\infty}\frac{(-\alpha^2 x^2/4)^m}{m!\,\Gamma(m \pm s + 1)}, \qquad s \neq n$$

For $s = n$, the above series are still valid but now $J_{-n} = (-1)^n J_n$ and the two solutions are not independent. J_s is the Bessel function of the first kind of order s.

3. The Neumann function (Bessel function of the second kind of order s)

$$N_s(\alpha x) = \frac{\cos s\pi \, J_s(\alpha x) - J_{-s}(\alpha x)}{\sin s\pi}, \qquad s \neq n$$

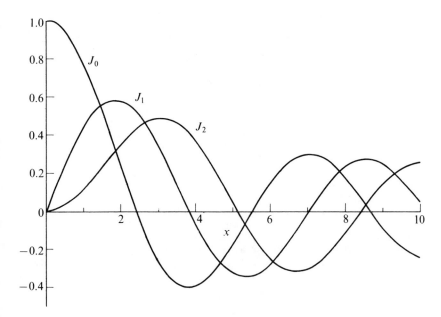

Fig. C-1   Cylindrical Bessel functions of the first kind.

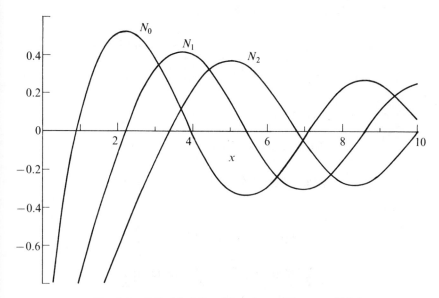

Fig. C-2   Cylindrical Bessel functions of the second kind.

is a second solution that is linearly independent of J_s and J_{-s}. The expression for N_{-s} can be obtained from that for N_s by replacing s by $-s$. Thus, we obtain

$$N_{-s}(\alpha x) = \cos s\pi \, N_s(\alpha x) + \sin s\pi \, J_s(\alpha x), \qquad s \neq n$$

When s becomes an integer n, the two are not independent since $N_{-n} = (-1)^n N_n$. Curves of $J_n(x)$ and $N_n(x)$ are shown in Fig. C-1 and Fig. C-2, respectively, for $n = 0, 1, 2, 3$.

4. The Hankel functions are defined as

$$\left. \begin{aligned} H_s^{(1)}(\alpha x) &= J_s(\alpha x) + j N_s(\alpha x) \\ H_s^{(2)}(\alpha x) &= J_s(\alpha x) - j N_s(ax) \end{aligned} \right\}, \qquad s = n$$

or

$$\left. \begin{aligned} H_s^{(1)}(\alpha x) &= \frac{e^{-js\pi} J_s(\alpha x) - J_{-s}(\alpha x)}{-j \sin s\pi} \\ H_s^{(2)}(\alpha x) &= \frac{e^{js\pi} J_s(\alpha x) - J_{-s}(\alpha x)}{j \sin s\pi} \end{aligned} \right\}, \qquad s \neq n$$

$H_n^{(1)}$ is bounded when the imaginary part of its argument is positive and $H_n^{(2)}$ is bounded when the imaginary part of its argument is negative.

5. Useful recurrence relations are

$$\left. \begin{aligned} \frac{2s}{\alpha x} B_s(\alpha x) &= B_{s-1}(\alpha x + B_{s+1}(\alpha x) \\ \frac{2}{\alpha} \frac{d}{dx} B_s(\alpha x) &\equiv 2B'_s(\alpha x) = B_{s-1}(\alpha x) - B_{s+1}(\alpha x) \end{aligned} \right\}, \qquad s = n$$

where $B_s(\alpha x)$ is any combination of linearly independent solutions. Also,

$$\frac{d}{dx}[x^n B_n(x)] = x^n B_{n-1}(x), \qquad \frac{d}{dx}[x^{-n} B_n(x)] = -x^{-n} B_{n+1}(x), \qquad n \neq s$$

6. Two normalization integrals are

$$\int_a^b x B_s(\alpha x) B_r(\alpha x) \, dx = \delta_{rs} N, \qquad r = m, s = n$$

$$\int_a^b x B_s(\alpha_i x) B_s(\alpha_j x) \, dx = \delta_{ij} N, \qquad s = n$$

where

$$N = \left[\frac{x^2}{2} \{ B_s^2(\alpha x) - B_{s-1}(\alpha x) B_{s+1}(\alpha x) \} \right]_a^b$$

$$= \left[\frac{x^2}{2} \left\{ B_s'^2(\alpha x) + \left(1 - \frac{s^2}{\alpha^2 x^2} \right) B_s^2(\alpha x) \right\} \right]_a^b$$

7. For small values of the argument, i.e., $\alpha x \longrightarrow 0$,

$$J_0(\alpha x) \longrightarrow 1$$

$$J_s(\alpha x) \longrightarrow \frac{1}{\Gamma(s+1)}\left(\frac{\alpha x}{2}\right)^s, \qquad s = n$$

$$N_0(\alpha x) \longrightarrow \frac{2}{\pi}\ln\frac{\alpha x}{2}$$

$$N_s(\alpha x) \longrightarrow -\frac{\Gamma(s)}{\pi}\left(\frac{2}{\alpha x}\right)^s, \qquad s = n$$

where $\Gamma(s) = (s-1)!$, etc.

8. For large values of the argument, i.e., $\alpha x \longrightarrow \infty$, $|\alpha x| \gg |s|$,

$$J_s(\alpha x) \longrightarrow \left(\frac{2}{\pi\alpha x}\right)^{1/2}\cos\left[\alpha x - \frac{\pi}{2}\left(s+\frac{1}{2}\right)\right]$$

$$N_s(\alpha x) \longrightarrow \left(\frac{2}{\pi\alpha x}\right)^{1/2}\sin\left[\alpha x - \frac{\pi}{2}\left(s+\frac{1}{2}\right)\right]$$

$$H_s^{(1)}(\alpha x) \longrightarrow \left(\frac{2}{\pi\alpha x}\right)^{1/2} e^{j\{\alpha x - (\pi/2)[s+(1/2)]\}}$$

$$H_s^{(2)}(\alpha x) \longrightarrow \left(\frac{2}{\pi\alpha x}\right)^{1/2} e^{-j\{\alpha x - (\pi/2)[s+(1/2)]\}}$$

$$, \qquad s = n$$

9. Useful miscellaneous relations are

$$H_{-s}^{(1)}(\alpha x) = e^{js\pi} H_s^{(1)}(\alpha x)$$

$$H_{-s}^{(2)}(\alpha x) = e^{-js\pi} H_s^{(2)}(\alpha x)$$

$$H_s^{(1)}(z^*) = [H_s^{(2)}(z)]^*$$

$$H_s^{(2)}(z^*) = [H_s^{(1)}(z)]^*$$

$$, \qquad s = n$$

s real, z complex or real, α and x real.

$$J_s(-x) = e^{js\pi} J_s(x)$$

$$J_{-s}(-x) = e^{-js\pi} J_{-s}(x)$$

$$N_s(-x) = e^{js\pi} N_s(x) + 2j J_{-s}(x) = e^{-js\pi} N_s(x) + 2j\cos s\pi\, J_s(x)$$

$$H_s^{(1)}(-x) = -e^{js\pi} H_s^{(2)}(x)$$

$$H_s^{(2)}(-x) = e^{js\pi} J_s(x) + e^{-js\pi} H_s^{(2)}(x) = e^{js\pi} H_s^{(1)}(x) + 2\cos s\pi\, H_s^{(2)}(x)$$

10. A variety of Wronskians are

$$
W[J_s(\alpha x), J_{-s}(\alpha x)] \equiv J_s(\alpha x)\frac{d}{d(\alpha x)}J_{-s}(\alpha x) - J_{-s}(\alpha x)\frac{d}{d(\alpha x)}J_s(\alpha x)
$$

$$
= J_s(\alpha x)J'_{-s}(\alpha x) - Jt'_s(\alpha x)J_{-s}(\alpha x)
$$

$$
= -\frac{2\sin s\pi}{\pi\alpha x}, \qquad s \neq n
$$

$$
W[J_s(\alpha x), N_s(\alpha x)] = \frac{2}{\pi\alpha x}, \qquad s = n
$$

$$
W[J_s(\alpha x), H_s^{(2)}(\alpha x)] = \frac{2}{j\pi\alpha x}, \qquad s = n
$$

$$
W[J_s(\alpha x), H_s^{(1)}(\alpha x)] = \frac{2j}{\pi\alpha x}, \qquad s = n
$$

$$
W[N_s(\alpha x), H_s^{(2)}(\alpha x)] = -\frac{2}{\pi\alpha x}, \qquad s = n
$$

11. The generating function for Bessel functions of the first kind is

$$
e^{(x/2)[t-(1/t)]} = \sum_{n=-\infty}^{\infty} J_n(x)t^n
$$

where x may be real or complex, n is a real integer, and $0 < |t| < 1$.

12. For the Laplacian in cylindrical coordinates

$$
\nabla^2 B_s(\alpha\rho)g(\phi)h(z) = B_s(\alpha\rho)g(\phi)\left(\frac{d^2}{dz^2} - \alpha^2\right)h(z), \qquad s = n
$$

provided $g(\phi)$ is a linear combination of $e^{\pm js\phi}$ or cos $s\phi$ and sin $s\phi$. $B_s(\alpha\rho)$ is any linear combination of Bessel functions.

13. Roots of the Bessel function of the first kind are given in Table C-1. The roots are ordered and designated by x_{np}. The first subscript indicates the order of the Bessel function and the second indicates the order of the zero.

TABLE C-1 Ordered Zeros x_{np} of $J_n(x)$

p \ n	0	1	2	3	4
1	2.405	3.832	5.136	6.380	7.588
2	5.520	7.016	8.417	9.761	11.065
3	8.654	10.173	11.620	13.015	14.372
4	11.792	13.324	14.796	16.223	17.616

14. Roots of the first derivative of the Bessel function of the first kind are given in Table C-2. The roots are ordered and designated by x'_{np}. Additional roots of J_n and J'_n may be found in C.L. Beattie, "Tables of First 700 Zeros of Bessel Functions $-J_\ell(x)$ and $J'_\ell(x)$," *Bell System Tech. J.*, Vol. 37, pp. 689–99, May 1958 (also Bell Monograph 3055). Also, most of the special functions listed here are tabulated for integer orders in M. Abramowitz and I. Stegun, eds., *Handbook of Mathematical Functions, with Formulas, Graphs, and Mathematical Tables*, Dover Publications, New York, 1965.

TABLE C-2   Ordered Zeros x'_{np} of $J'_n(x)$

p \ n	0	1	2	3	4
1	3.832	1.841	3.054	4.201	5.317
2	7.016	5.331	6.706	8.015	9.282
3	10.173	8.536	9.969	11.346	12.682
4	13.324	11.706	13.170	14.586	15.964

C.2 Modified Cylindrical Bessel Functions

1. The modified Bessel's equation is

$$\frac{d^2g}{dx^2} + \frac{1}{x}\frac{dg}{dx} - \left(\alpha^2 + \frac{s^2}{x^2}\right)g = \frac{1}{x}\frac{d}{dx}\left(x\frac{dg}{dx}\right) - \left(\alpha^2 + \frac{s^2}{x^2}\right)g = 0, \qquad s = n$$

where $g = C_s(\alpha x)$ and $C_s(\alpha x)$ is any linear combination of the modified Bessel functions $I_s(\alpha x)$, $I_{-s}(\alpha x)$, or $K_s(\alpha x)$.

2. Two linearly independent solutions of the modified Bessel's equation in series form are

$$I_{\pm s}(\alpha x) = \left(\frac{\alpha x}{2}\right)^{\pm s} \sum_{m=0}^{\infty} \frac{(\alpha^2 x^2/4)^m}{m!\,\Gamma(m \pm s + 1)}, \qquad s \neq n$$

$$I_{\pm s}(\alpha x) = e^{\mp js\pi/2} J_{\pm s}(j\alpha x), \qquad s = n$$

For $s = n$, the above series are still valid but now $I_{-n} = I_n$ and one obtains only one solution, the modified Bessel function of the first kind. Curves of the modified Bessel functions of the first and second kinds are shown in Fig. C-3 for $n = 0, 1$.

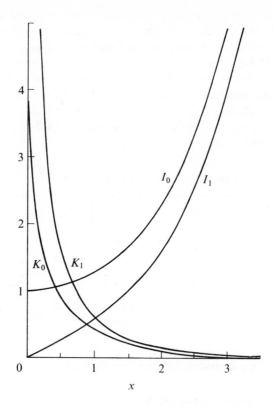

Fig. C-3 Modified cylindrical Bessel functions of the first
and second kinds.

3. The modified Bessel function of the second kind, $K_s(\alpha x)$, is linearly
independent of either $I_s(\alpha x)$ or $I_{-s}(\alpha x)$ and can be defined as

$$K_s(\alpha x) = \frac{\pi}{2 \sin s\pi}[I_{-s}(\alpha x) - I_s(\alpha x)], \qquad s = n$$

or

$$K_s(\alpha x) = \frac{\pi}{2}(j)^{s+1}H_s^{(1)}(j\alpha x), \qquad s = n$$

4. Useful miscellaneous relations are

$$I_s(-\alpha x) = e^{js\pi}I_s(\alpha x), \qquad s = n$$

$$K_s(-\alpha x) = e^{-js\pi}K_s(\alpha x) - j\pi I_s(\alpha x), \qquad s = n$$

$$I_{-n}(\alpha x) = I_n(\alpha x), \qquad n \neq s$$

$$K_{-s}(\alpha x) = K_s(\alpha x), \qquad s = n$$

$$I_s(-j\alpha x) = je^{-js\pi/2}J_s(\alpha x), \qquad s = n$$

5. The recurrence relations for $I_s(\alpha x)$ and $K_s(\alpha x)$ are not the same. Thus, we find the following:

$$\frac{2s}{\alpha x}I_s(\alpha x) = I_{s-1}(\alpha x) - I_{s+1}(\alpha x)$$

$$\frac{2}{\alpha}\frac{d}{dx}I_s(\alpha x) \equiv 2I'_s(\alpha x) = I_{s-1}(\alpha x) + I_{s+1}(\alpha x)$$
$$\qquad\qquad\qquad , \qquad s = n$$

$$-\frac{2s}{\alpha x}K_s(\alpha x) = K_{s-1}(\alpha x) - K_{s+1}(\alpha x)$$

$$-2K'_s(\alpha x) = K_{s-1}(\alpha x) + K_{s+1}(\alpha x)$$

6. Several Wronskians are the following:

$$W[I_s(\alpha x), K_s(\alpha x)] = -\frac{1}{\alpha x}, \qquad s = n$$

$$W[I_s(\alpha x), I_{-s}(\alpha x)] = -\frac{2}{\pi\alpha x}\sin s\pi, \qquad s \neq n$$

7. For large values of the argument, i.e., $\alpha x \longrightarrow \infty, |\alpha x| \gg |s|$,

$$I_s(\alpha x) \longrightarrow (2\pi\alpha x)^{-1/2}e^{\alpha x}$$
$$K_s(\alpha x) \longrightarrow (\pi/2\alpha x)^{-1/2}e^{-\alpha x} \qquad , \qquad s = n$$

8. For small values of the argument, i.e., $\alpha x \longrightarrow 0$,

$$I_0(\alpha x) \longrightarrow 1$$

$$I_s(\alpha x) \longrightarrow \frac{1}{\Gamma(s+1)}\frac{(\alpha x)^s}{2}$$

$$K_0(\alpha x) \longrightarrow -\ln\frac{\alpha x}{2} \qquad , \qquad s = n$$

$$K_s(\alpha x) \longrightarrow \frac{\Gamma(s)}{2}\left(\frac{2}{\alpha x}\right)^s$$

where $\Gamma(s) = (s-1)!$, etc.

9. The generating function for modified Bessel functions of the first kind is

$$e^{(x/2)[t+(1/t)]} = \sum_{n=-\infty}^{\infty} I_n(x)t^n$$

x real or complex, n a real integer, and $0 < |t| < 1$.

10. For the Laplacian in cylindrical coordinates

$$\nabla^2 C_s(\alpha\rho)g(\phi)h(z) = C_s(\alpha\rho)g(\phi)\Big(\frac{d^2}{dz^2} + \alpha^2\Big)h(z), \qquad s = n$$

provided $g(\phi)$ is a linear combination of $e^{\pm js\phi}$ or $\cos s\phi$ and $\sin s\phi$. $C_s(\alpha\rho)$ is any linear combination of $I_{\pm s}(\alpha\rho)$ and $K_s(\alpha\rho)$.

C.3 Spherical Bessel Functions

1. The spherical Bessel's equation is

$$x^2\frac{d^2h}{dx^2} + 2x\frac{dh}{dx} + [(\alpha x)^2 - s(s+1)]h$$

$$= \frac{d}{dx}\Big(x^2\frac{dh}{dx}\Big) + [(\alpha x)^2 - s(s+1)]h = 0, \qquad s = n$$

where $h = b_s(\alpha x)$ and $b_s(\alpha x)$ is any linear combination of spherical Bessel functions of the first and second kinds, including Hankel functions of the first and second kinds.

2. The solution to the spherical Bessel's equation can be expressed in terms of the usual Bessel functions by the relation

$$b_s(\alpha x) = \Big(\frac{\pi}{2\alpha x}\Big)^{1/2} B_{s+1/2}(\alpha x), \qquad s = n$$

3. Another form of spherical Bessel functions was used by Debye and Schelkunoff and can be defined as

$$\hat{B}_s(\alpha x) = \alpha x b_s(\alpha x) = \Big(\frac{\pi\alpha x}{2}\Big)^{1/2} B_{s+1/2}(\alpha x), \qquad s = n$$

and satisfy the differential equation

$$\Big[\frac{d^2}{dx^2} + \alpha^2 - \frac{s(s+1)}{x^2}\Big]\hat{B}_s(\alpha x) = 0, \qquad s = n$$

4. Two pairs of recurrence relations that the spherical Bessel function satisfies are

$$\frac{2s+1}{x}b_s(\alpha x) = b_{s-1}(\alpha x) + b_{s+1}(\alpha x)$$

$$(2s+1)b'_s(\alpha x) = sb_{s-1}(\alpha x) - (s+1)b_{s+1}(\alpha x)$$

$$\frac{2s+1}{x}\hat{B}_s(\alpha x) = \hat{B}_{s-1}(\alpha x) + \hat{B}_{s+1}(\alpha x)$$

$$, \qquad s = n$$

$$(2s + 1)\hat{B}'_s(\alpha x) = (s + 1)\hat{B}_{s-1}(\alpha x) - s\hat{B}_{s+1}(\alpha x)$$

$$\frac{d}{dx}[x^{n+1}b_n(x)] = x^{n+1}b_{n-1}(x), \qquad \frac{d}{dx}[x^{-n}b_n(x)] = -x^{-n}b_{n+1}(x)$$

$$\frac{d}{dx}[x^n\hat{B}_n(x)] = x^n\hat{B}_{n-1}(x), \qquad \frac{d}{dx}[x^{-(n+1)}\hat{B}_n(x)] = -x^{-(n+1)}\hat{B}_{n+1}(x), \qquad n = s$$

$$\frac{d}{dx}[xb_n(x)] = xb_{n-1}(x) - nb_n(x) = xb_{n+1}(x) + (n + 1)b_n(x)$$

The spherical Bessel functions are particularly useful when s in an integer n since then the $B_{n+1/2}(\alpha x)$ reduce to combinations of elementary functions. For the noninteger cases, one can replace the $b_s(\alpha x)$ or $\hat{B}_s(\alpha x)$ functions by the usual $B_{s+1/2}(\alpha x)$ functions and proceed accordingly. Some of the spherical Bessel functions of integer order are as follows:

$$\hat{J}_0(x) = xj_0(x) = \sin x$$

$$\hat{J}_1(x) = xj_1(x) = \frac{\sin x}{x} - \cos x$$

$$\hat{J}_2(x) = xj_2(x) = \left(\frac{3}{x^2} - 1\right)\sin x - \frac{3}{x}\cos x$$

$$\hat{H}_0^{(1)}(x) = xh_0^{(1)}(x) = -je^{jx}$$

$$\hat{H}_1^{(1)}(x) = xh_1^{(1)}(x) = -\left(1 + \frac{j}{x}\right)e^{jx}$$

$$\hat{N}_0(x) = xn_0(x) = -\cos x$$

$$\hat{N}_1(x) = xn_1(x) = -\sin x + \frac{\cos x}{x}$$

$$\hat{N}_2(x) = xn_2(x) = -\left[\frac{3}{x}\sin x + \left(\frac{3}{x^2} - 1\right)\cos x\right]$$

$$\hat{H}_0^{(2)}(x) = xh_0^{(2)}(x) = je^{-jx}$$

$$\hat{H}_1^{(2)}(x) = xh_1^{(2)}(x) = -\left(1 - \frac{j}{x}\right)e^{-jx}$$

Curves of $j_n(x)$ and $n_n(x)$ are shown in Figs. C-4 and C-5, respectively, for $n = 0, 1, 2$.

5. Useful normalization integrals are the following:

$$\int_a^b x^2 b_s(\alpha x)b_r(\alpha x)\, dx = \delta_{sr}N, \qquad s = n, r = m$$

$$\int_a^b \hat{B}_s(\alpha x)\hat{B}_r(\alpha x)\, dx = \delta_{sr}\hat{N}, \qquad s = n, r = m$$

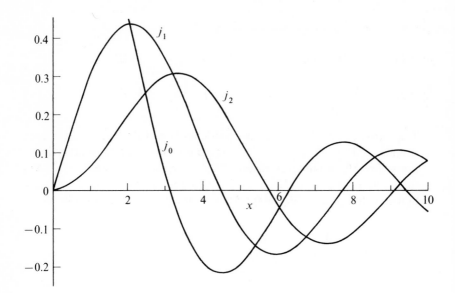

Fig. C-4   Spherical Bessel functions of the first kind.

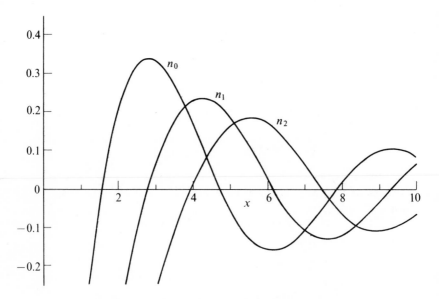

Fig. C-5   Spherical Bessel functions of the second kind.

where

$$N = \left[\frac{x^3}{2}\{b_s^2(\alpha x) - b_{s-1}(\alpha x)b_{s+1}(\alpha x)\}\right]_a^b$$

$$= \left[\frac{x^3}{2}\left\{b_s'^2(\alpha x) + \left[1 - \frac{s(s+1)}{\alpha^2 x^2}\right]b_s^2(\alpha x) + \frac{1}{\alpha x}b_s(\alpha x)b_s'(\alpha x)\right\}\right]_a^b$$

and

$$\hat{N} = \left[\frac{x}{2}\{\hat{B}_s^2(\alpha x) - \hat{B}_{s-1}(\alpha x)\hat{B}_{s+1}(\alpha x)\}\right]_a^b$$

$$= \left[\frac{x}{2}\left\{\hat{B}_s'^2(\alpha x) + \left[1 - \frac{s(s+1)}{\alpha^2 x^2}\right]\hat{B}_s^2(\alpha x) - \frac{1}{\alpha x}\hat{B}_s(\alpha x)\hat{B}_s'(\alpha x)\right\}\right]_a^b$$

6. For small values of the argument, i.e., $\alpha x \rightarrow 0$,

$$j_n(\alpha x) \longrightarrow \frac{n!}{(2n+1)!}(2\alpha x)^n, \qquad n = s$$

$$n_n(\alpha x) \longrightarrow -(2n-1)!\,(\alpha x)^{-(n+1)}, \qquad n \neq s$$

7. For large values of the argument, i.e., $\alpha x \rightarrow \infty$, $|\alpha x| \gg n$,

$$j_n(\alpha x) \longrightarrow \frac{1}{\alpha x}\cos\left[\alpha x - \frac{\pi}{2}(n+1)\right]$$

$$n_n(\alpha x) \longrightarrow \frac{1}{\alpha x}\sin\left[\alpha x - \frac{\pi}{2}(n+1)\right]$$

$$\qquad\qquad\qquad\qquad\qquad , \quad n \neq s$$

$$h_n^{(1)}(\alpha x) \longrightarrow \frac{1}{\alpha x}e^{j[\alpha x - (\pi/2)(n+1)]}$$

$$h_n^{(2)}(\alpha x) \longrightarrow \frac{1}{\alpha x}e^{-j[\alpha x - (\pi/2)(n+1)]}$$

8. Useful Wronskians are

$$W[j_n(\alpha x), n_n(\alpha x)] \equiv j_n(\alpha x)\frac{d}{d(\alpha x)}n_n(\alpha x) - n_n(\alpha x)\frac{d}{d(\alpha x)}j_n(\alpha x)$$

$$= j_n(\alpha x)n_n'(\alpha x) - j_n'(\alpha x)n_n(\alpha x) = \frac{1}{(\alpha x)^2}$$

$$W[j_n(\alpha x), h_n^{(2)}(\alpha x)] = -\frac{j}{(\alpha x)^2}$$

$$\qquad\qquad\qquad\qquad\qquad , \quad n = s$$

$$W[\hat{J}_n(\alpha x), \hat{N}_n(\alpha x)] = 1$$

$$W[\hat{J}_n(\alpha x), \hat{H}_n^{(2)}(\alpha x)] = -j$$

$$W[\hat{H}_n^{(2)}(\alpha x), \hat{N}_n(\alpha x)] = 1$$

$$W[r^n, r^{-(n+1)}] = -(2n+1)r^{-2}$$

9. Other useful relations are

$$j_n(-\alpha x) = (-1)^n j_n(\alpha x), \qquad \hat{J}_n(-\alpha x) = (-1)^{n+1} \hat{J}_n(\alpha x)$$

$$n_n(-\alpha x) = (-1)^{n+1} n_n(\alpha x), \qquad \hat{N}_n(-\alpha x) = (-1)^n \hat{N}_n(\alpha x)$$

$$h_n^{(1)}(-\alpha x) = (-1)^n h_n^{(2)}(\alpha x), \qquad \hat{H}_n^{(1)}(-\alpha x) = (-1)^{n+1} \hat{H}_n^{(2)}(\alpha x) , \qquad n \neq s$$

$$h_n^{(2)}(-\alpha x) = (-1)^n h_n^{(1)}(\alpha x), \qquad \hat{H}_n^{(2)}(-\alpha x) = (-1)^{n+1} \hat{H}_n^{(1)}(\alpha x)$$

10. The generating function for spherical Bessel functions of the first kind is

$$e^{\pm jx \cos\theta} = \sum_{n=0}^{\infty} (\pm j)^n (2n + 1) j_n(x) P_n(\cos\theta), \qquad n \neq s$$

11. For the Helmholtz equation in spherical coordinates

$$(\nabla^2 + k^2) b_s(kr) g(\theta) h(\phi)$$

$$= \frac{b_s(kr) h(\phi)}{r^2} \left[\frac{d^2}{d\theta^2} + \frac{1}{\tan\theta} \frac{d}{d\theta} + s(s+1) - \frac{t^2}{\sin^2\theta} \right] g(\theta),$$

$$s = n, t = m$$

provided $b_s(kr)$ is a linear combination of spherical Bessel functions and $h(\phi)$ is a linear combination of $e^{\pm jt\phi}$ or $\cos t\phi$ and $\sin t\phi$. The ordinary differential equation for $g(\theta)$ is the associated Legendre's equation. When $k = 0$, Helmholtz's equation reduces to Laplace's equation and the r-dependent functions change.

12. Roots of the spherical Bessel function of the first kind are given in Table C-3. The roots are ordered and designated by u_{np}, where the first subscript indicates the order of the spherical Bessel function and the second indicates the order of the zero.

TABLE C-3   Ordered Zeros u_{np} of $j_n(x)$

p \ n	1	2	3	4
1	4.493	5.763	6.988	8.183
2	7.725	9.095	10.417	11.705
3	10.904	12.323	13.698	15.040
4	14.066	15.515	16.924	18.301

Note: The zeros of $\hat{J}_n(x)$ are the same as those above for $j_n(x)$. Also, the zeros of $j_0(x)$ are omitted since they are just the zeros of $\sin x$.

13. Roots of the first derivative of the spherical Bessel function of the first kind are given in Table C-4. The roots are ordered in the same manner as above and are designated by u'_{np}.

TABLE C-4   Ordered Zeros u'_{np} of $j'_n(x)$

p \ n	0	1	2	3	4
1	0	2.0816	3.342	4.514	5.646
2	4.493	5.940	7.290	8.583	9.840
3	7.725	9.205	10.613	11.972	13.295
4	10.904	12.404	13.846	15.244	16.609

14. Roots of the first derivative of the spherical Debye-Schelkunoff Bessel function of the first kind are given in Table C-5. The roots are designated by $\hat{u}'_{np}$.

TABLE C-5   Ordered Zeros $\hat{u}'_{np}$ of $\hat{J}'_n(x)$

p \ n	1	2	3	4
1	2.744	3.870	4.973	6.062
2	6.117	7.443	8.722	9.968
3	9.317	10.713	12.064	13.380
4	12.486	13.921	15.314	16.674

Note: The zeros of $\hat{J}'_0(x)$ are omitted since they are just the zeros of $\cos x$.

C.4 Legendre Functions

1. The associated Legendre equation of degree s and order r is

$$(1 - x^2)\frac{d^2h}{dx^2} - 2x\frac{dh}{dx} + \left[s(s+1) - \frac{r^2}{1-x^2}\right]h = 0, \qquad s = n, r = m$$

or, for $x = \cos\theta$,

$$\frac{1}{\sin\theta}\frac{d}{d\theta}\left(\sin\theta\frac{dh}{d\theta}\right) + \left[s(s+1) - \frac{r^2}{\sin^2\theta}\right]h$$

$$= \frac{d^2h}{d\theta^2} + \frac{1}{\tan\theta}\frac{dh}{d\theta} + \left[s(s+1) - \frac{r^2}{\sin^2\theta}\right]h = 0$$

where h is any linear combination of the associated Legendre functions of the first and second kinds, $P_s^r(x)$ and $Q_s^r(x)$, respectively, or any linear combination of the associated Legendre functions $P_s^r(x)$ and $P_s^r(-x)$, hereafter designated $L_s^r(x)$ or $L_s^r(\cos\theta)$.

2. The function $P_s^r(x)$ is singular at $x = -1$ and the function $P_s^r(-x)$ is singular at $x = 1$. The function $Q_s^r(x)$ can be defined as

$$Q_s^r(x) = \frac{\pi}{2 \sin (s + r)\pi}[\cos (s + r)\pi P_s^r(x) - P_s^r(-x)], \qquad s \neq n, r = m$$

which indicates that it is singular at $x = \pm 1$.

3. Since the associated Legendre equation depends on r^2, another valid solution may be obtained for negative values of r. This solution is useful when one chooses (in a wave equation) exponential dependence in ϕ of the form $e^{\pm jr\phi}$ instead of the real form involving $\cos r\phi$ or $\sin r\phi$. In this case, r is an integer m, and these solutions may be defined[1] as

$$L_s^m(x) = (-1)^m(1 - x^2)^{m/2}\frac{d^m}{dx^m}L_s(x)$$

$$L_s^{-m}(x) = (-1)^m\frac{(s - m)!}{(s + m)!}L_s^m(x) \qquad , \qquad s = n$$

where L_s^m may be a combination of P_s^m and Q_s^m. In general, the associated Legendre equation has solutions $P_s^{\pm r}(\pm x)$, $Q_s^{\pm r}(\pm x)$, $P_{-s-1}^{\pm r}(\pm x)$, $Q_{-s-1}^{\pm r}(\pm x)$.

4. The functions $P_s(\pm x)$ can be expressed as the series

$$P_s(x) = \sum_{t=0}^{T} \frac{(-1)^t(s + t)!}{(t!)^2(s - t)!}\left(\frac{1 - x}{2}\right)^t$$

$$- \frac{\sin s\pi}{\pi} \sum_{t=T+1}^{\infty} \frac{(t - 1 - s)!(s + t)!}{(t!)^2}\left(\frac{1 - x}{2}\right)^t, \qquad s \neq n$$

where T is the nearest integer $T \leq s$. When s becomes an integer, the series has a finite number of terms and one obtains Legendre polynomials of degree n. Thus, one can use Rodrique's formula to obtain $P_n(\pm x)$, i.e.,

$$P_n(x) = \frac{1}{2^n n!}\frac{d^n}{dx^n}(x^2 - 1)^n$$

and

$$P_n(-x) = (-1)^n P_n(x)$$

---

[1] R. Harrington, *Time-Harmonic Electromagnetic Fields*, McGraw-Hill Book Company, Inc., New York, 1961, p. 468. The $(-1)^m$ is omitted by W. Smythe, *Static and Dynamic Electricity*, 2nd ed., McGraw-Hill Book Company, Inc., New York, 1950, p. 148, and others.

5. The Legendre functions of the second kind for s an integer n also become polynomials and can be expressed as

$$Q_n(x) = P_n(x)\left[\frac{1}{2}\ln\frac{1+x}{1-x}\right] - \frac{2n-1}{1-n}P_{n-1}(x) - \frac{2n-5}{3(n-1)}P_{n-3}(x) - \cdots$$

6. Some of the lower-degree polynomials are

$$P_0 = 1, \qquad P_1 = x = \cos\theta, \qquad P_2 = \tfrac{1}{2}(3x^2 - 1) = \tfrac{1}{4}(3\cos 2\theta + 1)$$

$$P_3 = \tfrac{1}{2}(5x^3 - 3x) = \tfrac{1}{8}(5\cos 3\theta + 3\cos\theta),$$

$$P_4 = \tfrac{1}{8}(35x^4 - 30x^2 + 3) = \tfrac{1}{64}(35\cos 4\theta + 20\cos 2\theta + 9)$$

$$Q_0 = \frac{1}{2}\ln\frac{1+x}{1-x} = \ln\cot\frac{\theta}{2},$$

$$Q_1 = \frac{x}{2}\ln\frac{1+x}{1-x} - 1 = \cos\theta\ln\cot\frac{\theta}{2} - 1$$

$$Q_2 = \frac{3x^2 - 1}{4}\ln\frac{1+x}{1-x} - \frac{3x}{2} = \frac{1}{2}(3\cos^2\theta - 1)\ln\cot\frac{\theta}{2} - \frac{3}{2}\cos\theta$$

and are shown in Figs. C-6 and C-7 for $n = 0, 1, 2, 3$.

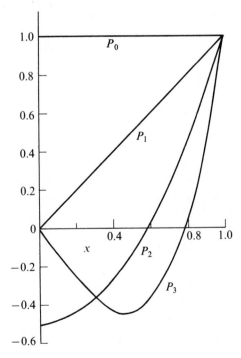

Fig. C-6 Legendre functions of the first kind.

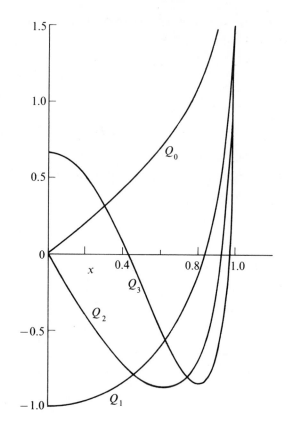

Fig. C-7 Legendre functions of the second kind.

7. Some of the lower-order associated Legendre polynomials are

$$P_1^1 = -(1 - x^2)^{1/2}, \qquad P_2^1 = 3x(1 - x^2)^{1/2}, \qquad P_2^2 = 3(1 - x^2),$$

$$P_3^1 = \tfrac{3}{2}(1 - x^2)^{1/2}(1 - 5x^2), \qquad P_3^2 = 15x(1 - x^2), \qquad P_3^3 = -15(1 - x^2)^{3/2}$$

$$Q_1^1 = -(1 - x^2)\left(\frac{1}{2}\ln\frac{1 + x}{1 - x} + \frac{x}{1 - x^2}\right),$$

$$Q_2^1 = -(1 - x^2)^{1/2}\left(\frac{3x}{2}\ln\frac{1 + x}{1 - x} + \frac{3x^2 - 2}{1 - x^2}\right)$$

$$Q_2^2 = (1 - x^2)^{1/2}\left[\frac{3}{2}\ln\frac{1 + x}{1 - x} + \frac{5x - 3x^3}{(1 - x^2)^2}\right]$$

8. Other useful relations are

$$P_n(1) = 1, \qquad P_n(-1) = (-1)^n$$

$$P_n^m(\pm 1) = 0, \qquad m > 0$$

$$P_n^m(0) = \begin{cases} (-1)^{(n-m)/2}\dfrac{1\cdot3\cdot5\cdots(n+m-1)}{2\cdot4\cdot6\cdots(n-m)}, & n+m \text{ even} \\ 0, & n+m \text{ odd} \end{cases}$$

$$Q_n^m(0) = \begin{cases} (-1)^{(n-m-1)/2}\dfrac{2\cdot4\cdot6\cdots(n+m-1)}{1\cdot3\cdot5\cdots(n-m)}, & n+m \text{ odd} \\ 0, & n+m \text{ even} \end{cases}$$

$$\left.\frac{d^r}{dx^r}P_n^m(x)\right]_{x=0} = (-1)^r P_n^{m+r}(0)$$

$$\left.\frac{d^r}{dx^r}Q_n^m(x)\right]_{x=0} = (-1)^r Q_n^{m+r}(0)$$
, r an integer

$$P_n^m(-x) = (-1)^{m+n} P_n^m(x)$$

$$Q_n^m(-x) = (-1)^{m+n+1} Q_n^m(x)$$

$$P_s^r(x) = P_{-s-1}^r(x), \qquad s=n, r=m$$

9. Useful recurrence relations are

$$L_s^{r+1}(x) = -2rx(1-x^2)^{-1/2}L_s^r(x) - (r+s)(s-r+1)L_s^{r-1}(x)$$

$$(1-x^2)\frac{d}{dx}L_s^r(x) = (r+s)(s-r+1)(1-x^2)^{1/2}L_s^{r-1}(x) + rxL_s^r(x)$$

$$\frac{d}{dx}L_s^r(x) = -rx(1-x^2)^{-1}L_s^r(x) - (1-x^2)^{-1/2}L_s^{r+1}(x)$$

$$(s-r+1)L_{s+1}^r(x) = (2s+1)xL_s^r(x) - (r+s)L_{s-1}^r(x)$$

$$(1-x^2)\frac{d}{dx}L_s^r(x) = -sxL_s^r(x) + (s+r)L_{s-1}^r(x)$$

$$\frac{d}{dx}L_s(x) = -(1-x^2)^{-1/2}L_s^1(x)$$

$$\frac{d}{d\theta}P_s(\cos\theta) = P_s^1(\cos\theta) \qquad\qquad s=n, r=m$$

The $L_s^r(x)$ above are any linear combination of $P_s^r(\pm x)$, $P_s^r(x)$ and $Q_s^r(x)$, or $P_s^{-r}(x)$ and $Q_s^{-r}(x)$.

10. A variety of orthogonality integrals are

$$\int_{-1}^1 \frac{P_n^s(x)P_n^t(x)\,dx}{1-x^2} = \frac{1}{s}\frac{(n+s)!}{(n-s)!}\delta_{st}, \qquad s>0;\ n,s,t \text{ integers}$$

$$\int_{-1}^1 P_s^m(x)P_t^m(x)\,dx = \int_0^\pi P_s^m(\cos\theta)P_t^m(\cos\theta)\sin\theta\,d\theta$$

$$= \frac{2}{2s+1}\frac{(s+m)!}{(s-m)!}\delta_{st}, \qquad s\geq m;\ m,s,t \text{ integers}$$

$$\int_{-1}^1 P_s^m(x)P_t^{-m}(x)\,dx = \frac{2}{2s+1}(-1)^m\,\delta_{st}, \qquad s\geq m;\ m,s,t \text{ integers}$$

$$\int_0^\pi \left(\frac{dP_s^m}{d\theta}\frac{dP_t^m}{d\theta} + \frac{m^2}{\sin^2\theta}P_s^m P_t^m\right)\sin\theta\, d\theta = \frac{2s(s+1)}{2s+1}\frac{(s+m)!}{(s-m)!}\delta_{st},$$

$$s, t, m \text{ integers}$$

$$= \frac{2[s(s+1)]^2}{2s+1}\delta_{st},$$

$$s, t \text{ integers}, m = 1$$

$$\int_0^\pi \left(\frac{P_n^1}{\sin\theta}\frac{dP_s^1}{d\theta} + \frac{P_s^1}{\sin\theta}\frac{dP_n^1}{d\theta}\right)\sin\theta\, d\theta = 0, \qquad n, s \text{ integers}$$

$$\int_{x_1}^{x_2} L_s^r(x)L_t^r(x)\, dx = \int_{\theta_2}^{\theta_1} L_s^r(\cos\theta)L_t^r(\cos\theta)\sin\theta\, d\theta = N_{1,2}\,\delta_{st},$$

$$\int_{x_1}^{x_2}\left[(1-x^2)\frac{dL_s^r(x)}{dx}\frac{dL_t^r(x)}{dx} + \frac{r^2}{1-x^2}L_s^r(x)L_t^r(x)\right]dx$$

$$= \int_{\theta_2}^{\theta_1}\left[\frac{dL_s^r(\cos\theta)}{d\theta}\frac{dL_t^r(\cos\theta)}{d\theta} + \frac{r^2}{\sin^2\theta}L_s^r(\cos\theta)L_t^r(\cos\theta)\right]\sin\theta\, d\theta$$

$$= s(s+1)N_{1,2}\,\delta_{st}, \qquad |x_1| \neq 1, |x_2| \neq 1, x_2 > x_1, \theta_1 > \theta_2,$$

$$\theta_{1,2} \neq 0, \pi \qquad r = m;\ s, t \text{ nonintegers}$$

and where L_s^r is any linear combination of $P_s^r(\pm x)$ or $P_s^r(x)$ and $Q_s^r(x)$. The norms are

$$N_1 = \frac{1}{2s+1}\left[(1-x^2)\frac{\partial L_s^r(x)}{\partial x}\frac{\partial L_s^r(x)}{\partial s}\right]_{x_1}^{x_2}$$

$$= \frac{1}{2s+1}\left[\sin\theta\,\frac{\partial L_s^r(\cos\theta)}{\partial\theta}\frac{\partial L_s^r(\cos\theta)}{\partial s}\right]_{\theta_2}^{\theta_1}$$

for

$$L_s^r(x_1) = L_s^r(x_2) = 0, \quad \text{or} \quad L_s^r(\cos\theta_1) = L_s^r(\cos\theta_2) = 0$$

and

$$N_2 = -\frac{1}{2s+1}\left[(1-x^2)L_s^r(x)\frac{\partial^2 L_s^r(x)}{\partial x\,\partial s}\right]_{x_1}^{x_2}$$

$$= -\frac{1}{2s+1}\left[\sin\theta L_s^r(\cos\theta)\frac{\partial^2 L_s^r(\cos\theta)}{\partial\theta\,\partial s}\right]_{\theta_2}^{\theta_1}$$

for

$$\frac{d}{dx}L_s^r(x)\bigg]_{x_1} = \frac{d}{dx}L_s^r(x)\bigg]_{x_2} = 0 \quad \text{or} \quad \frac{d}{d\theta}L_s^r(\cos\theta)\bigg]_{\theta_1} = \frac{d}{d\theta}L_s^r(\cos\theta)\bigg]_{\theta_2} = 0$$

The integrals and norms are also valid when $x_1 = -1$ provided $L_s^r(x) = P_s^r(-x)$ and when $x_2 = 1$ provided $L_s^r(x) = P_s^r(x)$.

11. Wronskians of Legendre functions are

$$W[P_s^r(x), Q_s^r(x)] = \frac{e^{j\pi r}2^{2r}\Gamma[1 + (r + s)/2]\Gamma[(1 + r + s)/2]}{(1 - x^2)\Gamma[1 + (s - r)/2]\Gamma[(1 + s - r)/2]},$$

$$s = n, r = m$$

$$W[P_s(x), Q_s(x)] = (1 - x^2)^{-1}, \qquad s = n$$

$$W[P_s(x), P_s(-x)] = -\frac{2}{\pi}\frac{\sin s\pi}{1 - x^2}, \qquad s \neq n$$

12. For the Helmholtz equation in spherical coordinates

$$(\nabla^2 + k^2)f(r)L_s^t(\cos\theta)h(\phi) = \frac{L_s^t(\cos\theta)h(\phi)}{r^2}$$

$$\times \left[r^2\frac{d^2}{dr^2} + 2r\frac{d}{dr} + (kr)^2 - s(s + 1) \right]f(r)$$

where $L_s^t(\cos\theta)$ is a linear combination of associated Legendre functions or polynomials of order t and degree s and $h(\phi)$ is a linear combination of $e^{\pm jt\phi}$ or $\cos t\phi$ and $\sin t\phi$. The ordinary differential equation for $f(r)$ is the spherical Bessel's equation. When $k = 0$, Helmholtz's equation reduces to Laplace's equation and the spherical Bessel's equation reduces to an equation with solutions of the form r^s and $r^{-(s+1)}$.

13. For unrestricted values of r, s, and x, the Legendre functions of the first and second kinds may be defined in terms of hypergeometric functions. (The same statement can be made about most of the useful functions in engineering and the applied sciences.) Thus, we may define[2]

$$P_s^r(x) = \frac{1}{\Gamma(1 - r)}\left(\frac{x + 1}{x - 1}\right)^{r/2} F\left(-s, s + 1; 1 - r; \frac{1 - x}{2}\right), \qquad |1 - x| < 2$$

$$Q_s^r(x) = e^{j\pi r}\sqrt{\pi}\,2^{-s-1}\frac{\Gamma(s + r + 1)}{\Gamma(s + \frac{3}{2})}x^{-r-s-1}(x^2 - 1)^{r/2}$$

$$\times F\left(1 + \frac{r + s}{2}, \frac{1 + r + s}{2}; s + \frac{3}{2}; x^{-2}\right)$$

where F is a hypergeometric function,

$$F(a, b; c; x) \equiv \sum_{n=0}^{\infty} \frac{\Gamma(a + n)}{\Gamma(a)}\frac{\Gamma(b + n)}{\Gamma(b)}\frac{\Gamma(c)}{\Gamma(c + n)}\frac{x^n}{n!},$$

$$c \neq 0, -1, -2, \ldots$$

[2]A. Erdélyi, W. Magnus, F. Oberhettinger, F. Tricomi, *Higher Transcendental Functions*, Vol. I, McGraw-Hill Book Company, Inc., New York, 1955.

14. Useful relations involving gamma functions and factorials are

$$\Gamma(z)\Gamma(1 - z) = \frac{\pi}{\sin \pi z}, \qquad z \text{ complex}$$

$$\Gamma(z + 1) = z\Gamma(z) = z!$$

$$\Gamma(1) = 0! = 1$$

$$(\tfrac{1}{2})! = (\tfrac{1}{2})\Gamma(\tfrac{1}{2})$$

$$\Gamma(\tfrac{1}{2}) = \sqrt{\pi} = (-\tfrac{1}{2})!$$

$$n! = 1 \cdot 2 \cdot 3 \cdot 4 \cdots n, \qquad n \text{ an integer}$$

$$n! = \pm\infty \quad \text{for} \quad n < 0, \qquad n \text{ an integer}$$

$\Gamma(z)$ may be defined as

$$\Gamma(z) \equiv \operatorname*{Lim}_{n \to \infty} \frac{1 \cdot 2 \cdot 3 \cdots n}{z(z + 1)(z + 2) \ldots (z + n)} n^z$$

z real or complex, or

$$\Gamma(z) \equiv \int_0^\infty e^{-t} t^{z-1} \, dt$$

real part of $z > 0$.

Index

284 Index